Wound Healing Biomaterials – Volume 1

Related titles

Nanomaterials in Tissue Engineering
(ISBN 978-0-85709-596-1)

Biomaterials for Cancer Therapeutics
(ISBN 978-0-85709-664-7)

Biomaterials and Medical Device – Associated Infections
(ISBN 978-0-85709-597-8)

Woodhead Publishing Series in Biomaterials:
Number 114

Wound Healing Biomaterials

Volume 1: Therapies and Regeneration

Edited by

Magnus S. Ågren

AMSTERDAM • BOSTON • CAMBRIDGE • HEIDELBERG
LONDON • NEW YORK • OXFORD • PARIS • SAN DIEGO
SAN FRANCISCO • SINGAPORE • SYDNEY • TOKYO
Woodhead Publishing is an imprint of Elsevier

Woodhead Publishing is an imprint of Elsevier
The Officers' Mess Business Centre, Royston Road, Duxford, CB22 4QH, UK
50 Hampshire Street, 5th Floor, Cambridge, MA 02139, USA
The Boulevard, Langford Lane, Kidlington, OX5 1GB, UK

British Library Cataloguing-in-Publication Data
A catalogue record for this book is available from the British Library

Library of Congress Cataloging-in-Publication Data
A catalog record for this book is available from the Library of Congress

ISBN: 978-1-78242-455-0 (print)
ISBN: 978-0-08-100605-4 (online)

For information on all Woodhead Publishing publications visit our website at https://www.elsevier.com/

Publisher: Matthew Deans
Acquisition Editor: Laura Overend
Editorial Project Manager: Lucy Beg
Production Project Manager: Poulouse Joseph
Designer: Matthew Limbert

Contents

List of contributors

A. Abdullahi University of Toronto, Toronto, ON, Canada

S. Amini-Nik University of Toronto, Toronto, ON, Canada

B. Azzimonti Department of Health Sciences, University of Piemonte Orientale, UPO, Alessandria, Novara Vercelli, Italy

A. Barbul Vanderbilt University Medical Center, Nashville, TN, United States

B.K.H.L. Boekema Association of Dutch Burn Centers, Beverwijk, The Netherlands

M. Cannas Department of Health Sciences, University of Piemonte Orientale, UPO, Alessandria, Novara Vercelli, Italy

A.L. Clement Worcester Polytechnic Institute, Worcester, MA, United States

B. Hinz University of Toronto, Toronto, ON, Canada

M.G. Jeschke University of Toronto, Toronto, ON, Canada

S.L. Kavalukas Vanderbilt University Medical Center, Nashville, TN, United States

D. Kletsas Laboratory of Cell Proliferation & Ageing, Institute of Biosciences & Applications, National Center for Scientific Research "Demokritos," Athens, Greece

T.J. Koh University of Illinois at Chicago, Chicago, IL, United States

F.V. Lali Blond McIndoe Research Foundation, Queen Victoria Hospital, East Grinstead, United Kingdom; The University of Brighton, Brighton, United Kingdom

R. Lundquist Reapplix ApS, Birkerød, Denmark

Y.H. Martin Blond McIndoe Research Foundation, Queen Victoria Hospital, East Grinstead, United Kingdom; The University of Brighton, Brighton, United Kingdom

A.D. Metcalfe Blond McIndoe Research Foundation, Queen Victoria Hospital, East Grinstead, United Kingdom; The University of Brighton, Brighton, United Kingdom

G. Mulder University of California at San Diego Medical Center, San Diego, CA, United States

G.D. Pins Worcester Polytechnic Institute, Worcester, MA, United States

H. Pratsinis Laboratory of Cell Proliferation & Ageing, Institute of Biosciences & Applications, National Center for Scientific Research "Demokritos," Athens, Greece

H.O. Rennekampff Universitätsklinikum der RWTH Aachen, Aachen, Germany

L. Rimondini Department of Health Sciences, University of Piemonte Orientale, UPO, Alessandria, Novara Vercelli, Italy

M. Sabbatini Department of Science and Innovation Technology (DISIT), University of Piemonte Orientale, UPO, Alessandria, Novara Vercelli, Italy

J.W. Shupp MedStar Washington Hospital Center, Washington, DC, United States

S. Tejiram MedStar Washington Hospital Center, Washington, DC, United States

M. Tenenhaus University of California at San Diego Medical Center, San Diego, CA, United States

M.M.W. Ulrich Association of Dutch Burn Centers, Beverwijk, The Netherlands; VU University Medical Center, Amsterdam, The Netherlands

N. Urao University of Illinois at Chicago, Chicago, IL, United States

M. Vlig Association of Dutch Burn Centers, Beverwijk, The Netherlands

Woodhead Publishing Series in Biomaterials

1 **Sterilisation of tissues using ionising radiations**
Edited by J. F. Kennedy, G. O. Phillips and P. A. Williams
2 **Surfaces and interfaces for biomaterials**
Edited by P. Vadgama
3 **Molecular interfacial phenomena of polymers and biopolymers**
Edited by C. Chen
4 **Biomaterials, artificial organs and tissue engineering**
Edited by L. Hench and J. Jones
5 **Medical modelling**
R. Bibb
6 **Artificial cells, cell engineering and therapy**
Edited by S. Prakash
7 **Biomedical polymers**
Edited by M. Jenkins
8 **Tissue engineering using ceramics and polymers**
Edited by A. R. Boccaccini and J. Gough
9 **Bioceramics and their clinical applications**
Edited by T. Kokubo
10 **Dental biomaterials**
Edited by R. V. Curtis and T. F. Watson
11 **Joint replacement technology**
Edited by P. A. Revell
12 **Natural-based polymers for biomedical applications**
Edited by R. L. Reiss et al
13 **Degradation rate of bioresorbable materials**
Edited by F. J. Buchanan
14 **Orthopaedic bone cements**
Edited by S. Deb
15 **Shape memory alloys for biomedical applications**
Edited by T. Yoneyama and S. Miyazaki
16 **Cellular response to biomaterials**
Edited by L. Di Silvio
17 **Biomaterials for treating skin loss**
Edited by D. P. Orgill and C. Blanco
18 **Biomaterials and tissue engineering in urology**
Edited by J. Denstedt and A. Atala

19 **Materials science for dentistry**
B. W. Darvell
20 **Bone repair biomaterials**
Edited by J. A. Planell, S. M. Best, D. Lacroix and A. Merolli
21 **Biomedical composites**
Edited by L. Ambrosio
22 **Drug–device combination products**
Edited by A. Lewis
23 **Biomaterials and regenerative medicine in ophthalmology**
Edited by T. V. Chirila
24 **Regenerative medicine and biomaterials for the repair of connective tissues**
Edited by C. Archer and J. Ralphs
25 **Metals for biomedical devices**
Edited by M. Niinomi
26 **Biointegration of medical implant materials: Science and design**
Edited by C. P. Sharma
27 **Biomaterials and devices for the circulatory system**
Edited by T. Gourlay and R. Black
28 **Surface modification of biomaterials: Methods analysis and applications**
Edited by R. Williams
29 **Biomaterials for artificial organs**
Edited by M. Lysaght and T. Webster
30 **Injectable biomaterials: Science and applications**
Edited by B. Vernon
31 **Biomedical hydrogels: Biochemistry, manufacture and medical applications**
Edited by S. Rimmer
32 **Preprosthetic and maxillofacial surgery: Biomaterials, bone grafting and tissue engineering**
Edited by J. Ferri and E. Hunziker
33 **Bioactive materials in medicine: Design and applications**
Edited by X. Zhao, J. M. Courtney and H. Qian
34 **Advanced wound repair therapies**
Edited by D. Farrar
35 **Electrospinning for tissue regeneration**
Edited by L. Bosworth and S. Downes
36 **Bioactive glasses: Materials, properties and applications**
Edited by H. O. Ylänen
37 **Coatings for biomedical applications**
Edited by M. Driver
38 **Progenitor and stem cell technologies and therapies**
Edited by A. Atala
39 **Biomaterials for spinal surgery**
Edited by L. Ambrosio and E. Tanner
40 **Minimized cardiopulmonary bypass techniques and technologies**
Edited by T. Gourlay and S. Gunaydin
41 **Wear of orthopaedic implants and artificial joints**
Edited by S. Affatato
42 **Biomaterials in plastic surgery: Breast implants**
Edited by W. Peters, H. Brandon, K. L. Jerina, C. Wolf and V. L. Young

43 **MEMS for biomedical applications**
Edited by S. Bhansali and A. Vasudev
44 **Durability and reliability of medical polymers**
Edited by M. Jenkins and A. Stamboulis
45 **Biosensors for medical applications**
Edited by S. Higson
46 **Sterilisation of biomaterials and medical devices**
Edited by S. Lerouge and A. Simmons
47 **The hip resurfacing handbook: A practical guide to the use and management of modern hip resurfacings**
Edited by K. De Smet, P. Campbell and C. Van Der Straeten
48 **Developments in tissue engineered and regenerative medicine products**
J. Basu and J. W. Ludlow
49 **Nanomedicine: Technologies and applications**
Edited by T. J. Webster
50 **Biocompatibility and performance of medical devices**
Edited by J.-P. Boutrand
51 **Medical robotics: Minimally invasive surgery**
Edited by P. Gomes
52 **Implantable sensor systems for medical applications**
Edited by A. Inmann and D. Hodgins
53 **Non-metallic biomaterials for tooth repair and replacement**
Edited by P. Vallittu
54 **Joining and assembly of medical materials and devices**
Edited by Y. (Norman) Zhou and M. D. Breyen
55 **Diamond-based materials for biomedical applications**
Edited by R. Narayan
56 **Nanomaterials in tissue engineering: Fabrication and applications**
Edited by A. K. Gaharwar, S. Sant, M. J. Hancock and S. A. Hacking
57 **Biomimetic biomaterials: Structure and applications**
Edited by A. J. Ruys
58 **Standardisation in cell and tissue engineering: Methods and protocols**
Edited by V. Salih
59 **Inhaler devices: Fundamentals, design and drug delivery**
Edited by P. Prokopovich
60 **Bio-tribocorrosion in biomaterials and medical implants**
Edited by Y. Yan
61 **Microfluidic devices for biomedical applications**
Edited by X.-J. James Li and Y. Zhou
62 **Decontamination in hospitals and healthcare**
Edited by J. T. Walker
63 **Biomedical imaging: Applications and advances**
Edited by P. Morris
64 **Characterization of biomaterials**
Edited by M. Jaffe, W. Hammond, P. Tolias and T. Arinzeh
65 **Biomaterials and medical tribology**
Edited by J. Paolo Davim
66 **Biomaterials for cancer therapeutics: Diagnosis, prevention and therapy**
Edited by K. Park

67 **New functional biomaterials for medicine and healthcare**
E.P. Ivanova, K. Bazaka and R. J. Crawford
68 **Porous silicon for biomedical applications**
Edited by H. A. Santos
69 **A practical approach to spinal trauma**
Edited by H. N. Bajaj and S. Katoch
70 **Rapid prototyping of biomaterials: Principles and applications**
Edited by R. Narayan
71 **Cardiac regeneration and repair Volume 1: Pathology and therapies**
Edited by R.-K. Li and R. D. Weisel
72 **Cardiac regeneration and repair Volume 2: Biomaterials and tissue engineering**
Edited by R.-K. Li and R. D. Weisel
73 **Semiconducting silicon nanowires for biomedical applications**
Edited by J. L. Coffer
74 **Silk biomaterials for tissue engineering and regenerative medicine**
Edited by S. Kundu
75 **Biomaterials for bone regeneration: Novel techniques and applications**
Edited by P. Dubruel and S. Van Vlierberghe
76 **Biomedical foams for tissue engineering applications**
Edited by P. Netti
77 **Precious metals for biomedical applications**
Edited by N. Baltzer and T. Copponnex
78 **Bone substitute biomaterials**
Edited by K. Mallick
79 **Regulatory affairs for biomaterials and medical devices**
Edited by S. F. Amato and R. Ezzell
80 **Joint replacement technology Second edition**
Edited by P. A. Revell
81 **Computational modelling of biomechanics and biotribology in the musculoskeletal system: Biomaterials and tissues**
Edited by Z. Jin
82 **Biophotonics for medical applications**
Edited by I. Meglinski
83 **Modelling degradation of bioresorbable polymeric medical devices**
Edited by J. Pan
84 **Perspectives in total hip arthroplasty: Advances in biomaterials and their tribological interactions**
S. Affatato
85 **Tissue engineering using ceramics and polymers Second edition**
Edited by A. R. Boccaccini and P. X. Ma
86 **Biomaterials and medical-device associated infections**
Edited by L. Barnes and I. R. Cooper
87 **Surgical techniques in total knee arthroplasty (TKA) and alternative procedures**
Edited by S. Affatato
88 **Lanthanide oxide nanoparticles for molecular imaging and therapeutics**
G. H. Lee
89 **Surface modification of magnesium and its alloys for biomedical applications Volume 1: Biological interactions, mechanical properties and testing**
Edited by T. S. N. Sankara Narayanan, I. S. Park and M. H. Lee

90 **Surface modification of magnesium and its alloys for biomedical applications Volume 2: Modification and coating techniques**
Edited by T. S. N. Sankara Narayanan, I. S. Park and M. H. Lee
91 **Medical modelling: the application of advanced design and rapid prototyping techniques in medicine Second Edition**
Edited by R. Bibb, D. Eggbeer and A. Paterson
92 **Switchable and responsive surfaces and materials for biomedical applications**
Edited by Z. Zhang
93 **Biomedical textiles for orthopaedic and surgical applications: fundamentals, applications and tissue engineering**
Edited by T. Blair
94 **Surface coating and modification of metallic biomaterials**
Edited by C. Wen
95 **Hydroxyapatite (HAP) for biomedical applications**
Edited by M. Mucalo
96 **Implantable neuroprostheses for restoring function**
Edited by K. Kilgore
97 **Shape memory polymers for biomedical applications**
Edited by L. Yahia
98 **Regenerative engineering of musculoskeletal tissues and interfaces**
Edited by S. P. Nukavarapu, J. W. Freeman and C. T. Laurencin
99 **Advanced cardiac imaging**
Edited by K. Nieman, O. Gaemperli, P. Lancellotti and S. Plein
100 **Functional marine biomaterials: Properties and applications**
Edited by S. K. Kim
101 **Shoulder and elbow trauma and its complications Volume 1: The shoulder**
Edited by R. M. Greiwe
102 **Nanotechnology-enhanced orthopedic materials: Fabrications, applications and future trends**
Edited by L. Yang
103 **Medical devices: Regulations, standards and practices**
Edited by S. Ramakrishna, L. Tian, C. Wang, S. L. and T. Wee Eong
104 **Biomineralisation and biomaterials: Fundamentals and applications**
Edited by C. Aparicio and M. Ginebra
105 **Shoulder and elbow trauma and its complications Volume 2: The Elbow**
Edited by R. M. Greiwe
106 **Characterisation and design of tissue scaffolds**
Edited by P. Tomlins
107 **Biosynthetic polymers for medical applications**
Edited by L. Poole-Warren, P. Martens and R.Green
108 **Advances in polyurethane biomaterials**
Edited by S. L. Cooper
109 **Nanocomposites for musculoskeletal tissue regeneration**
Edited by H. Liu
110 **Thin film coatings for biomaterials and biomedical applications**
Edited by H. J. Griesser
111 **Laser surface modification of biomaterials**
Edited by R. Vilar
112 **Biomaterials and regenerative medicine in ophthalmology Second edition**
Edited by T. V. Chirila and D. Harkin

113 **Extracellular matrix-derived medical implants in clinical medicine**
Edited by D. Mooradian
114 **Wound healing biomaterials Volume 1: Therapies and regeneration**
Edited by M. S. Ågren
115 **Wound healing biomaterials Volume 2: Functional biomaterials**
Edited by M. S. Ågren

Part One

Fundamentals and strategies for wound healing

Wound healing

1

S. Tejiram[1], *S.L. Kavalukas*[2], *J.W. Shupp*[1], *A. Barbul*[2]
[1]MedStar Washington Hospital Center, Washington, DC, United States; [2]Vanderbilt University Medical Center, Nashville, TN, United States

1.1 Introduction

Mammals are generally incapable of tissue regeneration, so repair of an injured organ depends upon a complex process that involves inflammatory cell migration and cytokine actions, collagen and extracellular matrix deposition, and scar remodeling. With few exceptions, all wound healing results in the formation of scar. The process of wound repair differs little from one tissue to another and is generally independent of the nature of the injury. A number of elaborate and redundant mechanisms that complement one another facilitate healing and drive the wound healing process forward.

1.2 Skin layers

The skin represents one of the largest organs of the body whose key roles include protection against outside pathogens and excessive water loss. The skin is comprised of three layers: the epidermis, dermis, and subcutaneous layer (hypodermis) (Fig. 1.1). The epidermis is the top layer of the skin and produces keratin that makes up the outermost physical barrier and the pigment melanin for ultraviolet (UV) radiation protection. It is the foremost layer involved in barrier protection against the outside environment through tight junctions and Langerhans cells. The dermis contains the papillary and reticular dermis and houses the skin appendages hair follicles, sebaceous glands and sweat glands necessary for tissue regeneration. The hypodermis is the bottom layer of skin that serves to attach the dermis to muscles and bones, houses blood vessels and nerves, and helps control body temperature.

1.3 Phases of wound healing

Injury starts a complex cascade of cellular and biochemical events, resulting in a healed wound. The wound healing response can be divided into four separate, but overlapping, phases: hemostasis and inflammation, proliferation, maturation and remodeling, and wound contraction.

Wound Healing Biomaterials - Volume 1. http://dx.doi.org/10.1016/B978-1-78242-455-0.00001-X

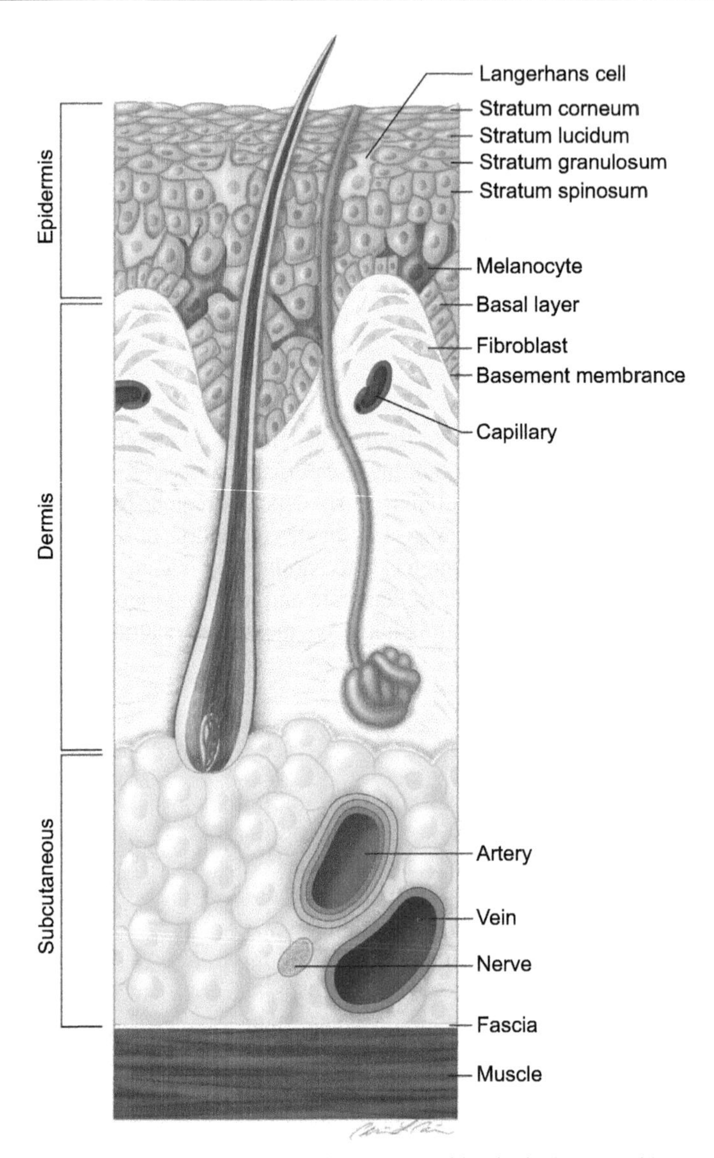

Figure 1.1 Anatomy of the skin depicting the important histological zones, skin structures, dermal appendages, and cell types comprising the layers of skin.
Adapted from Shupp JW, et al. A review of the local pathophysiologic bases of burn wound progression. J Burn Care Res 2010;31(6):849–73.

1.3.1 Hemostasis and inflammation

Hemostasis precedes inflammation. The rupture of vessels that accompanies injury exposes the subendothelial collagen to platelets, which leads to the aggregation and activation of the coagulation cascade. The contact between collagen and platelets as

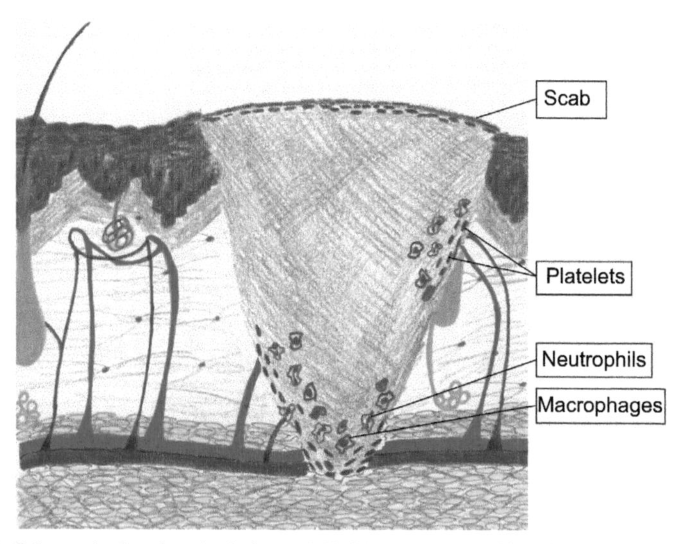

Figure 1.2 Schematic drawing depicting a full-thickness wound into hypodermis. Note the initial arrival of platelets to achieve hemostasis as well as neutrophils and macrophages to the healing-wound bed. The contact between collagen and platelets results in the release of cytokines and growth factors that ultimately provide chemotaxis for neutrophils, monocytes, fibroblasts, and more.

well as the presence of thrombin, fibronectin, and their fragments results in the release of cytokines and growth factors from platelet α-granules, such as platelet-derived growth factor (PDGF), transforming growth factor (TGF)-β, platelet activating factor, fibronectin, and serotonin. The locally formed fibrin clot serves as a scaffold for migrating cells such as neutrophils, monocytes, fibroblasts, and endothelial cells (Fig. 1.2).

The inflammatory phase is characterized by increased vascular permeability, chemotaxis of cells from the circulation into the wound milieu, the local release of cytokines and growth factors, and the activation of migrating cells. Within 6 h of injury, circulating immune cells migrate into the wound. Neutrophils are the first blood leukocytes to enter the wound site, and their numbers peak at 24–48 h after injury. Increased vascular permeability caused by inflammation and the release of prostaglandin, together with a concentration gradient of chemotactic substances such as complement factors, interleukin (IL)-1, tumor necrosis factor (TNF)-α, TGF-β, platelet factor-4, and bacterial products, stimulate neutrophil migration. The main function of the neutrophils is phagocytosis of bacteria and tissue debris. The presence of neutrophils is not essential for healing to take place, provided that bacterial contamination has not occurred [1]. Macrophages, derived from circulating monocytes, migrate next within the wound and peak by day 3 after injury. Wound macrophages have a much longer life span than the neutrophils and persist in the wound until healing is complete. Macrophages play a central role in the orchestration of the wound healing cascade. Their appearance is followed by lymphocytes, which occur in significant numbers around the fifth day after injury. In contrast to neutrophils, the presence and activation of both macrophages and lymphocytes in the wound is critical to the progression of the normal healing process (Fig. 1.3(a)) [2–4].

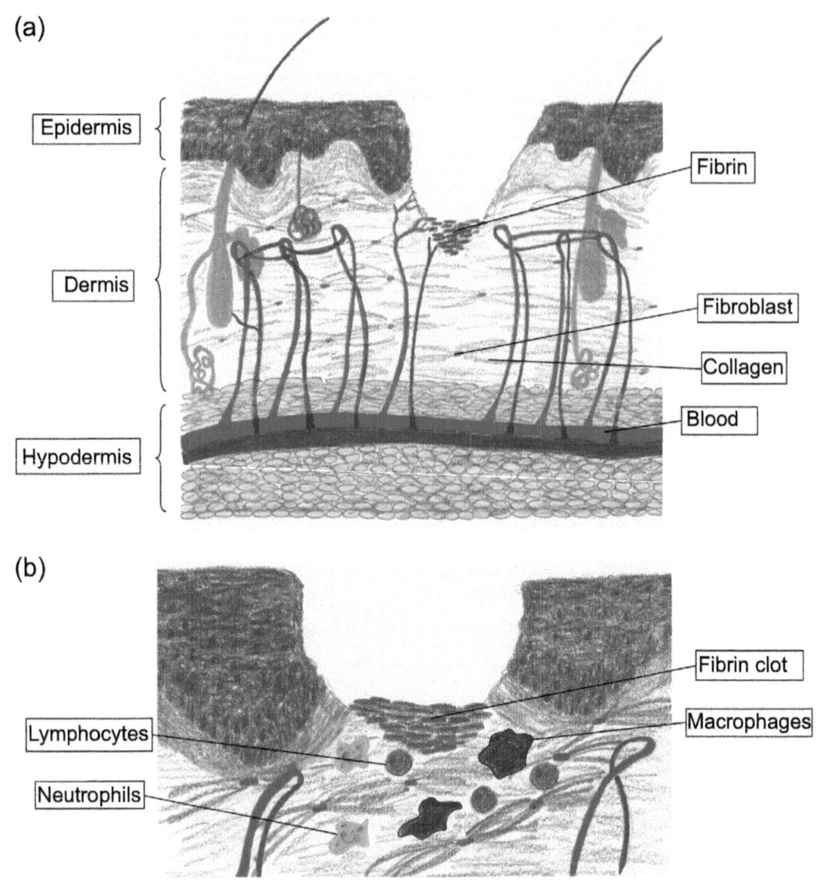

Figure 1.3 Schematic drawing of a partial-thickness wound in the inflammatory phase. (a) Each layer of the skin contributes to the wound healing process and associated extracellular matrix production. (b) Close-up of the wound bed depicting the arrival of neutrophils, macrophages, and lymphocytes to the wound bed. The fibrin clot helps achieve hemostasis while providing chemotaxis for the cells necessary to begin the wound healing process.

Chemotaxis of cells into the wound milieu is followed by cellular activation, which signifies the phenotypic alteration of cellular, biochemical, and functional properties induced by local mediators (Fig. 1.3(b)). Although neutrophils, macrophages, and lymphocytes predominate during inflammation, the contribution of each cell population to successful wound healing is variable.

The activation of macrophages has fundamental implications in several aspects of wound healing, such as debridement, matrix synthesis, and angiogenesis. The initial and brief release of factors from platelets is a first and strong stimulus of macrophage activation. Activated macrophages release cytokines that mediate angiogenesis, fibroplasia [5], and nitric oxide (NO) synthesis, which itself has antimicrobial properties and stimulates collagen synthesis [6,7]. Macrophages evolve phenotypically as healing progresses, switching from NO synthesis and inflammatory function to arginase expression and more synthetic functions [8]. The inflammatory phase of healing is vital to the proper evolution of subsequent wound healing phases. Reduced inflammatory responses profoundly affect subsequent healing, as observed clinically in patients who have diabetes mellitus or secondary to corticosteroid treatment [9]. On the other hand, persistence of the inflammatory phase is a hallmark of wound failure and/or excessive fibrosis.

1.3.2 Proliferation

Revascularization of the wound proceeds in parallel with fibroplasia. Angiogenesis occurs by a combination of proliferation and migration. Capillary buds sprout from blood vessels adjacent to the wound and extend into the wound space. Endothelial cells from the side of the venule closest to the wound begin to migrate in response to angiogenic stimuli. These capillary sprouts eventually branch at their tips and join to form capillary loops, through which blood begins to flow. New sprouts then extend from these loops to form a capillary plexus (Fig. 1.4) [10,11]. Mediators for endothelial cell growth and chemotaxis include cytokines produced by platelets, macrophages and lymphocytes in the wound [12,13], low oxygen tension [14], lactic acid [15], and biogenic amines [16]. Vascular endothelial growth factor is one of the most potent cytokines-stimulating angiogenesis. Basic fibroblast growth factor (bFGF), acidic

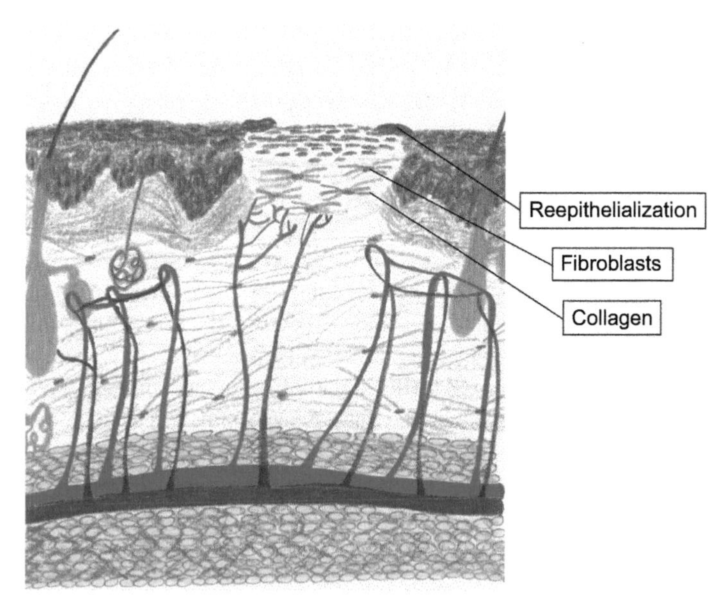

Figure 1.4 The proliferative phase is characterized by angiogenesis, fibroplasias, and collagen production. Reepithelialization of the wound begins from the wound margins or residual dermal epithelial appendages. Once epithelialization completes in 48 h in primary closed wounds, keratinization of the surface layer begins.

FGF, TGF-α, epidermal growth factor (EGF), and TGF-β have also been shown to play a role in stimulating new blood vessel formation [17–20].

Fibroblasts first appear in significant numbers in the wound on the third day after injury and achieve peak numbers around the seventh day [21]. Fibroblasts in the surrounding tissue need to become activated from their nonreplicative, quiescent state. Growth factors such as PDGF, bFGF, or EGF induce chemotaxis and proliferation of fibroblasts [22]. The rapid expansion of the fibroblast population at the wound site results from a combination of proliferation and migration [23]. Fibroblasts are derived from local mesenchymal cells, particularly those associated with blood vessel adventitia [24]. They are attracted to the wound and induced to proliferate by cytokines released initially from platelets and subsequently from macrophages and lymphocytes. Fibroblasts are the primary synthetic element in the repair process and are responsible for producing the majority of structural proteins necessary for tissue reconstruction. The main protein product of fibroblasts is collagen, a family of triple-chain glycoproteins that form the main constituent of the extracellular wound matrix. These are ultimately responsible for imparting tensile strength to the scar. Collagen is first detected in the wound around the third day after injury [25,26]. The levels then increase rapidly for approximately 3 weeks. It continues to accumulate at a more gradual pace for up to 3 months after wounding. A circulating population of fibrocytes (less than 0.5% of nonerythrocytic circulating cells) has been defined as contributing to the healing fibroblastic population and to collagen synthesis [27].

While granulation tissue and collagen synthesis are proceeding deep in the wound, epithelial integrity is being restored at the wound surface. Epithelialization of the wound begins within a few hours of injury. Epithelial cells arising from the wound margins and residual dermal epithelial appendages within the wound bed begin to migrate under the scab and over the underlying viable connective tissue. The epidermis immediately adjacent to the wound edge begins thickening within 24 h after injury. Marginal basal cells at the edge of the wound lose their firm attachment to the underlying dermis, enlarge, and begin to migrate across the surface of the provisional matrix, filling the wound. Fixed basal cells in a zone near the cut edge undergo a series of rapid mitotic divisions and migrate by moving over one another in a leapfrog fashion until the defect is covered [28,29]. Once the defect has been bridged, migrating epithelial cells lose their flattened appearance, become more columnar in shape, and increase their mitotic activity. Layering of the epithelium is reestablished, and the surface layer is eventually keratinized [30]. Epithelialization is complete in less than 48 h in approximated incised wounds but may take substantially longer in larger wounds in which there is a significant tissue defect.

If only the epithelium is damaged, such as occurs in split-thickness skin graft donor sites and superficial second-degree burns, repair consists primarily of epithelialization with minimal or absent fibroplasia and granulation tissue formation. The process is mediated by a combination of loss of contact inhibition, exposure of constituents of the extracellular matrix, particularly fibronectin [31], and cytokines produced by immune mononuclear cells such as macrophages and lymphocytes [32]. In particular, EGF, TGF-β, bFGF, PDGF, and insulin-like growth factor-1 have been shown to promote epithelialization.

Although much interest has been focused on the signals that "turn on" the proliferative phase, little is known about the signals that bring this phase to a controlled end. Stimuli for the activation of cells do not have to be long lasting [33], but the induced phenotypic alterations are often not maintained in vitro over time [34,35]. It is probable that negative feedback mechanisms play a role. The destiny of some cells remains unclear after completion of their function in the wound healing cascade. Neutrophils undergo apoptosis and are ingested by macrophages. Macrophages also undergo apoptosis and in the process release arginase, which influences wound metabolism [36]. Macrophages also play an important role in decreasing the cellularity of healed wounds by ingesting apoptotic fibroblasts [8].

1.3.3 Maturation and remodeling

Changes in the wound matrix composition follow a certain pattern over time. Initially, the wound matrix is mainly composed of fibrin and fibronectin that originate from hemostasis and macrophages [37]. Another early expressed protein is thrombospondin-1, which also supports cellular recruitment in the wound milieu [38]. Glycosaminoglycans, proteoglycans, and other proteins such as secreted protein acidic rich in cysteine are synthesized next and support further matrix deposition and remodeling [38,39]. Subsequently, collagen becomes the predominant scar protein. Collagen is initially deposited in a seemingly haphazard fashion. Thereafter, the individual collagen fibrils are reorganized by cross-linking into regularly aligned bundles oriented along the lines of stress in the healing wound. Fibroblasts are also responsible for the production of other matrix constituents, including fibronectin and the glycosaminoglycans (Fig. 1.5) [37].

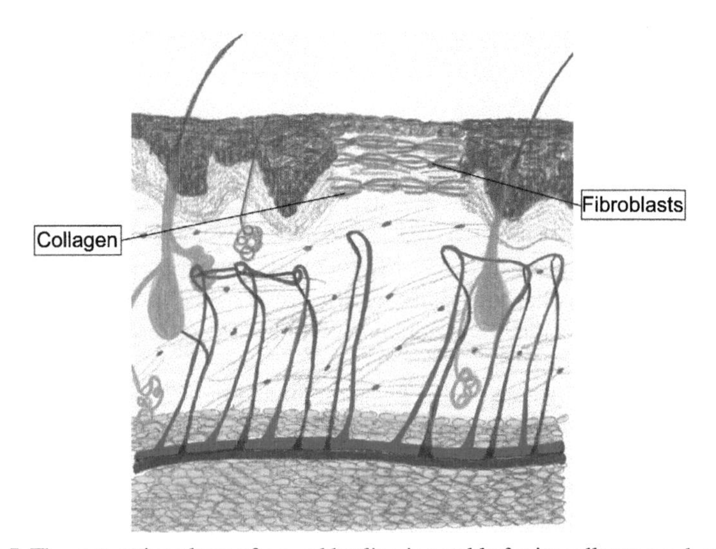

Figure 1.5 The maturation phase of wound healing is notable for its collagen production and remodeling. Compared to uninjured dermal collagen, which has a more basket weave-like pattern, scar collagen is typically thinner and arranged parallel to the skin.

The intact dermis is predominantly composed of collagen I (80–90%) and III (10–20%). In granulation tissue, the proportion of type III collagen is increased (30%), whereas in the mature scar the proportion of type III collagen is low (10%) [40–43]. The early appearance of type III collagen coincides with the presence of fibronectin; it has been proposed that the coating of denatured collagen with fibronectin facilitates its phagocytosis [37,44]. Nonetheless, the role of the early deposition of type III collagen, which does not significantly contribute to the strength of the wound, remains unclear [45,46].

Net collagen synthesis is increased for at least 4–5 weeks after wounding [25,26]. The increased rate of collagen synthesis during healing arises not only from the increased number of fibroblasts in the wound but also from a net increase in collagen production by each cell. The structure of the matrix changes with time as well. Normal dermis shows a basket weave-like pattern, while thinner collagen fibers are arranged parallel to the skin in a scar. These thinner collagen fibers gradually thicken after wounding and organize along the stress line of the wound. This is accompanied by increased scar tensile strength, which indicates a positive correlation between fiber thickness and orientation with tensile strength [47]. Biochemically, collagen derived from granulation tissue differs from collagen derived from nonwounded skin, with more hydroxylation and glycosylation of lysine residues that correlate to a thinner fiber diameter [48].

Despite a prolonged ongoing remodeling phase, the orientation of collagen fibers in the healed scar tissue does not become as organized as in the intact dermis. As a functional corollary, scar-breaking strength is always less than the strength of unwounded skin. A time course of the breaking strength shows that after 1 week the wound has only 3% and after 3 weeks 20% of its final strength [41]. After 3 months the scar has approximately 80% of the strength of unwounded skin, and no further increase occurs [49].

1.3.4 Wound contraction

Wound contraction is the approximation of the wound edges together, whereas wound contracture is the shortening of the scar itself. Healing by primary or secondary intention determines the role of wound contraction in the healing process. Several theories have been proposed for the mechanisms of wound contraction [50]. It has been postulated that a special cell, the myofibroblast, is responsible for contraction, whereas another theory suggests that the locomotion of all fibroblasts leads to a reorganization of the matrix and contraction.

The cytoskeletal structure of the myofibroblast differs from that of the normal fibroblast. Typically, the myofibroblast expresses α-smooth muscle actin in thick bundles called stress fibers that allow it to contract [51]. The α-smooth muscle actin is undetectable until day 6 and is then progressively expressed for the next 15 days of wound healing [52]. After 4 weeks, this expression fades, and the cell is believed to undergo apoptosis [53]. The appearance of the myofibroblast does not correspond perfectly with the time course of wound contraction, which starts almost immediately after wounding and continues for the next 2–3 weeks [52].

1.4 Growth factors and wound healing

The transition from the immediate postinjury chaos through the structured initial stages of hemostasis, early repair, and finally an organized stable scar is orchestrated by an ever more complex network of growth factors, cytokines, and chemokines. The immediate release of cell-signaling molecules following injury appears to trigger the inflammatory cascade with a subsequent recruitment of inflammatory type cells. Cell-mediated phagocytosis, epithelialization, granulation tissue formation, and angiogenesis are all heralded and signaled by changes in the local growth factor and cytokine milieu. With the increasing sophistication of molecular assays, a growing list of growth factors and their downstream cellular targets has been described.

Eight growth factors/growth factor families deserve special attention for their role in the wound healing process: EGF, TGF-β, FGF, VEGF, granulocyte macrophage-colony stimulating factor (GM-CSF), PDGF, connective tissue growth factor, and the wound associated proinflammatory cytokines IL family and TNF-α.

Three members of the epidermal growth factor family have roles in wound healing: EGF, heparin-binding EGF, and TGF-α [54]. These growth factors are released by a variety of cells found in the wound during early repair such as platelets, macrophages, lymphocytes, keratinocytes, fibroblasts, and vascular smooth muscle cells. Downstream effects include stimulation of keratinocyte migration during epithelialization, angiogenesis, fibroblast proliferation, and survival. The EGF family is also implicated in the induction of the early inflammatory response as well as its regulation through the generation of antiinflammatory mediators.

TGF-β is part of a large superfamily of growth factors involved in regulatory and modulatory roles during the healing process, specifically TGF-β1, TGF-β2, TGF-β3, the bone morphogenic proteins (BMP), and the activins [55]. TGF-β1 and TGF-β2, produced by platelets, macrophages, keratinocytes, and fibroblasts, serve to recruit inflammatory cells and fibroblasts and help transform fibroblasts into myofibroblasts. Interestingly, TGF-β3 inhibits scar formation and has likely a role in the counter-regulatory system that ensures appropriate but not excessive scar formation. BMPs are found throughout normal skin and influence keratinocyte differentiation. Activins, expressed by fibroblasts, endothelial cells, and keratinocytes, induce keratinocyte differentiation and stimulate fibroblasts to deposit matrix substances [56].

FGF-1 (aFGF), FGF-2 (bFGF), FGF-7 (KGF), FGF-10 (KGF-2), and FGF-22 have multiple roles in wound healing. Produced by inflammatory cells, vascular endothelial cells, fibroblasts, and keratinocytes, these factors are involved in epithelialization, angiogenesis, granulation tissue formation, and the production and remodeling of the extracellular matrix [57].

The VEGF family, composed of six known members, is primarily responsible for the angiogenesis seen during the early phases of wound healing. Not surprisingly, concentrations in normal skin are negligible, and expression is dramatically upregulated in response to hypoxia and other growth factors. VEGF is actively expressed by keratinocytes, macrophages, fibroblasts, and platelets with primarily paracrine-mediated action on endothelial cells. VEGF stimulation increases vascular permeability, basement membrane breakdown, migration of endothelial cells, and, ultimately, early vascular progenitor cell proliferation. Monocyte migration and activation as well as smooth muscle migration and proliferation have also been associated with VEGF [58].

GM-CSF is a specific cytokine increased in epidermal cells following injury. It is believed to increase the number and activity of neutrophils in the early inflammatory phase. Additionally, this cytokine is thought to upregulate keratinocyte proliferation and, via endothelial cell migration and proliferation, promote angiogenesis [55].

A four-member family, PDGF is one of the first factors expressed during wounding and remains constitutively expressed throughout the entire wound healing process. Consequently, PDGF is released by all cells involved in the wound healing process, from platelet to vascular endothelium. PDGF is chemotactic for early inflammatory cells (neutrophils and macrophages) as well as fibroblasts and smooth muscle cells. Furthermore, PDGF triggers the release of TGF-β from macrophages and an increased secretion of VEGF. PDGF has also been described as independently contributing to angiogenesis, epithelialization, and the synthesis of the extracellular matrix [54].

Connective tissue growth factor, an integrin-specific protein that is synthesized by fibroblasts, is increased following acute injury. Interestingly, the major target of this cytokine is the fibroblast itself, and this cytokine has been associated with increased granulation tissue formation, extracellular matrix formation and remodeling, and epithelialization. Knock-out studies have demonstrated that this growth factor is required for wound healing [55].

Inflammatory cytokines, responsible for the growth and differentiation of early inflammatory cells, are essential for successful wound healing. IL-1, IL-6, and TNF-α are actively secreted throughout the inflammatory phase. Produced by neutrophils, monocytes, macrophages, and keratinocytes, these factors serve to promote the early chemotaxis of neutrophils and the proliferation of keratinocytes as well as the activation of fibroblasts [55].

1.5 Acute and chronic wounds

Broadly speaking, wounds are either acute or chronic, differentiated by a lack of healing within 6 weeks.

1.5.1 Acute wounds

Acute wounds are classified into accidental wounds and surgical wounds. Acute accidental wounds can be further divided based on the mechanism of injury and/or the extent and type of tissue injury. Surgical wounds are classified as to their potential for infection into clean, clean-contaminated, contaminated, and dirty. Clean wounds are carried out into noninfected, noninflamed tissue without entry into a visceral organ and have an infection risk of 1–2% [59]. Clean-contaminated wounds are those carried out through a noninfected, noninflamed tissue plane with a controlled entry into a visceral organ without spillage or contamination of contents. Depending on the organ system entered, clean-contaminated wounds have a potential wound infection risk of up to 10%. Given the low incidence of surgical site infections (SSI), primary closure remains the recommended wound care strategy [59]. Contaminated wounds are defined as those in which unintentional hollow viscus penetration has occurred, a large break in sterile technique, incision into an area of purulence or inflammation, and operations in which gross spillage of enteric contents occurs. Contaminated wounds are associated with up to a 60% incidence of SSI, and frequently decisions on wound management are made on a case-by-case basis. Many surgeons will choose to close these wounds primarily with the express understanding that greater than 50% will require subsequent reopening and healing by secondary intention. Other surgeons are unwilling to risk SSI and allow all contaminated wounds to heal by secondary intention. One additional strategy is to allow the wound to heal by tertiary intention (delayed primary closure). In this technique, the sutures for closure are placed but not tied, and the wound remains open for the first 3–5 postoperative days. In the absence of visible infection, the sutures are tied, usually at the bedside, and the wound is allowed to heal [59]. The final category of operative wounds, dirty or infected, include wounds where an incision is made through devitalized or grossly infected tissue, wounds whereby the abdomen is entered with a perforated viscus or gross fecal contamination, or wounds that contain a foreign body. Given the variability of the contamination and bacterial burden, these wounds have a large (greater than 60%) incidence of subsequent SSI and are best closed by either secondary intention or delayed primary closure [59].

Acute accidental wounds can be categorized based on the mechanisms of occurrence and presentation. Major categories include abrasions, lacerations, and thermal injury. Despite different wound management techniques, the overall approach to wound care is similar. The fundamental goal is to foster rapid healing with an optimal cosmetic result and minimal functional impairment. To accomplish this goal, secondary complications (such as infection and hypertrophic scarring) should be minimized while optimizing the healing environment [60].

1.5.2 Chronic wounds

Chronic wounds and skin ulcers affect approximately 6 million patients in the United States and as many as 37 million globally [61]. In the United States alone, associated treatments cost as much as 10 billion dollars annually [62]. Chronic wounds are defined by the presence of a skin defect or lesion that persists longer than 6 weeks or has a frequent recurrence [62]. The prolonged inflammatory phase associated with chronic wounds involves the neutrophil-mediated destruction of extracellular matrix elements [63] that stimulate further production of proinflammatory cytokines, resulting in a cycle of sustained elevated inflammatory levels [62]. The prolonged inflammatory response favors an increased collagen deposition that often results in excessive and fibrotic scar formation [64]. Chronic wounds are characterized by less mitogenic activity [65,66], higher levels of proinflammatory cytokines [67], and elevated protease activity [68] as compared to normal wound healing. Pathologic wound healing places patients at risk for such complications as hernias, anastomotic leaks, wound dehiscence, infection, adhesions, hypertrophic scarring, and even sepsis [69]. A detailed and systematic evaluation of a nonhealing wound is therefore imperative to determine the appropriate therapeutic planning [62]. The etiology and maintenance of the chronic wound is multifactorial and involves bacterial contamination, ongoing tissue ischemia, inadequate inflammatory response, and failure of wound protein synthesis. Advanced age, immobility, and malnutrition are further contributors to this process.

1.5.2.1 Pressure ulcers

Pressure ulcers, commonly referred to as "bed sores," are the most common type of chronic wound with incidence rates in excess of 70% [70]. Associated with increased morbidity and mortality [70,71], pressure ulcers are a significant financial burden to the healthcare system [72] with the overall surgical, nursing, and medical care costs estimated as high as 5 billion dollars annually in the United States. They form after exposure to sustained mechanical loading or in combination with shear forces on tissues overlying bony prominences. Patients with the inability to reposition themselves, such as those who are paralyzed, debilitated, unconscious, or elderly, are particularly at risk [73]. These patients are further burdened by malnutrition or disuse atrophy that further increases their risk for pressure ulcer development. Patients with neurologic injuries have the added difficulty of sensory deficits that prevent them from recognizing chronic pressure.

Pressure ulcers are classified into four stages (Table 1.1) [74,75]. The most common anatomical location for pressure ulcers is the sacrum and heels, but other locations involve bony prominences such as the ischium, greater trochanters, elbows, knees, and posterior scalp [76]. Continuous pressure impairs flow through tissue microvasculature, and tissue loss begins at pressures in excess of 25 mmHg. Therapeutic and preventative measures include the identification of high-risk patients, frequent assessments, repositioning, pressure reduction, wound care, bacterial control, and surgical debridement [70,77].

1.5.2.2 Diabetic foot ulcers

In the United States, diabetic foot ulcers occur in 15–20% of all diabetics, resulting in 50,000–75,000 amputations each year and accounting for nearly 20 billion dollars in medical costs, rehabilitation expenses, and productivity loss.

Table 1.1 Staging of pressure ulcers

Stage	Description
Stage I	Erythematous, nonblanchable intact skin, usually over bony prominences
Stage II	Shallow open ulcer with partial-thickness loss of dermis
Stage III	Full-thickness tissue loss and subcutaneous fat exposure
Stage IV	Full-thickness tissue loss with exposed periosteum bone, tendon, or muscle

Contrary to notions popular in the medical community, no studies have convincingly demonstrated that diabetic foot ulcers result from obliteration of the microvasculature. Instead, tissue loss arises from decreased oxygenation caused by glycosylation of hemoglobin and erythrocyte membranes and obstructed infrapopliteal arteries. Glycosylation of hemoglobin results in an increased oxygen affinity while glycosylation of cell membranes decreases erythrocyte deformation, increases blood viscosity, and decreases flow. Simultaneously, diabetic patients develop peripheral neuropathies that render their feet insensate. The combination of undetected prolonged pressure and inadequate tissue oxygenation leads to tissue loss, most typically over the clawed metatarsal heads.

Studies show that meticulous regulation of blood glucose levels can decrease some of the end-organ complications of diabetes. Diabetic patients must be taught meticulous foot care, including daily inspection, avoidance of tight shoes, and frequent callus removal. Patients who have severe peripheral neuropathy can be treated with shoe orthotics that transfer pressure away from the metatarsal heads.

Revascularization can be helpful in some cases. Magnetic resonance angiography has become increasingly useful in identifying appropriate vessels for distal bypass. Local debridement and dressing changes can lead to healing of these wounds; however, many patients proceed to loss of digits, feet, and limbs if local measures are unsuccessful.

Only one growth factor, PDGF-BB, has been approved by FDA for the treatment of diabetic ulcers. Unfortunately, there is a noted dose-dependent increase in neoplasm risk that has hindered widespread adaptation [78]. Several studies of chronic, nonhealing wounds have demonstrated correlations with insufficient levels of FGF, VEGF modulators, EGF, and TGF-β [55]. While animal and some early phase I trials have shown promising results, these were not confirmed in phase II trials. The failure of these individual growth factors is most likely due to the coexistence of many other factors that interact with and modulate the response of other growth factors. It is not known how and when such interactions occur during the wound healing cascade. Although the concept of an exogenously applied growth factor to reverse a nonhealing process is highly attractive, this approach alone would likely not be successful. Another therapeutic approach is the use of concentrated platelet formulations that deliver multiple growth factors at superphysiological levels [79]. Nevertheless, success is more likely to be achieved by a combination of interventions that would correct the local and systemic imbalance in factors that led to impaired wound healing.

1.5.2.3 Arterial occlusive

An adequate supply of oxygen and nutrients is required for normal wound healing and is dependent on an intact vascular system. Inadequate blood supply is a major cause of lower extremity ulceration and can worsen other chronic wound etiologies [80]. Wounds caused by arterial insufficiency are typically characterized by pain and appear pale with an overlying eschar and little to no granulation tissue. The surrounding skin is generally atrophic and hairless with associated symptoms such as claudication and, in extreme cases, rest pain [81]. A histologic examination will show sparse wound granulation tissue with little neovasculogenesis [82]. Oxygen is necessary for the aerobic metabolism and energy production required for wound healing as well as for infection control [80]. Treatment focuses on limb preservation strategies of reestablishing vascular flow while maintaining adequate wound care [83].

1.5.2.4 *Venous stasis*

Up to 85% of lower extremity ulcers are caused by chronic venous insufficiency [84] and affect as much as 1% of the US population, causing substantial morbidity and associated healthcare expenditures [85,86]. Venous stasis ulcers manifest from valvular incompetence-induced venous reflux that leads to vessel wall weakness and dilation [87]. The resulting increased hypertension leads to extremity swelling, edema, and decreased transcapillary flow that blocks oxygen and nutrient delivery to the skin [77]. The gold standard for venous stasis ulcer management involves compression therapy, topical antimicrobial treatment, wound care, and vascular surgery [88].

1.5.2.5 *Lymphedema*

Though less frequent than venous stasis ulcers, ulcer development from chronic lymphedema still represents a serious complication [89]. In developed countries, lymph node resection, chronic inflammation, and obesity are the most common etiologies leading to lymphedema whereas parasite infections are more commonly seen in developing countries [90]. In tropical and subtropical regions, the World Health Organization has identified 120 million filarial lymphedema cases distributed among 81 countries [91]. Affected lymphatic vessels are typically dilated and filled with inflammatory infiltrates. The occlusion of lymphatic drainage causes interstitial fluid accumulation that eventually impedes skin perfusion [92]. While wound care in these ulcers is similar to other chronic wound management strategies, there is a specific focus on limb decongestion [90]. General conservative strategies range from therapeutic exercise and mobilization, skin hygiene, compression bandages, topical antimicrobial control, and surgical intervention in those who have failed conservative management [90,92].

1.5.2.6 *Calciphylaxis*

Calciphylaxis, or calcific uremic arteriolopathy, is a rare and dangerous condition that primarily occurs in patients with end-stage renal disease [93]. It is characterized by the calcification of small- and medium-sized blood vessels that leads to tissue necrosis [94]. It is notable for painful, purpuric, violaceous plaques that progress to nonhealing ulcers, tissue necrosis, and gangrene [93]. Skin changes typically present in areas of increased subcutaneous fat such as the breast, buttocks, and thighs.

A skin biopsy will confirm the presence of calcium deposits in the intima of affected arterioles. Associated risk factors include female gender, obesity, hyperphosphatemia, hypercalcemia, end-stage renal disease, hyperparathyroidism, and oral anticoagulation [93,94]. Due to an inadequate understanding of its pathophysiology, treatment strategies are limited but focus on correcting disturbances of calcium, phosphorus, and parathyroid hormone metabolism [93]; hemodialysis in renal failure patients; parathyroidectomy in those with hyperparathyroidism; and surgical debridement of associated ulcers [95].

1.5.2.7 Warfarin-induced skin necrosis

Skin necrosis from warfarin is an uncommon condition that occurs in 1 in 10,000 patients. Warfarin is a vitamin K antagonist that inactivates vitamin K-dependent clotting factors II, VII, IX, and X as well as proteins C and S [96]. Protein C falls more rapidly than other factors due to its shorter half-life. This causes a paradoxical hypercoagulable state that places patients at risk for microthrombi development in cutaneous and subcutaneous venules. The resulting cutaneous lesions are similar to those in calciphylaxis and appear 3–10 days following warfarin initiation. A histologic examination will show widespread thrombosis without arteriolar calcification [97].

Associated risk factors include obesity, perimenopausal age, viral infections, hepatic disease, deficiency of protein C or S, Factor V Leiden, antithrombin III deficiency, hyperhomocysteinemia, and antiphospholipid antibodies [98]. Areas of risk are those with reduced blood supply to adipose tissue such as the breasts, buttocks, abdomen, thighs, and calves.

Treatment starts with the immediate cessation of warfarin and reversal of its effects with vitamin K and fresh frozen plasma. Anticoagulation is then continued using heparin. Prevention focuses on starting heparin before warfarin treatment [99]. Wound treatment includes local wound care, ulcer debridement, and in severe cases, mastectomy or amputation [96].

1.5.2.8 Infection

Bacterial burden is thought to play a major role in impaired chronic wound healing and its associated complications. The microbial community, or microbiome, presents a microbial load, diversity, and interaction that significantly impacts a chronic wound more than any single organism alone [100]. Chronic wounds refractory to conventional antimicrobial treatment often has colonization by multiple species that results in biofilm formation. A biofilm refers to the community of microorganisms that interact synergistically to increase their pathogenicity, antimicrobial resistance, and protection. Biofilms have life cycles comprised of stages that include attachment, growth, maturation, and detachment. In the attachment stage, bacteria adhere to an environment guided by receptors on its own surface. In the growth stage, bacteria encase themselves in a carbohydrate-rich matrix that shields them from inflammatory and immune cells. Dying inflammatory cells outside of the matrix still secrete enzymes and cytokines that further drive inflammation and prolong wound healing. Biofilm development has prompted investigations into alternative interventions such as stronger and targeted systemic or topical antimicrobial therapies, local wound care, and surgical debridement [101].

1.6 Excessive scarring

Excessive scarring is considered to be an overaggressive wound healing response following cutaneous injury from surgery, trauma, burns, or inflammation [102–104]. The process of epithelialization and maturation is impaired in chronic wounds and represents a major source of abnormal scar formation [105]. Cutaneous scars are present in up to 16% of the population and particularly affect darkly pigmented individuals of African, Hispanic, or Asian descent [106,107]. Long-term sequelae of excessive scarring include physical, esthetic, and psychosocial impairments such as joint contractures, chronic pain, pruritus, anxiety, and depression [108]. Surveys have indicated that as high as 91% of respondents undergoing surgical procedures would value any improvement in scarring [109]. Some reports estimate hypertrophic scar formation in 39–86% of patients undergoing surgery and as high as 91% in burn-injured patients [106,110,111]. Excessive scarring affects all tissue types and organs, ranging from cutaneous fibroproliferative disorders like hypertrophic and keloid scarring to endogenous intra-abdominal scarring. Each scar type is characteristically different from the other and may require different treatment strategies (Fig. 1.6(a)–(e)).

Excessive collagen deposition has been implicated in the pathogenesis of scar formation. Collagen synthesis is as much as 3 times higher in hypertrophic scars and 20 times higher in keloids [112]. Regulation of collagen type I synthesis is also impaired and results in a higher ratio of type I to type III collagen in hypertrophic scars [113,114]. Increased fibroblast activity and disturbed apoptotic mechanisms in scars have further implicated the role of collagen in scar formation as well [115,116].

1.6.1 Hypertrophic scar

Hypertrophic scars are characterized by the rapid formation of excess cutaneous tissue following injury. Hypertrophic scars are characteristically elevated, erythematous, and contained within the original site of injury. Most scars appear within 4–8 weeks after initial injury, and a large percentage will gradually regress to a flat scar over a period

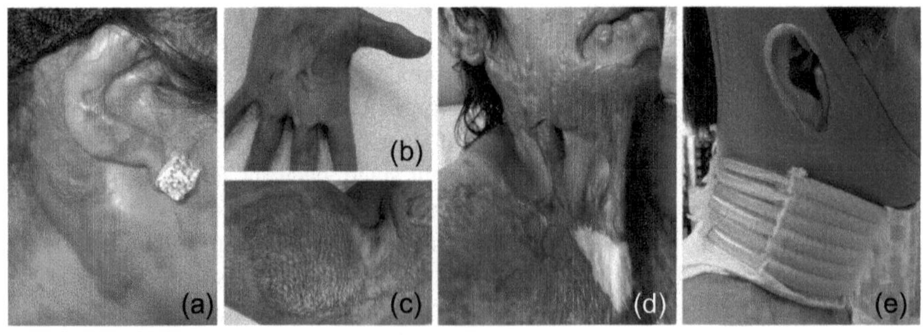

Figure 1.6 Different forms of hypertrophic scars formed following thermal injury (a–c). Significant contractures can develop that impair activities of daily living (d). Major rehabilitation and assist devices must be employed to maximize the recovery of thermally injured patients (e).

of years. This scar type often occurs following a major dermal trauma such as a burn injury as well as in areas closed under high tension, such as the neck, joint flexor surfaces, and skin creases. Histologically, it is characterized by an increased fibroblast density and activity [106] with an abundance of collagen, specifically type III, that are oriented parallel to the surface of the epithelium in wavy bundles. Hypertrophic scars also contain small nodular structures composed of myofibroblasts, small vessels, and fine collagen fibers [117].

1.6.2 Keloid

Compared to hypertrophic scars, keloid scars spread beyond the confines of a wound and can develop months to years after injury. Keloid scars present with elevated, erythematous, and shiny surfaces beyond the margins of the original wounds. There seems to be no defined relationship between the extent of the initial injury and the appearance of keloids. Interestingly, keloids are highly associated with dark skin pigmentation and have been described in high-tension areas not previously wounded such as the earlobes, upper back region, and deltoids. They rarely occur on the eyelids, genitalia, palms, or soles. Keloids are considered permanent and do not regress. Keloids similarly have increased fibroblast density and proliferation rates [106] but histologically are devoid of organized collagen bundles and instead are composed of disorganized sheets of type I and type III collagen haphazardly arranged with respect to the epithelial surface [104,117]. There is a strong familial association with the disease. Keloid formation appears to have a predilection for autosomal dominance with incomplete penetration and variable expression [118]. Surgical intervention usually leads to a recurrence of the keloid scar, often with worse results.

1.6.3 Adhesions

An intra-abdominal adhesion is the result of injury to the peritoneal lining that results in fibrin deposition, a localized inflammatory process, and ultimately intra-abdominal scar formation. In most cases, the early adhesive band is reabsorbed within 10 days. Often, however, these scars can become organized, thicker, innervated, and fibrotic. Once organized, they are unlikely to resolve. Intra-abdominal adhesions are frequently responsible for disease states, such as small bowel obstructions, and pose technical challenges for any subsequent operation.

Like scarring in any other part of the body, intra-abdominal scarring is principally mediated by inflammation, infection, ischemia, and foreign bodies. As there are no effective treatment modalities for adhesions, prevention is paramount. Careful attention must be paid while operating within the peritoneal cavity to avoid excessive tissue tension, tissue devitalization, and ischemia. Furthermore, foreign bodies prone to inflammation, such as silk sutures, should be used sparingly and judiciously. Gloves with talcum powder should be avoided; talcum powder is an effective agent for pleurodesis because of its inflammatory response. In addition, Seprafilm, a sodium hyaluronate-based bioresorbable membrane, has been approved by FDA for the prevention of intra-abdominal adhesions. Widespread adaptation of this product is limited as there is an increased incidence of anastomotic breakdown with its use [119]. Laparoscopic surgery results in less adhesion formation and its wider use should ultimately result in a decreased number of operations for adhesion-related pathology [120].

1.6.4 Pathophysiology

There is substantial evidence to suggest that scarring is associated with a prolonged inflammatory response with a greater number of inflammatory cells present. Aberrances in wound healing include increased cell migration and proliferation, cytokine synthesis and secretion, and extracellular matrix production and deposition. Prolongation of this inflammatory response in large wounds such as burn injury or wound infection exaggerates the healing response and increases the fibrogenic activity central to scar development [103]. Both keloid and hypertrophic scarring are associated with an increased and aberrant fibroblast response. Keloid fibroblasts have upregulated growth factor receptors, respond more intensely to growth factors, and a have reduced expression of downregulatory proteins [104,117]. The prolonged deficiency of extracellular matrix (ECM) degradation and regulation ultimately leads to the development of a scar [121].

1.6.5 Skin and scar evaluation

Numerous assessment scales have been developed to standardize scar evaluation. These tools range from clinical to patient observation, such as the Vancouver Scar Scale (VSS) or the Patient and Observer Scar Assessment Scale. The VSS is a clinically accepted tool used to standardize and quantify scar assessment between observers [122]. In this scale, scoring is given to clinical features seen in abnormal scars and includes skin pigmentation, vascularity, pliability, and height. Scores are greater in scars that are more hypertrophic [123]. Some variations of the VSS and other scoring systems additionally incorporate patient specific feedback such as the degree of pain or pruritus associated with a given scar [102].

The Fitzpatrick Scale was originally developed as a method to classify a patient's ability to burn and tan when exposed to UV radiation. With emphasis placed on skin and eye color, a questionnaire asks a patient to grade his or her tendency to burn and tan at 24 h and 7 days following unprotected sun exposure. The results are correlated to a six-point scale with lighter skin pigmentation comprising Fitzpatrick skin types I–IV, brown pigmented skin classified as skin type V, and black skin classified as type VI [124]. Despite its common use as a classification system, however, this score does little to predict skin behavior to cutaneous trauma or procedures. Furthermore, while the scale does incorporate the effects of UV radiation exposure into its system, it should be noted that all skin types remain susceptible to UV radiation [125].

1.6.6 Available treatment methods

There is a wide variety of invasive or noninvasive treatments for excessive or unesthetic cutaneous scarring. These include pressure and compression therapy, silicone gels and sheeting, intralesional injections, chemoradiation, laser therapy, and surgical scar correction [109,126]. Each of these modalities is not without limitations, however. For example, pressure and compression therapy is hampered by its inability to consistently deliver precise or quantifiable therapy while surgical management has reported relatively high recurrence rates and treatment delays [126]. Despite the variety of treatment modalities, there remains a paucity of evidence and no established gold standard for scar therapy [106].

1.6.6.1 Intralesional injections

Intralesional injections involve the direct application of pharmacotherapy to scars. Corticosteroid applications suppress proinflammatory mediators, cells, and ECM production involved in scar formation [127]. Corticosteroids can soften and flatten aberrant scars but are unable to narrow or completely eliminate scars [104]. Serial injections are required for a therapeutic response and treatment typically continues until scar stability or side effects such as hypopigmentation or telangiectasias arise. Another intralesional injection used is the pyrimidine analog 5-flurouracil, which targets the rapidly proliferating fibroblasts responsible for collagen production. While shown to be effective in hypertrophic scars, studies are mixed on its efficacy in keloid scars [128]. The disadvantages of injection use include the serial nature of the regimen, associated pain, signs of purpura and ulceration at scars and injection sites, skin atrophy, hypopigmentation, and even skin necrosis [129].

1.6.6.2 Silicone gel

Silicone sheets were introduced in the early 1980s as a method for improving scar appearance. It is thought that the occlusive therapy affects local keratinocytes and alters the growth factor secretion involved in fibroblast regulation [104,130]. They additionally decrease capillary permeability, inflammatory mediators, and collagen synthesis. Application is recommended as soon as epithelialization occurs with daily applications of at least a 12-h duration for a minimum of 2 months to prevent rebound hypertrophy [131].

1.6.6.3 Radiation

Radiation for the treatment of keloids was first described in 1906 [104] and is suggested to directly affect fibroblast proliferation and induce apoptosis. The total recommended dosage for the treatment of keloids varies between 15 and 20 Gy fractionated over 5–6 treatments [128]. Drawbacks include hyperpigmentation at the radiated site and a risk of radiation-induced malignancy in potentially high carcinogenic areas, such as the breast and thyroid, and radiation is contraindicated in children [104]. Furthermore, it has variable results with rates of 10–100% recurrence when used alone.

1.6.6.4 Laser therapy

Laser revision has existed for over 30 years and has grown considerably in popularity. Stimulating tissue remodeling by producing microscopic patterns of thermal injury in the dermis, laser therapy is used as a tool to improve the tone, texture, and pigmentation of skin [126]. Symptom improvements include a reduction in erythema, pruritus, and pain as well as skin flattening with improved skin textures [132]. Laser types range from carbon dioxide to the pulsed dye and have been tested in the setting of hypertrophic and keloid scars with varying results. Laser therapy is thought to work by altering the signaling pathways that favor collagen degradation and fibroblast apoptosis [133,134]. Current laser-based methods of scar management can be grouped as ablative or nonablative as well as fractionated and nonfractionated [135,136]. Ablative lasers typically produce dramatic outcomes due to their mechanism of direct tissue vaporization. Nonablative lasers in turn are more moderate in improvement while minimizing the possible side effects of laser usage. Fractional therapy generates columns of thermally denatured skin of controlled width and depth to allow for significant clinical results with minimal post-treatment recovery [137].

1.6.6.5 Pressure therapy

The use of pressure therapy in scars was first described in 1835 [127]. However, it was not popularized until the 1970s when physicians noted improved scars following lower extremity pressure garment use on healing burns [138]. Pressure therapy is thought to work by decreasing the blood flow to the scar, resulting in collagenase-mediated collagen breakdown, hypoxia-induced fibroblast and collagen degradation, and decreased scar hydration [104,127]. Pressure therapy aids collagen maturation, flattens scars, and improves thinning and pliability. While studies have examined its effects on components of scar formation, such as myofibroblasts [139], matrix metalloproteinases [140], and TNF-α [141], a full understanding of its pathophysiology remains limited. The lack of standardized protocols and a validated animal model have further stymied efforts to understand scar behavior in pressure treatment [142,143]. Pressure therapy is recommended to start immediately after complete wound epithelialization with applications of 8–24 h in length per day to maximize success rates and prevent rebound hypertrophy. A pressure between 24 and 30 mm Hg must be achieved to exceed capillary pressures but preserve peripheral blood circulation. Success is largely dependent on patient compliance [104,128].

1.6.6.6 Surgery

Scar revision as a treatment focuses on two goals: excision and narrowing of scars or changing the direction of scars [129]. Z-plasty focuses on the extension of a reduced distance and turns the main axis of the scar parallel to skin creases [104]. Z-plasty is typically used in patients with hypertrophic scars crossing joints or wrinkle creases at a right angle. The technique brings the new scar in line with relaxed skin tension lines to improve healing [129]. W-plasty causes scar disruption and allows every other scar shank to be positioned within the skin creases [104]. W-plasty is optimal for the correction of facial scars.

Hypertrophic scars are more amenable to surgical correction. In comparison, the total excision of keloids stimulates additional collagen synthesis that may cause a rapid recurrence of keloids larger than their predecessors [104]. The recurrence rates of keloid surgical resections are high and range from 45% to 100% [144]. As a result, a combination adjuvant treatment with other modalities such as pressure, corticosteroids, and radiotherapy has been advocated [104].

1.7 Burns

Approximately 3.5 million people are burned annually in the United States [145]. Thermal wounds are among the most devastating of all traumatic injuries [146–148] and are the fourth most common type of trauma worldwide with the most vulnerable groups being children, women, and the elderly [149]. Up to 67% of individuals surviving a burn injury will develop hypertrophic scars [150] that can cause serious physical impairments, body dysmorphic issues, and social reintegration difficulties [151,152]. Scars formed after a burn injury can present with concomitant pruritus, pain [153], and, in severe scar contracture, significantly impaired activities of daily living [154].

Advances in surgical technique, critical care, and antimicrobial therapy have decreased the mortality seen in severe burn injuries [155,156]. The management of burn injuries requires a multidisciplinary care team and focused resources to allow for a lengthy treatment ranging from ambulatory to intensive-care therapies [157]. This has led to the establishment of specialized burn centers that focus on the continued care of the burned population [149]. The American Burn Association has emphasized referrals to verified burn centers to ensure that patients receive established standards of care [158,159].

1.7.1 Burn severity

The French surgeon Ambroise Pare was the first to describe burns by different degrees in the 1500s, and in 1832, Dupuytren expanded this classification into six severities based on depth [160,161]. Burn severity is still estimated based on a clinical assessment of the involved skin depth and is now categorized as superficial, partial, or full-thickness [162,163]. The assessment of burn depth includes the evaluation of parameters such as wound appearance, capillary refill, and sensation [164]. The surface area and depth of the burn injury are key determinants in patient mortality and morbidity. This is because the keratinocyte reservoirs and growth factor-producing dermal macrophages and fibroblasts that drive successful wound healing are located in the papillary dermis (Fig. 1.7) [165].

1.7.1.1 Superficial burns

Superficial, or first-degree, burns affect only the epidermis and have no dermal tissue loss. These burns are characterized by pain and erythema that does not blister. This usually resolves in 3–5 days without scarring [165]. The most common manifestation of a first-degree burn is a sunburn caused by long exposure to the sun [163]. These burns typically require little management beyond supportive care and observation.

1.7.1.2 Superficial partial-thickness burns

The term "superficial partial-thickness" is synonymous with "superficial second degree" and describes burns that involve the epidermis and the papillary dermis. These burns

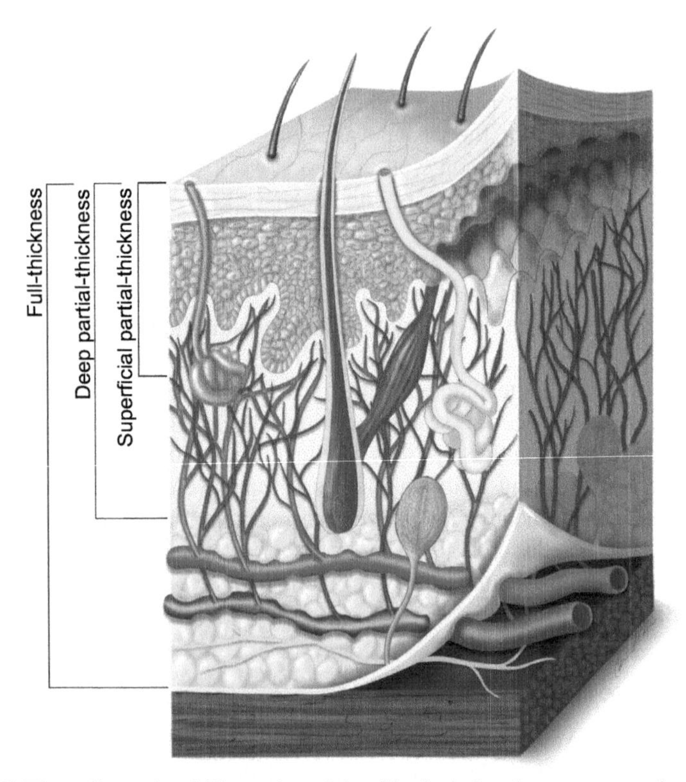

Figure 1.7 Three-dimensional illustration of the skin depicting the anatomy and associated skin structures affected by the thickness of the thermal injury. Deeper burns affect more skin structures and result in worse outcomes.
Adapted from Shupp JW, et al. A review of the local pathophysiologic bases of burn wound progression. J Burn Care Res 2010;31(6):849–73.

are more painful and involve the formation of blisters with copious exudate production [163]. On examination, the dermis blanches and rebounds with a brisk capillary refill [165]. Since the remainder of the dermis is not involved, these burn wounds heal within 14 days without surgical intervention and usually have minimal to no scarring [164].

1.7.1.3 Deep partial-thickness burns

Deep dermal, or deep second degree, burns involve skin layers beyond the papillary dermis into the reticular dermis [163]. Burn wounds of this severity have derangements in sensation due to the affected nervous tissue in the deeper dermal layers. The dermis appears paler with an increased or nonexistent capillary refill [165]. Hypertrophic scar formation is more likely at this depth with the severity of scarring thought to be dependent on the length of time allowed for wound closure. If a deep partial-thickness burn takes longer than 2 weeks to heal following injury, the likelihood of scarring increases significantly [164]. Burns of this size are clinically relevant such that they typically require excision and skin grafting.

1.7.1.4 Full-thickness burns

Full-thickness, or third degree, burns extend through all layers of the skin and involve the underlying subcutaneous tissue [163]. These burns are characterized by a complete anesthesia of the involved skin as well as a leathery, nonblanching appearance. This depth of burn requires aggressive intervention that necessitates early surgical excision and grafting to optimize outcomes [164]. The dermal appendages that house reservoirs of progenitor cells are destroyed, eliminating keratinocyte reservoirs, cytokine-producing macrophages, and extracellular matrix-producing fibroblasts [166,167]. Circumferential wounds with a rigid eschar can produce a tourniquet effect that compromises the venous outflow and eventually the arterial inflow. This can occur in extremities and lead to compartment syndrome, but abdominal and thoracic compartment syndromes may also occur. Early excision with subsequent grafting is critical to this depth of burn and has resulted in both improved survival as well as decreased morbidity, hospital length of stay, and associated hospital expenditures [168].

1.7.1.5 Jackson zones and burn depth progression

In 1947, Jackson originally described three concentric areas of tissue injury following a burn injury [165,169,170]. The *zone of hyperemia* is the outermost area of injured tissue characterized by increased perfusion that will eventually recover unless a significant insult, such as severe sepsis or prolonged hypoperfusion, influences it otherwise [169]. The next concentric area, the *zone of stasis*, describes the intermediate, potentially salvageable tissue that can be influenced by insults such as prolonged hypotension, infection, or edema into unsalvageable necrotic tissue. Management strategies aim to maximize resuscitation and tissue perfusion to salvage as much tissue as possible. In the central *zone of coagulation*, irreversible tissue loss and protein denaturation results in tissue necrosis. The white tissue in this zone represents the most direct damage from thermal trauma with a complete destruction of the subpapillary vasculature seen microscopically [165].

The conversion of burn wound depth and surface area remains a central tenet of burn-wound management. As Jackson realized in 1969 when he observed the ability of the zone of stasis to accept autografts and bleed upon excision, certain factors can prevent or contribute to the progression of tissue destruction following a burn injury [171]. Several studies have looked at the mechanisms of cell death in the vicinity of the burn wound. Gravante et al. [172] suggested the role of apoptosis in burn-wound progression when they found higher rates of apoptotic dermal cells in deep partial-thickness burns as compared to normal skin and other burn depths. They additionally noted that evidence of apoptosis continued for at least 23 days after the initial burn insult [173]. Singer et al. [174] reported on the additional contribution of oncosis with apoptosis in the zone of stasis.

Other factors contribute to burn-wound progression. The prolonged inflammatory state following a burn injury can lead to collagen degradation, keratinocyte apoptosis, and impaired neutrophil apoptosis [175–177]. Adherent neutrophils further exacerbate damage through cell degranulation, free radical production, and venous congestion [178]. Bacterial infection and endotoxin production increase burn-wound infiltrates and prolong the inflammatory response. Topical antibiotic use has demonstrated decreased rates of burn-wound progression [179]. Massive fluid shifts from the intravascular space to interstitium also contribute due to wound edema that increases tissue pressure and impairs circulation [180]. Thrombus formation [181] and a concomitant reduction in plasmin-mediated fibrinolysis [182,183] additionally impairs perfusion and risks further burn-wound progression.

1.7.2 Types of burns

1.7.2.1 Scald

Scald injuries occur when a hot liquid contacts the skin and causes thermal injury. A significant portion of scald injuries are experienced by children or the elderly. The most common mechanism is due to spillage of hot drinks or exposure to hot bathing water. Scald injuries typically cause either first or superficial second-degree dermal burns [169].

1.7.2.2 Grease

Grease burns are a common cause of injury in both the home and workplace. Due to the viscosity, high boiling point, and lower specific heat of oil, grease burns are typically deeper than scald [184]. Patient populations commonly affected include young females and children and usually occur in the home related to attempts to extinguish a frying pan fire. A majority of grease fire-injured patients require excision and grafting [185]. Due to operative requirements and prolonged hospital stays, education and prevention are advocated, specifically as they pertain to proper methods of extinguishing grease fires [186].

1.7.2.3 Contact

Contact injuries are caused when the skin is touched by an extremely hot object or from abnormally long contact with an object. Patients with an altered mental status, as in substance abuse, or mobility-challenged, like the elderly, typically experience prolonged contact injuries. Brief contact burns from extremely hot objects are usually due to industrial accidents. Contact burns are typically deep second-degree or third-degree burns [169].

1.7.2.4 Flame

Flame injuries comprise the majority of all adult burns and are associated with concomitant morbidities such as related trauma injury or inhalation injury. Flame injuries tend to be either deep secondary or third degree in depth [169].

1.7.2.5 Electrical contact

Electrical injuries represent 3–4% of burn unit admissions and are caused by the passage of electrical current through the body. Mortality is reported between 3% and 15%

with the most common cause of death being cardiac arrest after acute arrhythmias. Those affected are primarily young, working men such as power company linemen or electricians [187]. Electrical injuries usually present with no signs other than a small eschar at the entry and exit points in which electricity traveled. The amount of heat generated is proportional to the associated voltage at the time of injury. Low-voltage injuries cause small, deep contact burns at the entry and exit sites and can interfere with the cardiac cycle. Voltages in excess of 1000 V can cause extensive tissue damage, necrosis, or even limb loss. Associated muscle damage can cause a rhabdomyolysis that leads to renal damage. These injuries require aggressive resuscitation, debridement, and close observation in a critical care setting. The risk of mortality increases greatly in electrical injuries in excess of 70,000 V [169].

1.7.2.6 Electrothermal (arc)

Arc or "flash" injuries occur when exposed skin is injured by the heat generated from an arc of current traveling from a voltage source. This commonly occurs in electrical or converter boxes and, as a result, the face and hands are typically involved in this injury. Exposed clothing is similarly at risk and can result in deeper burns if caught on fire. No current passes through the patient's body with this injury [169].

1.7.2.7 Chemical

Chemical injuries are usually the result of industrial accidents but can occur by exposure to household chemical products or even acts of terrorism. Contact with skin by these chemical agents causes coagulative necrosis and associated deep tissue injury until removed completely. Alkali burns (cement) tend to penetrate deeper and cause worse burns than acids. Management includes the immediate removal of affected skin and irrigation of all involved areas [169].

1.8 Animal models

Understanding the mechanisms of tissue repair can maximize therapeutic interventions that promote wound healing and minimize scarring. Studying human tissue can be impractical due to factors such as the timing, depth, and location of injury [188]. Animal models provide the opportunity to examine factors that influence healing in a complex biologic environment [189].

Developing a reproducible wound comparable to human wound healing is important to the study of wound physiology. Every animal is different and must be appropriately considered for the type of wound to be studied. For example, despite the large number of publications using mice for wound healing study, human skin is thicker, relatively hairless, and uses less contraction to heal as compared to mice [189]. Rodents in particular heal greater than 90% of their wounds by contraction, and this must be considered when used for wound healing study. Larger animals more closely resemble human skin [189]. Studies of hypertrophic scars in red Duroc swine have increasingly documented similarities to human wound healing and scar formation at the molecular, cellular, and gross levels [188,190]. Duroc swine demonstrate a sparse hair coat, thick epidermis, well-differentiated dermal papillary body, and abundant elastic and adipose tissue similar to human skin [188]. Additional advantages include their comparable size to adult humans, the ability to create larger wounds, and flatter surfaces. It must be noted that patients typically present with comorbidities such as malnutrition, infection, diabetes, or even steroid use that further impair wound healing. As a result, no single model can fully correlate to human physiology [191].

1.8.1 Injury types

1.8.1.1 Granuloma models

Granuloma models are designed to test the ability to induce an inflammatory response and wall off foreign material. Devices that collect cells, fluid, and ECM are placed into subcutaneous tissue and analyzed over time. These models can then study the degree of inflammation and collagen deposition that occurs in response to a treatment or stimuli [192]. For example, collagen deposition can be determined by the hydroxyproline level assay of fluid collected from this device [37,189].

1.8.1.2 Incision models

Incision models are used to test the strength of healing incisions over time as well as agents that stimulate or impair incisional healing [49,193]. An incision made through the skin is closed with sutures or staples or is allowed to heal spontaneously and is then examined by a tensiometer to determine the biomechanical strength over time. This model determines the breaking strength, or the force required to break skin, as well as the tensile strength, which is the breaking strength per area of incision [191]. A "bursting strength," designed to simulate increased intra-abdominal pressure involved in wound dehiscence, can also be determined when a balloon placed inside a peritoneal cavity is inflated until disruption of the incision occurs. Other information, such as incisional extensibility or collagen cross-linking, can also be examined in this model [194].

1.8.1.3 Open-wound models

In this model, an open wound is created and its rate of closure is studied in the context of agents or environments that may increase or impair wound healing [195]. Additional insights into wound healing processes, such as granulation tissue formation, collagen deposition, epithelialization rates, wound contraction, and hypertrophic scar formation, can also be studied at specific wound depths. For example, partial-thickness wounds in porcine models have been used to examine the dermal response, epithelialization rates, and subsequent hypertrophic scar formation that occurs following wound creation [196,197]. Full-thickness wound models further examine tissue repair, skin graft usage, and the success of dermal skin substitutes [198]. Techniques such as basic histology, immunohistochemistry, and in situ hybridization are used to assess wounds. Molecular biologic analysis such as mRNA, protein, and apoptosis assays can also be easily performed [199].

1.8.1.4 Burn models

Models examining the influence of a burn injury on wound healing and its associated systemic response are numerous. Burn depth progression has been studied using parallel heated metal plates to burn skin and examine the progression of burn depth in surrounding skin over time. Individual characteristics shown to influence the degree of burn include both the temperature and duration of the contacting agent, the thickness of the skin, and the adequacy of skin perfusion [191]. Smaller, partial-thickness burns can be created using lower temperatures and shorter durations via scald or contact burns. To create and study larger burns, animals are dipped into a hot or inflammable liquid while a template is used to limit the size of injury.

1.8.2 Impaired healing models

1.8.2.1 Malnutrition

Malnutrition is a major factor that influences tissue repair [200,201]. Models examine either protein only or combination protein and caloric deficiencies. Protein only malnutrition is implemented using calorie adequate but protein poor diets while combination deficiency studies limit overall food intake [191]. Both wound healing and breaking strength have been demonstrated to be proportionally reduced in these models [202]. In addition to malnutrition, other studies have focused on specific nutritional deficiencies such as fat-soluble vitamins, trace elements, eg, zinc, and amino acids [191].

1.8.2.2 Infection

Initial studies of bacteria interestingly exhibited accelerated tissue repair [203,204]. Studies showed that a mild inoculum of bacteria could stimulate an inflammatory response and actually accelerate tissue repair [205]. The introduction of large, virulent organisms in contrast causes a massive production of destructive bacterial enzymes that impair tissue repair [191]. Infections at sites other than the wound have also been shown to alter wound healing [206]. Due to the increased burden infection brings to these animals and the possibility of anorexia, attention must be paid to nutritional intake [191].

1.8.2.3 Ischemia

The inadequate delivery of blood is a known factor in impaired wound healing, particularly in those with lower extremity ulcers [80], and clinically relevant animals have been sought as a result. In one model, flaps of skin are created on the backs of rodents with the goal of longer flaps having less perfusion. The healing of these flaps is tested for incisional strength [207]. Variations on this technique include clipping vessels to reduce the blood supply or using bipedical or musculocutaneous flaps

[191]. Another model uses the skin of rabbit ears to create a wound at the level of the cartilage. As the cartilage is relatively avascular, wound closure is delayed, allowing for the examination of granulation tissue formation and epithelialization rates over time [208].

1.9 Conclusion

Wound healing is a complex physiologic process impacted by behavioral, environmental, and injury types. There is a wide variety of treatment modalities available for all wound types and sizes. Animal and wound models serve to further understand wound healing behavior while new and improved treatment modalities are sought.

Conflict of interest

The authors report no proprietary or commercial interest in any product mentioned or concept discussed in this chapter.

References

[1] Simpson DM, Ross R. The neutrophilic leukocyte in wound repair. A study with antineutrophil serum. J Clin Invest 1972;51(8):2009–23.
[2] Leibovich SJ, Ross R. The role of the macrophage in wound repair. A study with hydrocortisone and antimacrophage serum. Am J Pathol 1975;78(1):71–100.
[3] Barbul A, et al. The effect of in vivo T helper and T suppressor lymphocyte depletion on wound healing. Ann Surg 1989;209(4):479–83.
[4] Mirza R, et al. Selective and specific macrophage ablation is detrimental to wound healing in mice. Am J Pathol 2009;75(6):2454–62.
[5] Polverini PJ, et al. Activated macrophages induce vascular proliferation. Nature 1977; 269(5631):804–6.
[6] Malawista SE, Montgomery RR, van Blaricom G. Evidence for reactive nitrogen intermediates in killing of staphylococci by human neutrophil cytoplasts. A new microbicidal pathway for polymorphonuclear leukocytes. J Clin Invest 1992;90(2):631–6.
[7] Schäffer MR, et al. Nitric oxide, an autocrine regulator of wound fibroblast synthetic function. J Immunol 1997;158(5):2375–81.
[8] Brancato SK, Albina JE. Wound macrophages as key regulators of repair: origin, phenotype, and function. Am J Pathol 2011;178(1):19–25.
[9] Fahey 3rd TJ, et al. Diabetes impairs the late inflammatory response to wound healing. J Surg Res 1991;50(4):308–13.
[10] Ausprunk DH, Folkman J. Migration and proliferation of endothelial cells in preformed and newly formed blood vessels during tumor angiogenesis. Microvasc Res 1977;14(1):53–65.
[11] Burger PC, Chandler DB, Klintworth GK. Corneal neovascularization as studied by scanning electron microscopy of vascular casts. Lab Invest 1983;48(2):169–80.

[12] Knighton DR, Silver IA, Hunt TK. Regulation of wound-healing angiogenesis-effect of oxygen gradients and inspired oxygen concentration. Surgery 1981;90(2):262–70.
[13] Harlan JM. Consequences of leukocyte-vessel wall interactions in inflammatory and immune reactions. Semin Thromb Hemost 1987;13(4):434–44.
[14] Remensnyder JP, Majno G. Oxygen gradients in healing wounds. Am J Pathol 1968; 52(2):301–23.
[15] Imre G. Studies on the mechanism of retinal neovascularization. Role of lactic acid. Br J Ophthalmol 1964;48:75–82.
[16] Zauberman H, et al. Stimulation of neovascularization of the cornea by biogenic amines. Exp Eye Res 1969;8(1):77–83.
[17] Schreiber AB, Winkler ME, Derynck R. Transforming growth factor-alpha: a more potent angiogenic mediator than epidermal growth factor. Science 1986;232(4755):1250–3.
[18] Lynch SE, Colvin RB, Antoniades HN. Growth factors in wound healing. Single and synergistic effects on partial thickness porcine skin wounds. J Clin Invest 1989;84(2):640–6.
[19] Gospodarowicz D, et al. Corpus luteum angiogenic factor is related to fibroblast growth factor. Endocrinology 1985;117(6):2383–91.
[20] Gospodarowicz D, et al. Structural characterization and biological functions of fibroblast growth factor. Endocr Rev 1987;8(2):95–114.
[21] Ross R, Benditt EP. Wound healing and collagen formation: I. Sequential changes in components of guinea pig skin wounds observed in the electron microscope. J Biophys Biochem Cytol 1961;11:677–700.
[22] Ågren MS, Steenfos HH, Dabelsteen S, Hansen JB, Dabelsteen E. Proliferation and mitogenic response to PDGF-BB of fibroblasts isolated from venous leg ulcers is ulcer-age-dependent. J Invest Dermatol 1999;112(4):463–9.
[23] Clark RAF. Overview and general considerations of wound repair. In: The molecular and cellular biology of wound repair. New York: Plenum Press; 1988. p. 3–50.
[24] Ross R, Everett NB, Tyler R. Wound healing and collagen formation: VI. The origin of the wound fibroblast studied in parabiosis. J Cell Biol 1970;44(3):645–54.
[25] Diegelmann RF, Rothkopf LC, Cohen IK. Measurement of collagen biosynthesis during wound healing. J Surg Res 1975;19(4):239–43.
[26] Madden JW, Peacock Jr EE. Studies on the biology of collagen during wound healing. I. Rate of collagen synthesis and deposition in cutaneous wounds of the rat. Surgery 1968;64(1):288–94.
[27] Blakaj A, Bucala R. Fibrocytes in health and disease. Fibrogen Tissue Repair 2012; 5(Suppl. 1):S6.
[28] Winter GD. Formation of the scab and the rate of epithelization of superficial wounds in the skin of the young domestic pig. Nature 1962;193:293–4.
[29] Usui ML, et al. Morphological evidence for the role of suprabasal keratinocytes in wound reepithelialization. Wound Repair Regen 2005;13(5):468–79.
[30] Johnson FR, McMinn RMH. The cytology of wound healing of body surfaces in mammals. Biol Rev Camb Philos Soc 1960;35(3):364–410.
[31] Woodley DT, Bachmann PM, O'Keefe EJ. The role of matrix components in human keratinocyte re-epithelialization. Prog Clin Biol Res 1991;365:129–40.
[32] Lynch SE. Interactions of growth factors in tissue repair. Prog Clin Biol Res 1991; 365:341–57.
[33] Korn JH. Fibroblast prostaglandin E2 synthesis. Persistence of an abnormal phenotype after short-term exposure to mononuclear cell products. J Clin Invest 1983;71(5):1240–6.
[34] Regan MC, et al. The wound environment as a regulator of fibroblast phenotype. J Surg Res 1991;50(5):442–8.

[35] Vande Berg JS, Rudolph R, Woodward M. Comparative growth dynamics and morphology between cultured myofibroblasts from granulating wounds and dermal fibroblasts. Am J Pathol 1984;114(2):187–200.

[36] Albina JE, et al. Temporal expression of different pathways of 1-arginine metabolism in healing wounds. J Immunol 1990;144(10):3877–80.

[37] Kurkinen M, et al. Sequential appearance of fibronectin and collagen in experimental granulation tissue. Lab Invest 1980;43(1):47–51.

[38] Reed MJ, et al. Differential expression of SPARC and thrombospondin 1 in wound repair: immunolocalization and in situ hybridization. J Histochem Cytochem 1993; 41(10):1467–77.

[39] Bentley JP. Rate of chondroitin sulfate formation in wound healing. Ann Surg 1967; 165(2):186–91.

[40] Bailey AJ, et al. Collagen polymorphism in experimental granulation tissue. Biochem Biophys Res Commun 1975;66(4):1160–5.

[41] Ehrlich HP, Krummel TM. Regulation of wound healing from a connective tissue perspective. Wound Repair Regen 1996;4(2):203–10.

[42] Miller EJ. Biochemical characteristics and biological significance of the genetically-distinct collagens. Mol Cell Biochem 1976;13(3):165–92.

[43] Nyström A. Collagens in wound healing. In: Ågren MS, editor. Wound healing biomaterials, vol. 2. Cambridge, UK: Woodhead Publishing; 2016, p. 171–201.

[44] Grinnell F, Billingham RE, Burgess L. Distribution of fibronectin during wound healing in vivo. J Invest Dermatol 1981;76(3):181–9.

[45] Clore JN, Cohen IK, Diegelmann RF. Quantitation of collagen types I and III during wound healing in rat skin. Proc Soc Exp Biol Med 1979;161(3):337–40.

[46] Gay S, et al. Collagen types in early phases of wound healing in children. Acta Chir Scand 1978;144(4):205–11.

[47] Doillon CJ, et al. Collagen fiber formation in repair tissue: development of strength and toughness. Coll Relat Res 1985;5(6):481–92.

[48] Forrest L. Current concepts in soft connective tissue wound healing. Br J Surg 1983; 70(3):133–40.

[49] Levenson SM, et al. The healing of rat skin wounds. Ann Surg 1965;161:293–308.

[50] Rudolph R, et al. Wound contraction and scar contracture. In: Cohen IK, Diegelmann RF, Lindblad WS, editors. Wound healing: biochemical & clinical aspects. Philadelphia: W.B. Saunders Co.; 1992, p. 96–114.

[51] Schmitt-Gräff A, et al. Heterogeneity of myofibroblast phenotypic features: an example of fibroblastic cell plasticity. Vichows Archiv 1994;425(1):3–24.

[52] Darby I, Skalli O, Gabbiani G. Alpha-smooth muscle actin is transiently expressed by myofibroblasts during experimental wound healing. Lab Invest 1990;63(1):21–9.

[53] Desmoulière A, et al. Apoptosis mediates the decrease in cellularity during the transition between granulation tissue and scar. Am J Pathol 1995;146(1):56.

[54] Werner S, Grose R. Regulation of wound healing by growth factors and cytokines. Physiol Rev 2003;83(3):835–70.

[55] Barrientos S, et al. Growth factors and cytokines in wound healing. Wound Repair Regen 2008;16(5):585–601.

[56] McLean CA, et al. Temporal expression of activin in acute burn wounds–from inflammatory cells to fibroblasts. Burns 2008;34(1):50–5.

[57] Singer AJ, Clark RA. Cutaneous wound healing. N Engl J Med 1999;341(10):738–46.

[58] Kumar I, et al. Angiogenesis, vascular endothelial growth factor and its receptors in human surgical wounds. Br J Surg 2009;96(12):1484–91.

[59] Reichman DE, Greenberg JA. Reducing surgical site infections: a review. Rev Obstet Gynecol 2009;2(4):212–21.
[60] Singer AJ, Hollander JE, Quinn JV. Evaluation and management of traumatic lacerations. N Engl J Med 1997;337(16):1142–8.
[61] Kalorama and Information. World Wound Care Markets 2008. May 2008. New York.
[62] Wild T, et al. Basics in nutrition and wound healing. Nutrition 2010;26(9):862–6.
[63] Diegelmann RF. Excessive neutrophils characterize chronic pressure ulcers. Wound Repair Regen 2003;11(6):490–5.
[64] Mast BA, Schultz GS. Interactions of cytokines, growth factors, and proteases in acute and chronic wounds. Wound Repair Regen 1996;4(4):411–20.
[65] Bucalo B, Eaglstein WH, Falanga V. Inhibition of cell proliferation by chronic wound fluid. Wound Repair Regen 1993;1(3):181–6.
[66] Zillmer R, Trøstrup H, Karlsmark T, Ifversen P, Ågren MS. Duration of wound fluid secretion from chronic venous leg ulcers is critical for interleukin-1α, interleukin-1β, interleukin-8 levels and fibroblast activation. Arch Dermatol Res 2011;303(8):601–6.
[67] Yager DR, Nwomeh BC. The proteolytic environment of chronic wounds. Wound Repair Regen 1999;7(6):433–41.
[68] Yager DR, et al. Wound fluids from human pressure ulcers contain elevated matrix metalloproteinase levels and activity compared to surgical wound fluids. J Invest Dermatol 1996;107(5):743–8.
[69] Goldberg SR, Diegelmann RF. Wound healing primer. Crit Care Nurs Clin North Am 2012;24(2):165–78.
[70] Moore ZE, Cowman S. Repositioning for treating pressure ulcers. Cochrane Database Syst Rev 2015;1:CD006898.
[71] Kroger K, et al. Prevalence of pressure ulcers in hospitalized patients in Germany in 2005: data from the Federal Statistical Office. Gerontology 2009;55(3):281–7.
[72] Bennett G, Dealey C, Posnett J. The cost of pressure ulcers in the UK. Age Ageing 2004;33(3):230–5.
[73] Moore Z, Cowman S, Conroy RM. A randomised controlled clinical trial of repositioning, using the 30 degrees tilt, for the prevention of pressure ulcers. J Clin Nurs 2011;20(17-18):2633–44.
[74] Tew C, et al. Recurring pressure ulcers: identifying the definitions. A National Pressure Ulcer Advisory Panel white paper. Wound Repair Regen 2014;22(3):301–4.
[75] Pieper B, Langemo D, Cuddigan J. Pressure ulcer pain: a systematic literature review and national pressure ulcer advisory panel white paper. Ostomy Wound Manag 2009; 55(2):16–31.
[76] Gallagher P, et al. Prevalence of pressure ulcers in three university teaching hospitals in Ireland. J Tissue Viability 2008;17(4):103–9.
[77] Eaglstein WH, Falanga V, Barbul A. Wound healing. Philadelphia: W.B. Saunders Co.; 1997. p. 689–700.
[78] Wieman TJ, Smiell JM, Su Y. Efficacy and safety of a topical gel formulation of recombinant human platelet-derived growth factor-BB (becaplermin) in patients with chronic neuropathic diabetic ulcers. A phase III randomized placebo-controlled double-blind study. Diabetes Care 1998;21(5):822–7.
[79] Ågren MS, Rasmussen K, Pakkenberg B, Jørgensen B. Growth factor and proteinase profile of Vivostat® platelet-rich fibrin linked to tissue repair. Vox Sang 2014;107(1):37–43.
[80] Zhang Z, Lv L, Guan S. Wound bed preparation for ischemic diabetic foot ulcer. Int J Clin Exp Med 2015;8(1):897–903.

[81] Lawrence TW, et al. Clinical management of nonhealing wounds. In: Cohen IK, Diegelmann RF, Lindblad WS, editors. Wound healing: biochemical & clinical aspects. Philadelphia: W.B. Saunders Co.; 1992, p. 541–61.
[82] Nunan R, Harding KG, Martin P. Clinical challenges of chronic wounds: searching for an optimal animal model to recapitulate their complexity. Dis Model Mech 2014; 7(11):1205–13.
[83] Setacci C, et al. Chapter IV: Treatment of critical limb ischaemia. Eur J Vasc Endovasc Surg 2011;42(Suppl. 2):S43–59.
[84] Misciali C, et al. Vascular leg ulcers: histopathologic study of 293 patients. Am J Dermatopathol 2014;36(12):977–83.
[85] Crenshaw BH, Roberson KY, Stevermer JJ. PURLs: A new adjunctive Tx option for venous stasis ulcers. J Fam Pract 2015;64(3):182–4.
[86] Collins L, Seraj S. Diagnosis and treatment of venous ulcers. Am Fam Physician 2010;81(8):989–96.
[87] Meissner MH, et al. Primary chronic venous disorders. J Vasc Surg 2007;46 (Suppl. S): 54S–67S.
[88] Valencia IC, et al. Chronic venous insufficiency and venous leg ulceration. J Am Acad Dermatol 2001;44(3):401–21; quiz 422–4.
[89] Moffatt CJ, et al. Prevalence of leg ulceration in a London population. QJM 2004; 97(7):431–7.
[90] Lin CT, Ou KW, Chang SC. Diabetic foot ulcers combination with lower limb lymphedema treated by staged charles procedure: case report and literature review. Pak J Med Sci 2013;29(4):1062–4.
[91] Moffatt CJ, et al. Lymphoedema: an underestimated health problem. QJM 2003; 96(10):731–8.
[92] Karnasula VM. Management of ulcers in lymphoedematous limbs. Indian J Plast Surg 2012;45(2):261–5.
[93] Sprague SM. Painful skin ulcers in a hemodialysis patient. Clin J Am Soc Nephrol 2014;9(1):166–73.
[94] Zhou Q, et al. Calciphylaxis. Lancet 2014;383(9922):1067.
[95] Breakey W, et al. Warfarin-induced skin necrosis progressing to calciphylaxis. J Plast Reconstr Aesthet Surg 2014;67(10):e244–6.
[96] Thornsberry LA, LoSicco KI, English 3rd JC. The skin and hypercoagulable states. J Am Acad Dermatol 2013;69(3):450–62.
[97] Crotty R, et al. Heparin, warfarin, or calciphylaxis? Am J Hematol 2014;89(7):785–6.
[98] Kakagia DD, et al. Warfarin-induced skin necrosis. Ann Dermatol 2014;26(1):96–8.
[99] Nazarian RM, et al. Warfarin-induced skin necrosis. J Am Acad Dermatol 2009; 61(2):325–32.
[100] Misic AM, Gardner SE, Grice EA. The wound microbiome: modern approaches to examining the role of microorganisms in impaired chronic wound healing. Adv Wound Care (New Rochelle) 2014;3(7):502–10.
[101] Vyas KS, Wong LK. Detection of biofilm in wounds as an early indicator for risk for tissue infection and wound chronicity. Ann Plast Surg 2016;76(1):127–31.
[102] Brewin MP, Lister TS. Prevention or treatment of hypertrophic burn scarring: a review of when and how to treat with the pulsed dye laser. Burns 2014;40(5):797–804.
[103] Tredget EE, Levi B, Donelan MB. Biology and principles of scar management and burn reconstruction. Surg Clin North Am 2014;94(4):793–815.
[104] Wolfram D, et al. Hypertrophic scars and keloids–a review of their pathophysiology, risk factors, and therapeutic management. Dermatol Surg 2009;35(2):171–81.

[105] Pastar I, et al. Epithelialization in wound healing: a comprehensive review. Adv Wound Care (New Rochelle) 2014;3(7):445–64.
[106] Ud-Din S, Bayat A. New insights on keloids, hypertrophic scars, and striae. Dermatol Clin 2014;32(2):193–209.
[107] Brown JJ, Bayat A. Genetic susceptibility to raised dermal scarring. Br J Dermatol 2009;161(1):8–18.
[108] Lawrence JW, et al. Epidemiology and impact of scarring after burn injury: a systematic review of the literature. J Burn Care Res 2012;33(1):136–46.
[109] Monstrey S, et al. Updated scar management practical guidelines: non-invasive and invasive measures. J Plast Reconstr Aesthet Surg 2014;67(8):1017–25.
[110] Wittgruber G, Parvizi D, Koch H. Nonsurgical therapy for postoperative hypertrophic scars. Eur Surg 2012;44(2):74–8.
[111] Bouzari N, Davis SC, Nouri K. Laser treatment of keloids and hypertrophic scars. Int J Dermatol 2007;46(1):80–8.
[112] Rockwell WB, Cohen IK, Ehrlich HP. Keloids and hypertrophic scars: a comprehensive review. Plast Reconstr Surg 1989;84(5):827–37.
[113] Friedman DW, et al. Regulation of collagen gene expression in keloids and hypertrophic scars. J Surg Res 1993;55(2):214–22.
[114] Abergel RP, et al. Biochemical composition of the connective tissue in keloids and analysis of collagen metabolism in keloid fibroblast cultures. J Invest Dermatol 1985;84(5):384–90.
[115] Nakaoka H, Miyauchi S, Miki Y. Proliferating activity of dermal fibroblasts in keloids and hypertrophic scars. Acta Derm Venereol 1995;75(2):102–4.
[116] Luo S, et al. Abnormal balance between proliferation and apoptotic cell death in fibroblasts derived from keloid lesions. Plast Reconstr Surg 2001;107(1):87–96.
[117] Gauglitz GG, et al. Hypertrophic scarring and keloids: pathomechanisms and current and emerging treatment strategies. Mol Med 2011;17(1-2):113–25.
[118] Marneros AG, et al. Clinical genetics of familial keloids. Arch Dermatol 2001; 137(11):1429–34.
[119] Fazio VW, et al. Reduction in adhesive small-bowel obstruction by seprafilm adhesion barrier after intestinal resection. Dis Colon Rectum 2006;49(1):1–11.
[120] Ouaissi M, et al. Post-operative adhesions after digestive surgery: their incidence and prevention: review of the literature. J Visc Surg 2012;149(2):e104–14.
[121] Ghahary A, et al. Collagenase production is lower in post-burn hypertrophic scar fibroblasts than in normal fibroblasts and is reduced by insulin-like growth factor-1. J Invest Dermatol 1996;106(3):476–81.
[122] Baryza MJ, Baryza GA. The Vancouver scar scale: an administration tool and its interrater reliability. J Burn Care Rehabil 1995;16(5):535–8.
[123] Oliveira GV, et al. Hypertrophic versus non hypertrophic scars compared by immunohistochemistry and laser confocal microscopy: type I and III collagens. Int Wound J 2009;6(6):445–52.
[124] Fitzpatrick TB. The validity and practicality of sun-reactive skin types I through VI. Arch Dermatol 1988;124(6):869–71.
[125] Roberts WE. Skin type classification systems old and new. Dermatol Clin 2009;27(4): 529–33, viii.
[126] Anderson RR, et al. Laser treatment of traumatic scars with an emphasis on ablative fractional laser resurfacing: consensus report. JAMA Dermatol 2014;150(2):187–93.
[127] Urioste SS, Arndt KA, Dover JS. Keloids and hypertrophic scars: review and treatment strategies. Semin Cutan Med Surg 1999;18(2):159–71.

[128] Slemp AE, Kirschner RE. Keloids and scars: a review of keloids and scars, their pathogenesis, risk factors, and management. Curr Opin Pediatr 2006;18(4):396–402.
[129] Baisch A, Riedel F. Hyperplastic scars and keloids: part II: surgical and non-surgical treatment modalities. HNO 2006;54(12):981–92.
[130] Gallant-Behm CL, Mustoe TA. Occlusion regulates epidermal cytokine production and inhibits scar formation. Wound Repair Regen 2010;18(2):235–44.
[131] Brissett AE, Sherris DA. Scar contractures, hypertrophic scars, and keloids. Facial Plast Surg 2001;17(4):263–72.
[132] Chan HH, et al. The use of pulsed dye laser for the prevention and treatment of hypertrophic scars in Chinese persons. Dermatologic Surg 2004;30(7):987–94.
[133] Kuo YR, et al. Activation of ERK and p38 kinase mediated keloid fibroblast apoptosis after flashlamp pulsed-dye laser treatment. Lasers Surg Med 2005;36(1):31–7.
[134] Kuo YR, et al. Suppressed TGF-beta1 expression is correlated with up-regulation of matrix metalloproteinase-13 in keloid regression after flashlamp pulsed-dye laser treatment. Lasers Surg Med 2005;36(1):38–42.
[135] Mamalis AD, et al. Laser and light-based treatment of keloids–a review. J Eur Acad Dermatol Venereol 2014;28(6):689–99.
[136] Preissig J, Hamilton K, Markus R. Current laser resurfacing technologies: a review that Delves beneath the surface. Semin Plast Surg 2012;26(3):109–16.
[137] Metelitsa AI, Alster TS. Fractionated laser skin resurfacing treatment complications: a review. Dermatol Surg 2010;36(3):299–306.
[138] Linares HA, Larson DL, Willis-Galstaun BA. Historical notes on the use of pressure in the treatment of hypertrophic scars or keloids. Burns 1993;19(1):17–21.
[139] Costa AM, et al. Mechanical forces induce scar remodeling. Study in non-pressure-treated versus pressure-treated hypertrophic scars. Am J Pathol 1999;155(5):1671–9.
[140] Reno F, et al. Effect of in vitro mechanical compression on Epilysin (matrix metalloproteinase-28) expression in hypertrophic scars. Wound Repair Regen 2005;13(3):255–61.
[141] Reno F, et al. In vitro mechanical compression induces apoptosis and regulates cytokines release in hypertrophic scars. Wound Repair Regen 2003;11(5):331–6.
[142] Eming SA, Martin P, Tomic-Canic M. Wound repair and regeneration: mechanisms, signaling, and translation. Sci Transl Med 2014;6(265):265sr6.
[143] Honardoust D, et al. Novel methods for the investigation of human hypertrophic scarring and other dermal fibrosis. Methods Mol Biol 2013;1037:203–31.
[144] Shaffer JJ, Taylor SC, Cook-Bolden F. Keloidal scars: a review with a critical look at therapeutic options. J Am Acad Dermatol 2002;46(2 Suppl. Understanding):S63–97.
[145] Traber DL, et al. Surgical research. San Diego, CA: Academic Press; 2001.
[146] Peck MD, et al. Burns and injuries from non-electric-appliance fires in low- and middle-income countries Part II. A strategy for intervention using the Haddon matrix. Burns 2008; 34(3):312–9.
[147] Peck MD, et al. Burns and fires from non-electric domestic appliances in low and middle income countries Part I. The scope of the problem. Burns 2008;34(3):303–11.
[148] The global burden of disease. WHO Communication; 2004. p. 1–160.
[149] Bezuhly M, Fish JS. Acute burn care. Plast Reconstr Surg 2012;130(2):349e–58e.
[150] Arno AI, et al. Up-to-date approach to manage keloids and hypertrophic scars: a useful guide. Burns 2014;40(7):1255–66.
[151] Parks DH, Carvajal HF, Larson DL. Management of burns. Surg Clin North Am 1977;57(5):875–94.
[152] Esselman PC. Burn rehabilitation: an overview. Arch Phys Med Rehabil 2007; 88(12 Suppl. 2):S3–6.

[153] Bell L, et al. Pruritus in burns: a descriptive study. J Burn Care Rehabil 1988;9(3):305–8.
[154] Woo SH, Seul JH. Optimizing the correction of severe postburn hand deformities by using aggressive contracture releases and fasciocutaneous free-tissue transfers. Plast Reconstr Surg 2001;107(1):1–8.
[155] Baxter CR, Shires T. Physiological response to crystalloid resuscitation of severe burns. Ann N Y Acad Sci 1968;150(3):874–94.
[156] Janzekovic Z. A new concept in the early excision and immediate grafting of burns. J Trauma 1970;10(12):1103–8.
[157] Hettiaratchy S, Dziewulski P. ABC of burns. Introduction. BMJ 2004;328(7452):1366–8.
[158] Gibran NS, Committee on Organization and Delivery of Burn Care, American Burn Association. Practice guidelines for burn care, 2006. J Burn Care Res 2006; 27(4):437–8.
[159] Supple KG, Fiala SM, Gamelli RL. Preparation for burn center verification. J Burn Care Rehabil 1997;18(1 Pt 1):58–60.
[160] Dupuytren G, Doane AS. Clinical lectures on surgery, delivered at Hôtel Dieu in 1832. New York: Collins & Hannay; 1833.
[161] Song D, et al. Plastic surgery. 4, 4. London [u.a.]: Elsevier, Saunders; 2013.
[162] Watts AM, et al. Burn depth and its histological measurement. Burns 2001;27(2):154–60.
[163] Hettiaratchy S, Papini R. Initial management of a major burn: II–assessment and resuscitation. BMJ 2004;329(7457):101–3.
[164] Devgan L, et al. Modalities for the assessment of burn wound depth. J Burns Wounds 2006;5:e2.
[165] Shupp JW, et al. A review of the local pathophysiologic bases of burn wound progression. J Burn Care Res 2010;31(6):849–73.
[166] Shakespeare P. Burn wound healing and skin substitutes. Burns 2001;27(5):517–22.
[167] Singh V, et al. The pathogenesis of burn wound conversion. Ann Plast Surg 2007; 59(1):109–15.
[168] Thompson P, et al. Effect of early excision on patients with major thermal injury. J Trauma 1987;27(2):205–7.
[169] Hettiaratchy S, Dziewulski P. ABC of burns: pathophysiology and types of burns. BMJ 2004;328(7453):1427–9.
[170] Jackson DM. The diagnosis of the depth of burning. Br J Surg 1953;40(164):588–96.
[171] Jackson DM. Second thoughts on the burn wound. J Trauma 1969;9(10):839–62.
[172] Gravante G, et al. Apoptotic death in deep partial thickness burns vs. normal skin of burned patients. J Surg Res 2007;141(2):141–5.
[173] Gravante G, et al. Inverse relationship between the apoptotic rate and the time elapsed from thermal injuries in deep partial thickness burns. Burns 2008;34(2):228–33.
[174] Singer AJ, et al. Apoptosis and necrosis in the ischemic zone adjacent to third degree burns. Acad Emerg Med 2008;15(6):549–54.
[175] Bucky LP, et al. Reduction of burn injury by inhibiting CD18-mediated leukocyte adherence in rabbits. Plast Reconstr Surg 1994;93(7):1473–80.
[176] Han YP, et al. TNF-alpha stimulates activation of pro-MMP2 in human skin through NF-(kappa)B mediated induction of MT1-MMP. J Cell Sci 2001;114(Pt 1):131–9.
[177] Parihar A, et al. Oxidative stress and anti-oxidative mobilization in burn injury. Burns 2008;34(1):6–17.
[178] Hu Z, Sayeed MM. Activation of PI3-kinase/PKB contributes to delay in neutrophil apoptosis after thermal injury. Am J Physiol Cell Physiol 2005;288(5):C1171–8.
[179] Sawhney CP, et al. Long-term experience with 1 per cent topical silver sulphadiazine cream in the management of burn wounds. Burns 1989;15(6):403–6.

[180] Baskaran H, Yarmush ML, Berthiaume F. Dynamics of tissue neutrophil sequestration after cutaneous burns in rats. J Surg Res 2000;93(1):88–96.
[181] Cotran RS. The delayed and prolonged vascular leakage in inflammation. II. An electron microscopic study of the vascular response after thermal injury. Am J Pathol 1965; 46:589–620.
[182] Isik S, et al. Saving the zone of stasis in burns with recombinant tissue-type plasminogen activator (r-tPA): an experimental study in rats. Burns 1998;24(3):217–23.
[183] Rockwell WB, Ehrlich HP. Fibrinolysis inhibition in human burn blister fluid. J Burn Care Rehabil 1990;11(1):1–6.
[184] Klein MB, et al. Patterns of grease burn injury: development of a classification system. Burns 2005;31(6):765–7.
[185] Schubert W, Ahrenholz DH, Solem LD. Burns from hot oil and grease: a public health hazard. J Burn Care Rehabil 1990;11(6):558–62.
[186] Fiebiger B, et al. Causes and treatment of burns from grease. J Burn Care Rehabil 2004;25(4):374–6.
[187] Maghsoudi H, Adyani Y, Ahmadian N. Electrical and lightning injuries. J Burn Care Res 2007;28(2):255–61.
[188] Domergue S, Jorgensen C, Noel D. Advances in research in animal models of burn-related hypertrophic scarring. J Burn Care Res 2014;36(5):e259–66.
[189] Greenhalgh DG. Models of wound healing. J Burn Care Rehabil 2005;26(4):293–305.
[190] Silverstein P, et al. Hypertrophic scar in the experimental animal. In: International symposium and workshop on the relation of the ultrastructure of collagen to the healing of wounds and to the surgical management of hypertropic scar. Springfield, III: Thomas; 1973.
[191] Greenhalgh DG, et al. Surgical research. San Diego, CA: Academic Press; 2001.
[192] Roberts AB, et al. Transforming growth factor type beta: rapid induction of fibrosis and angiogenesis in vivo and stimulation of collagen formation in vitro. Proc Natl Acad Sci USA 1986;83(12):4167–71.
[193] Charles D, et al. An improved method of in vivo wound disruption and measurement. J Surg Res 1992;52(3):214–8.
[194] Haxton H. The influence of suture materials and methods on the healing of abdominal wounds. Br J Surg 1965;52(5):372–5.
[195] Svedman C, et al. Skin suction blister wound exposed to u.v. irradiation: a burn wound model for use in humans. Burns 1991;17(1):41–6.
[196] Greenhalgh DG. The healing of burn wounds. Dermatol Nurs 1996;8(1):13–23, 66.
[197] Roesel JF, Nanney LB. Assessment of differential cytokine effects on angiogenesis using an in vivo model of cutaneous wound repair. J Surg Res 1995;58(5):449–59.
[198] Middelkoop E, van den Bogaerdt AJ, Lamme EN, Hoekstra MJ, Brandsma K, Ulrich MMW. Porcine wound models for skin substitution and burn treatment. Biomaterials 2004;25(9):1559–67.
[199] Brown DL, Kane CD, Chernausek SD, Greenhalgh DG. Differential expression and localization of insulin-like growth factors I and II in cutaneous wounds of diabetic and nondiabetic mice. Am J Pathol 1997;151(3):715–24.
[200] Howes EL, Briggs H, Shea R, Harvey SC. Effect of complete and partial starvation on the rate of fibroplasia in the healing wound. Arch Surg 1933;27(5):846–58.
[201] Rhoads JE, Fliegelman MT, Panzer LM. The mechanism of delayed wound healing in the presece of hypoproteinemia. JAMA 1942;118(1):21–5.
[202] Greenhalgh DG, Gamelli RL. Do nutritional alterations contribute to adriamycin-induced impaired wound healing? J Surg Res 1988;45(3):261–5.

[203] Tenorio A, et al. Accelerated healing in infected wounds. Surg Gynecol Obstet 1976; 142(4):537–43.
[204] Levenson SM, et al. Wound healing accelerated by *Staphylococcus aureus*. Arch Surg 1983;118(3):310–20.
[205] Hayward P, et al. Fibroblast growth factor reserves the bacterial retardation of wound contraction. Am J Surg 1992;163(3):288–93.
[206] Greenhalgh DG, Gamelli RL. Is impaired wound healing caused by infection or nutritional depletion? Surgery 1987;102(2):306–12.
[207] Chen C, et al. Molecular and mechanistic validation of delayed healing rat wounds as a model for human chronic wounds. Wound Repair Regen 1999;7(6):486–94.
[208] Mustoe TA, et al. Growth factor-induced acceleration of tissue repair through direct and inductive activities in a rabbit dermal ulcer model. J Clin Invest 1991;87(2):694–703.

Growth factors in fetal and adult wound healing

2

H. Pratsinis, D. Kletsas
Laboratory of Cell Proliferation & Ageing, Institute of Biosciences & Applications, National Center for Scientific Research "Demokritos," Athens, Greece

2.1 Introduction

2.1.1 Overview of the wound healing process in adults

Tissue injury in the adult organism activates a series of temporally overlapping processes, roughly categorized into three steps, ie, inflammation, tissue formation, and tissue remodeling, leading to repair (Clark, 1996; Werner and Grose, 2003). Immediately after injury and vascular rupture, blood coagulation follows, where the role of platelets is very important, since they aggregate and release several mediators of the healing process, among them being a series of growth factors (Rozman and Bolta, 2007; Stathakos et al., 1991). These factors attract in a chemotactic manner immune cells, such as monocytes/macrophages, producing more growth factors and cytokines in the region of the wound (Eming et al., 2007). In parallel, a provisional matrix is formed, necessary for the migration of cells into the wound area, among them being fibroblasts; this cell type is central in the healing process, especially due to the production of extracellular matrix (ECM) components, such as collagen, hyaluronic acid, and proteoglycans. Hence, the provisional matrix is replaced by a more permanent granulation tissue. Fibroblasts play an additional role in this phase as they can be converted to myofibroblasts, leading to wound contraction (Micallef et al., 2012). Two other important processes progressing alongside are the formation of a new epithelium covering the injured tissue, ie, reepithelialization, and angiogenesis in the newly formed tissue, termed neovascularization. Several weeks later, tissue remodeling is activated, aiming at the restoration of the initial tissue architecture. In this phase, catabolic and anabolic procedures are orchestrated. The balance between matrix metalloproteinases (MMPs) and their inhibitors, tissue inhibitors of metalloproteinases (TIMPs), is thus crucial. In extended wounds or in pathological cases, as in keloids, the initial tissue architecture is never achieved, and nonremodeled scars are observed. Other cases of imperfect healing are chronic wounds, featuring an intense and perpetuating inflammation, as well as cells incapable of responding to growth factors due to senescence (Ågren et al., 1999, 2000).

All of the events described above are controlled by a broad range of growth factors and cytokines mostly derived by the degranulating platelets and the immune cells attracted at the very beginning of the repair process (Werner and Grose, 2003).

Wound Healing Biomaterials - Volume 1. http://dx.doi.org/10.1016/B978-1-78242-455-0.00002-1

2.2 Growth factors implicated in wound healing

The main growth factors released from platelets are platelet-derived growth factor (PDGF), transforming growth factor-β (TGF-β), epidermal growth factor (EGF), fibroblast growth factor (FGF), insulin-like growth factor (IGF), and vascular endothelial growth factor (VEGF; Durante et al., 2013; Eppley et al., 2004; Jeong et al., 2014). They are shortly presented in this section.

2.2.1 Platelet-derived growth factors

PDGF was originally isolated from human platelets (Antoniades et al., 1975). The term collectively describes a family of dimeric disulfide–bound isoforms with a molecular weight of approximately 30 kDa: the homodimers PDGF-AA and PDGF-BB and the heterodimer PDGF-AB, the most common isoform in human platelets (Heldin and Westermark, 1999). The A and B polypeptide chains are synthesized as precursor molecules that undergo proteolytic processing before their secretion (Ostman et al., 1991). The mature A and B chains contain approximately 100 amino acid residues, and they have approximately 60% amino acid sequence identity (Fredriksson et al., 2004). Two new members of the PDGF family have been identified (Heldin et al., 2002), the homodimers of the PDGF-C and PDGF-D chains, which are activated after their secretion by cleavage of their amino-terminal domains (Bergsten et al., 2001; LaRochelle et al., 2001; Li et al., 2000). PDGFs act on target cells through the activation of two structurally related protein tyrosine–kinase receptors, α and β, with molecular weights of approximately 170 kDa and 180 kDa, respectively (Heldin and Westermark, 1999). Binding of PDGFs to their receptors triggers homo- or heterodimerization and cross-phosphorylation on specific tyrosine residues, consequently activating a panel of signaling cascades, including, among others, the PLCγ, the PI 3-K/Akt, and the Ras/Raf/MEK/ERK pathways (Levitzki, 2004).

PDGF is considered to be the most potent mitogen for cells of mesenchymal origin (Bonner, 2004), a fact attributed to the parallel activation of all of the abovementioned signaling pathways (Tallquist and Kazlauskas, 2004). Other biological effects of PDGFs include cell survival, actin cytoskeleton rearrangements, chemotaxis, stimulation of ECM-molecules' production, secretion of ECM-degrading enzymes, and induction of collagen matrix contraction (Heldin and Westermark, 1999). Therefore, PDGFs are implicated in the early phases of wound healing through the attraction of various cell types in the wound area and the promotion of their proliferation, during ECM deposition and remodeling, and, finally, during the contraction phase (Barrientos et al., 2008). Studies showing reduced and/or delayed expression of PDGFs and their receptors in various cases of impaired wound healing support their importance for this process (Ashcroft et al., 1997; Beer et al., 1997, 2000; Pierce et al., 1995).

2.2.2 Transforming growth factor-β superfamily

The TGF-β superfamily encompasses proteins encoded by 42 open reading frames in human, 9 in fly, and 6 in worm, having a common motif of 6 conserved cysteine residues (Shi and Massague, 2003). These proteins can be further grouped in two subfamilies, the TGF-β/Activin/Nodal subfamily and the bone morphogenetic protein (BMP)/growth and differentiation factor/Muellerian inhibiting substance subfamily, based on sequence similarities and the specific signaling pathways activated (Shi and Massague, 2003). Most of them encode dimeric polypeptides regulating cell proliferation, differentiation, adhesion, migration, and death in a developmental context-dependent and cell type–specific manner (Moustakas and Heldin, 2009). Among these members, the most important during the wound healing process in mammals are TGF-β1, TGF-β2, and TGF-β3, activin, and BMP-2, BMP-4, BMP-6, and BMP-7 (Werner and Grose, 2003). All TGF-β ligands act on cells through binding to type I and type II receptors that form heterotetrameric complexes in the presence of the ligand (Moustakas and Heldin, 2009). These receptors possess a cytoplasmic kinase domain that has a strong serine/threonine kinase activity and a weaker tyrosine kinase activity, hence they are considered to be dual specificity kinases. A common characteristic of all TGF-β superfamily members is that downstream signaling is primarily influenced by members of the Smad family. To this end, the receptor-activated Smads or R-Smads (SMAD2 and SMAD3 for the TGF-β subfamily and SMAD1, SMAD5, and SMAD8 for the BMP subfamily) are phosphorylated by type I receptors and associate with the common mediator Smad or co-Smad, ie, SMAD4. Then their complex is transferred to the nucleus to regulate gene transcription (Moustakas and Heldin, 2009). Moreover, regulation of the signaling strength and duration is accomplished by the inhibitory Smads or I-Smads (SMAD6 and SMAD7; Moustakas and Heldin, 2009; Park, 2005). Beyond this so-called "canonical" Smad pathway (Kardassis et al., 2009), members of the mitogen-activated protein kinases (MAPKs) ERK, Jun N-terminal kinase, p38, PI 3-K kinases, PP2A phosphatases, and Rho family members can also be involved in TGF-β superfamily members' signaling (Derynck and Zhang, 2003).

TGF-β is considered to be the prototype of the multifunctional growth factor (Mauviel, 2005). While it inhibits the division of epithelial and endothelial cells, it can be either inhibitory or stimulatory for fibroblastic cell proliferation (Massague, 1990). The members of the TGF-β-superfamily are regarded as the main regulators of ECM synthesis and degradation through the coordinated regulation of complex gene sets (Schiller et al., 2004). Furthermore, TGF-β regulates migration and collagen gel contraction and exhibits immunosuppressant activities, thus playing a central role in all phases of wound repair, from inflammation and granulation tissue formation to epithelialization and matrix formation and remodeling (Barrientos et al., 2008; Massague, 1990; Werner and Grose, 2003). Animal models knocking out or overexpressing members of the TGF-β superfamily or of the Smad pathway reveal their complex role in wound healing (Ashcroft et al., 1999; Brown et al., 1995; Koch et al., 2000; Munz et al., 1999).

Among the genes regulated by members of the TGF-β superfamily, the connective tissue growth factor (CTGF) or CCN family member 2 is worth mentioning in the context of wound repair. It was isolated from human endothelial cells, cloned, and sequenced in the laboratory of Gary Grotendorst (Bradham et al., 1991). CTGF contains six members of cysteine-rich, heparin-binding proteins, which integrate and modulate the signals of TGF-β superfamily members as well as other growth factors and integrins, further regulating numerous biological processes, such as differentiation, migration, proliferation, and cell adhesion (Zuo et al., 2010). CTGF acts not only on cells of mesenchymal origin (Frazier et al., 1996) being implicated in fibrotic responses (Ihn, 2002) but also on epithelial cells (Secker et al., 2008). Hence in the wound repair process CTGF is involved in granulation tissue formation, epithelialization, and matrix formation and remodeling (Barrientos et al., 2008).

2.2.3 Epidermal growth factor family

EGF, transforming growth factor-α, heparin-binding EGF (HB–EGF), amphiregulin, epiregulin, betacellulin, neuregulins, epigen, and proteins encoded by Vaccinia virus and other poxviruses are all the members of the EGF family (Nanney and King, 1996; Werner and Grose, 2003). In addition, more distantly related proteins known as neuregulins (heregulins, neu differentiation factors, NDF 1–4) also bind to some EGF receptor family members (Werner and Grose, 2003). All these growth factors serve as agonists of the four ErbB high-affinity receptors, ie, EGFR/ErbB1, HER2/ErbB2, HER3/ErbB3, and HER4/ErbB4, which possess tyrosine kinase activity and, upon ligand binding, form homo- or heterodimers, and undergo autophosphorylation. Substrates involved in the downstream transduction of the signal include the Ras/Raf/MEK/ERK axis, G-proteins, and the Jak/Stat pathway, among others (Nanney and King, 1996). Notably, different EGF family members may stimulate the same receptor yet stimulate divergent biological responses in cells in vitro and in vivo; this functional selectivity is the subject of intense study by the pharmaceutical industry, especially for cancer therapeutics (Wilson et al., 2009).

EGF family members promote the proliferation of most cell types in the wound area, such as keratinocytes, fibroblasts, macrophages, and vascular endothelial cells (Schultz et al., 1991). Furthermore, they regulate keratinocyte differentiation as well as the migration of fibroblasts and endothelial cells. Hence EGF family members play an important role during the wound healing process, mainly through the promotion of the epithelialization and neovascularization processes (Schultz et al., 1991; Barrientos et al., 2008). Knocking out an EGF family member was not reflected in a wound healing phenotype (Luetteke et al., 1993; Mann et al., 1993), possibly due to the compensatory effects of other family members. Despite these finding, a controlled delivery of EGF or HB-EGF seems to accelerate wound healing in organ cultures and mouse models (Hardwicke et al., 2008; Johnson and Wang, 2013; Ko et al., 2011; Tolino et al., 2011).

2.2.4 Fibroblast growth factors

The FGF family comprises 23 members (22 of them have been identified in humans) characterized by a central core of 140 amino acids exhibiting high homology throughout all family members as well as a strong affinity for heparin and heparin-like glycosaminoglycans (Itoh and Ornitz, 2011; Powers et al., 2000). Among them the three most important molecules involved in the wound repair process are FGF-2 or basic FGF, FGF-7 or keratinocyte growth factor-1 (KGF-1), and FGF-10 or keratinocyte growth factor-2 (KGF-2; Barrientos et al., 2008; Werner, 1998). FGFs mediate their cellular responses by binding to and activating a family of four receptor tyrosine kinases, designated the high-affinity FGF receptors (FGF-R1 to FGF-R4); the low-affinity ones are heparin and heparin sulfate proteoglycans, which do not transmit a signal but play accessory and regulatory roles (Eswarakumar et al., 2005). The diversity of FGFRs is further expanded due to the expression of numerous splice variants of each gene (Powers et al., 2000). The main downstream effectors of FGFR signaling include the Ras/Raf/MEK/ERK pathway, the PLCγ/DAG/PKC pathway, and the PI 3-K/Akt axis (Eswarakumar et al., 2005; Laestander and Engstrom, 2014).

FGF family members regulate the proliferation and motility of various cell types, such as fibroblasts, vascular endothelial cells, and keratinocytes as well as the synthesis and deposition of various ECM components, hence being implicated in granulation tissue formation, neovascularization, epithelialization, and tissue remodeling (Barrientos et al., 2008; Peng et al., 2011; Powers et al., 2000; Oladipupo et al., 2014). FGF-2 neutralization or knock out leads to impaired or delayed wound healing, respectively (Broadley et al., 1989; Ortega et al., 1998), while the expression of a dominant-negative mutant of FGF receptor delays wound epithelialization (Werner et al., 1994).

2.2.5 Insulin-like growth factors

The IGF family consists of two polypeptides: IGF-I (70kDa) and IGF-II (67kDa). They share 62% homology with proinsulin (Baserga et al., 1997). Insulin may also be included in the family as well as six IGF binding proteins interacting with IGF family members. Their signals are transmitted in the cells by four receptors: the type I insulin-like growth factor-receptor (IGF-IR), the insulin receptor (IR), the IGF-II/mannose 6-phosphate receptor, and hybrid IGF-IR/IR receptors (Ryan and Goss, 2008). IGF-IR is mainly responsible for most of IGFs' actions, upon which ligand binding on its extracellular α subunits undergoes conformational changes, leading to the activation of the intracellular tyrosine kinase domains of the β subunits and autophosphorylation (Jones and Clemmons, 1995). Downstream signaling proceeds through the direct phosphorylation of the insulin receptor substrates IRS-1 to –4 and Shc and subsequently the PI 3-K/Akt/mTOR and the Ras/Raf/MEK/ERK axis (Ryan and Goss, 2008).

IGFs exert proliferative and survival effects on a vast variety of target cell types, and they can affect differentiation, ECM regulation, chemotaxis, secretion of hormones, neurotransmitters, and other growth factors as well as the uptake of amino acids and glucose (Jones and Clemmons, 1995). They often operate as local mediators of the actions of hormones or other growth factors (Jones and Clemmons, 1995). IGFs can act in an autocrine or paracrine manner (Barreca et al., 1992), hence they affect the epithelialization of the wound. Generally, lower levels of IGFs are observed in various situations of impaired wound healing (Werner and Grose, 2003). For example, in one study, reduced endothelial insulin/IGF signaling was responsible for decreased angiogenesis and granulation tissue formation upon wounding in mice (Aghdam et al., 2012). This mechanism may contribute to diabetes-associated impaired healing (Aghdam et al., 2012). On the other hand, IGF overexpression increases collagen deposition that may lead to hypertrophic scarring (Khorramizadeh et al., 1999).

2.2.6 Vascular endothelial growth factors

The VEGF family currently includes VEGF-A, VEGF-B, VEGF-C, VEGF-D, VEGF-E, and placenta growth factor (PLGF; Tugues et al., 2011). These dimeric molecules are distant relatives of PDGF, since their monomers share a cysteine knot motif comprising an 8-residue ring formed by three disulphide bridges (Zachary, 1998; Heldin and Westermark, 1999). The variety of VEGF family increases further due to alternative splicing of the above genes. They bind to heparin, and, with high affinity, to three different transmembrane tyrosine kinase receptors, designated VEGFR1 (or Flt-1), VEGFR2 (or KDR), and VEGFR3 (Werner and Grose, 2003). VEGF receptors, like those of PDGF or EGF, undergo dimerization and autophosphorylation upon ligand binding and then transmit their signal through the activation of the MAPK cascade, PLCγ, PKC, FAK, and paxillin (Zachary, 1998).

VEGFs mainly regulate vasculogenesis and angiogenesis during development (Gale and Yancopoulos, 1999), while they play important roles in wound angiogenesis and granulation tissue formation (Barrientos et al., 2008; Werner and Grose, 2003). This has been shown in vivo, in various species by VEGF neutralization or retroviral delivery of a dominant-negative VEGF receptor (Howdieshell et al., 2001; Nissen et al., 1998; Tsou et al., 2002). Apart from their mitogenic effects, VEGF also increases vascular permeability (Ferrara et al., 2003).

2.3 Differences in wound repair between fetuses and adults

In contrast to the repair process in the adult organism, fetuses are able to heal their wounds in what seems to be perfect tissue regeneration, marked by the absence of scar formation. This is influenced by the size of the wound, whether it is incisional or excisional, and the age of the fetus (Cass et al., 1997). After the first two trimesters of gestation, a transition from scarless repair to adult-like scar formation is observed, which in humans happens after about 24 weeks of gestation (Lorenz et al., 1992). Tissue repair in the fetus is achieved in a sterile, fluid environment with a relative lack of inflammation (Eming et al., 2007). Accordingly, it has been hypothesized that these alterations are responsible for the differential repair process in fetuses and adults (McCallion and Ferguson, 1996). In this direction, the incorporation of bacteria in fetal wounds has been shown to trigger neovascularization and fibroplasia, characteristic responses of adult-like healing (Frantz et al., 1993). Furthermore, attempts have been made to identify the factors contained in amniotic fluid that could possibly support scarless healing (Burrington, 1971; Longaker et al., 1990). However, a series of experimental data support the importance of the intrinsic differences between fetal and adult tissue. Specifically, it has been shown that when grafts of human fetal skin were placed subcutaneously in adult athymic nude mice, they healed their experimental wounds in the absence of a scar, while cutaneous grafts healed with a scar (Lorenz et al., 1992). In addition, the wounds in full-thickness adult sheepskin transplanted onto the backs of early fetal lambs healed with scar formation (Longaker et al., 1994). Moreover, no significant differences between fetal and adult skin fibroblasts were observed in their response to amniotic fluid (Chrissouli et al., 2010). Finally, even embryos born at a very early developmental stage and with a poorly developed immune system, such as these of the opossum *Monodelphis domesticus*, can still heal in a fetal mode, characterized by rapid epithelialization, minimal inflammatory and angiogenic responses, and a lack of fibrosis (Armstrong and Ferguson, 1995). In conclusion, it seems that although the environment can affect scar formation, fetal skin healing properties are primarily intrinsic to the fetal tissue, most probably reflecting the unique features of its cells.

2.4 The interplay of fibroblasts and growth factors in fetal and adult wound healing

This section focuses on the characteristics of fetal cells, especially fibroblasts, and particularly on their response to growth factors, which represent an important determinant of tissue homeostasis and repair. The effects of growth factors on fetal and adult fibroblast functions are summarized in Table 2.1.

2.4.1 Expression of growth factors and their receptors

Normally, fibroblasts, central players of the repair process, express the respective receptors and are capable of responding to members of the PDGF, TGF-β, EGF, FGF, and IGF families (Bonner, 2004; Jones and Clemmons, 1995; Okada-Ban et al., 2000; Buckley-Sturrock et al., 1989). Even VEGF, which typically was considered to act on endothelial cells only, has been reported to directly stimulate fibroblasts (Wu et al., 2006), a fact further associated with fibrotic responses (Greaves et al., 2013). Furthermore, fibroblasts have been shown to express and produce PDGF (Rojas-Valencia et al., 1995), TGF-β (Pratsinis et al., 1997; Zeng et al., 1996), EGF (Kurobe et al., 1985), FGF (Artuc et al., 2002; Lonergan et al., 2003), IGF (Ankrapp and Bevan, 1993; Barreca et al., 1992), and VEGF (Coppe et al., 2006). Therefore, a variety of studies on the mechanisms underlying the intrinsic ability of the fetal tissue for complete tissue regeneration are concentrating on the differential expression of these growth factors and/or their receptors between fetal and adult cells. In an animal study using immunohistochemistry, TGF-β1, TGF-β2, and FGF-2 were present in neonatal and adult wounds but were not detected in the fetal wounds, while PDGF was present in fetal, neonatal, and adult wounds in mice (Whitby and Ferguson, 1991). Due to technical limitations, the cells producing the growth factors, besides the platelets, were not determined (Whitby and Ferguson, 1991).

In a similar study in rabbits, a higher expression of TGF-β1 and TGF-β2 in adult wounds as compared to fetal ones was observed, based on immunohistochemical localization, but it was attributed to macrophages and not to fibroblasts (Nath et al., 1994). Regarding EGF- and PDGF-B gene expression, no differences were observed among fibroblast cultures from the fetal rat skin of various gestational ages. In contrast, in whole rat skin both genes demonstrated a marked decrease in their expression with increasing

Table 2.1 Effects of growth factors on fetal versus adult fibroblasts

Function	Growth factor	Fetal	Adult	References
Fibroblast proliferation	PDGF	↑	↑	Betsholtz and Westermark (1984), Pratsinis et al. (2004)
	EGF	↑	↑	Betsholtz and Westermark (1984), Levine et al. (1992)
	FGF-2	↑	↑	Armatas et al. (2014), Giannouli and Kletsas (2006)
	IGF-I, IGF-II	↑	↑	Rolfe et al. (2007a)
	TGF-β1	↑	↑	Rolfe et al. (2007b)
	TGF-β1	ND	↑	Ishikawa et al. (1990), Soma and Grotendorst (1989), Yamakage et al. (1992)
	TGF-β1, TGF-β2, TGF-β3	↓	↑	Armatas et al. (2014), Giannouli and Kletsas (2006), Pratsinis et al. (2004)
Fibroblast migration	PDGF	↔	↑	Ellis et al. (1997)
	EGF	↔	↑	
	FGF-1, FGF-2	↔	↑	
	TGF-β1, TGF-β2 confluent	↓	↔	Ellis et al. (1997), Ellis and Schor (1998)
	TGF-β1, TGF-β2 subconfluent	↔	↓	Ellis and Schor (1998)
	TGF-β3 confluent	↓	↑	
	TGF-β3 subconfluent	↓	↓	
Collagen-I expression	TGF-β1	↓	↑	Carter et al. (2009)
Collagen-III expression	TGF-β1	↔	↓	Carter et al. (2009)
Collagen synthesis	IGF-I	↔	↑	Rolfe et al. (2007a)
	IGF-II	↑	↔	
Hyaluronic acid synthesis	TGF-β1, TGF-β2 confluent	↓	↔	Ellis and Schor (1998)
	TGF-β1, TGF-β2 subconfluent	↔	↓	
	TGF-β3 confluent	↓	↑	
	TGF-β3 subconfluent	↓	↓	
α1 integrin	TGF-β1, TGF-β2, TGF-β3	↓	↔	Moulin et al. (2001)
α2 integrin		↓	↔	
α3 integrin		↔	↑	
β1 integrin		↓	↑	
MMP-2	PDGF-AB	↓	↔	Cullen et al. (1997)
MMP-9		↑	↑	
MMP-2	TGF-β1	↑	↔	Cullen et al. (1997)
MMP-9		↑	↔	

Table 2.1 Continued

Function	Growth factor	Fetal	Adult	References
Fibroblast-populated collagen gel contraction	PDGF-AA, PDGF-BB	ND	↑	Tingström et al. (1992)
	TGF-β1	ND	↑	
	TGF-β1, TGF-β2, TGF-β3	↓	↑	Moulin et al. (2001)
	TGF-β1	↑	ND	Piscatelli et al. (1994)

↑, Upregulation; ↓, downregulation; ↔, no effect; *ND*, not detected.

gestational age, especially at the transition point from scarless to scar-forming repair (Peled et al., 2001). On the other hand, in a similar approach using primary mouse skin fibroblast cultures, it was observed that fetal cells express lower basal CTGF mRNA levels than adult ones, yet the induction of the CTGF gene by TGF-β1 and TGF-β3 was much more intense in fetal than in adult fibroblasts (Colwell et al., 2006). Furthermore, murine fetal skin fibroblasts were found to express higher levels of the TGF-β3 isoform than adult ones, while no differences were observed in the expression of TGF-β1 and TGF-β2 and of both TGF-βRI and TGF-βRII (Colwell et al., 2007), a very important observation given the antiscarring properties of TGF-β3 (Shah et al., 1995). In a preliminary report testing TGF-β1, TGF-β2, TGF-β3, FGF-1, FGF-2, KGF, PDGF-A, and PDGF-B mRNA expression in fetal and adult human skin fibroblasts, the main finding was that the expression level of TGF-β1, FGF-1, and FGF-2 was 2-fold higher in adult compared with fetal fibroblasts (Broker et al., 1999). This was proposed to contribute to suboptimal wound healing in adult wounds compared with the scarless healing of fetal wounds. In sharp contrast, the TGF-β1, FGF-1, and FGF-2 proteins were more abundantly expressed in fetal fibroblasts than in adult fibroblasts (Lee et al., 2000); this was attributed mainly to technical issues and/or differences in mRNA translation and degradation. The latter findings were supported by the observation that human fetal fibroblasts secrete higher TGF-β1 levels (assessed by enzyme-linked immunosorbent assay) in their conditioned medium than adult ones (Hanasono et al., 2003). However, the expression of a TGF-β activator, ie, latent TGF-β binding protein-1, and a TGF-β mediator, ie, IGF binding protein-3, has been shown to be significantly lower in fetal fibroblasts compared to adult ones (Gosiewska et al., 2001). In another study, no differences in the gene expression levels of all three TGF-β isoforms were reported between fetal and adult human skin fibroblasts by using quantitative real-time PCR (Rolfe et al., 2007b). Regarding TGF-β-receptor expression at the protein level, fetal fibroblasts were found to express higher TGF-βRI and lower TGF-βRII levels compared to adult ones; however, these variations did not cause any downstream difference in SMAD-2 and -3 phosphorylation (Armatas et al., 2014).

In conclusion, a series of studies have reported on differential growth factor and/or growth factor–receptor expression in fetal versus adult fibroblasts, yet these differences were not linked directly to functional alterations.

2.4.2 Fibroblast proliferation

PDGF is a potent mitogen for cells of mesenchymal origin and has been shown to stimulate the proliferation of both fetal and adult human skin fibroblasts (Betsholtz and Westermark, 1984). Moreover, the degree of PDGF-induced proliferative response is comparable in fetal and adult fibroblasts (Pratsinis et al., 2004). This is in accordance with observations regarding fetal and adult fibroblasts from different tissues, such as the lung (Clark et al., 1993). EGF is also mitogenic for human skin fibroblasts from both developmental stages (Betsholtz and Westermark, 1984) with comparable potency but on a lower level than that of PDGF (our unpublished data). Results obtained from cultures of fetal and adult prostatic fibroblasts suggest that this response to EGF could be independent of the tissue of origin (Levine et al., 1992). Similarly, FGF-2 is mitogenic for human skin fibroblasts originating from both fetal and adult donors, with potency between EGF and PDGF-BB (our unpublished observations). On the other hand, regarding IGFs, it was reported that human fetal skin fibroblasts exhibit a weaker mitogenic response to both IGF-I and IGF-II than postnatal ones, although they express comparable levels of IGF-IR, and that fetal cells, in contrast to postnatal ones, failed to phosphorylate ERK 1 and Shc (p46) in response to IGF-I and to IGF-II, respectively (Rolfe et al., 2007a).

As mentioned previously, TGF-β is the prototype of the multifunctional growth factor (Tian and Schiemann, 2009), a fact apparent also at the level of cell proliferation regulation, as exemplified by the early observations that this factor can either inhibit or stimulate proliferation, depending on the cellular context (Holley et al., 1985). With regard to human skin fibroblast proliferation, it had been reported initially that TGF-β1 does not stimulate DNA synthesis in neonatal fibroblasts and that it inhibits the mitogenic activity of PDGF in a density-dependent fashion (Paulsson et al., 1988). In contrast, several other studies reported a mitogenic action of TGF-β1 in confluent cultures of newborn and adult skin fibroblasts (Ishikawa et al., 1990; Soma and Grotendorst, 1989; Yamakage et al., 1992), while in another study a weak mitogenic response to this growth factor was described for both fetal and postnatal cells (Rolfe et al., 2007c). However, through a direct systematic comparison of different cell strains cultured under identical conditions, a clear-cut difference between fetal and adult human skin fibroblasts in their proliferative response to TGF-β was observed, ie, this growth factor is inhibitory for fetal cells, while it is stimulatory for adult ones (Pratsinis et al., 2004). This was observed for the three TGF-β isoforms, and, since neonatal fibroblasts were found to respond similarly to adult ones (Pratsinis et al., 2004), it seems that this change coincides with the transition from scarless repair to scar-forming healing. Regarding the mechanisms underlying these responses, it was shown that the inhibition of human fetal skin fibroblast proliferation by TGF-β is mediated through the activation of Protein Kinase A (PKA) and the subsequent induction of the cyclin-dependent kinase inhibitors $p21^{WAF1}$ and $p15^{INK4B}$ (Giannouli and Kletsas, 2006). In human adult skin fibroblasts, TGF-β does not activate PKA but induces an autocrine loop involving the upregulation of extracellular FGF-2 and, through FGF-R1, activation of the MEK/ERK pathway, thus leading to cell proliferation (Giannouli and Kletsas, 2006). In parallel, both SMAD2 and SMAD3 were found to be phosphorylated in response to TGF-β in both fetal and adult human skin fibroblasts (Giannouli and Kletsas, 2006; Rolfe et al., 2007c; Walraven et al., 2014). Although differences in Smad expression and duration of phosphorylation between fetal and adult fibroblasts have been reported (Walraven et al., 2014), completely blocking the Smad pathway using an siRNA approach against SMAD4 led to the abrogation of both the inhibitory and stimulatory effects of TGF-β on fetal and adult fibroblast proliferation, respectively (Armatas et al., 2014).

Most of these in vitro studies were performed with cultures growing on a plastic surface, while in vivo cells are surrounded by ECM components. Especially during wound healing, the extracellular space reflects the dynamic changes characterizing the repair phases, eg, during the early phase fibroblasts grow and migrate on a provisional matrix containing mainly fibrin and fibronectin, while later they are surrounded by polymerized collagen. Notably, fetal and adult human skin fibroblasts were shown to retain their differential proliferative response to TGF-β (inhibition versus proliferation, respectively) when cultured in the presence of fibronectin and unpolymerized or polymerized collagen (Armatas et al., 2014).

2.4.3 Fibroblast migration

Cell migration in the wound area is very important in the early healing process for populating the void with the cell types necessary to support the formation of the new tissue. Fibroblasts are the main producers of the ECM components that are necessary for granulation tissue formation. Significant differences exist between fetal and adult fibroblasts, since the former display an increased migratory activity due to intrinsic features and different responses to growth factors.

Fetal fibroblasts were reported to possess a superior migratory profile in three-dimensional collagen gel matrices as compared to adult ones (Schor et al., 1985). Furthermore, fetal cells were found to migrate irrespective of cell density, while adult fibroblasts migrate more easily when they are plated on the collagen matrix at low density (Schor et al., 1985). It was suggested that these differences were linked to the abundance of hyaluronic acid (HA), given that high HA levels enhance the migratory activity of cells, while treatment with hyaluronidase completely blocks cell migration (Schor et al., 1988, 1989). Fetal fibroblasts appear to have more receptors for HA than adult cells (Alaish et al., 1994). Furthermore, they were also reported to secrete a soluble factor (migration stimulation factor), stimulating cell migration through novel HA synthesis (Schor et al., 1988). Hence the differences in migratory activity between fetal and adult fibroblasts are partly due to the differential secretion of and response to HA. In addition, fetal skin fibroblasts were observed to migrate in the presence of a serum-depleted medium, while adult fibroblast migration required the presence of serum (Kondo and Yonezawa, 1995). Since an anti-FGF-2 antibody blocked fetal fibroblast migratory activity, this was linked to autocrine secretion of FGF-2 by fetal cells (Kondo et al., 1993).

The differential regulation of fetal and adult fibroblast migration by growth factors largely depends on the presence of ECM components as well as on culture density (Schor, 1994). On three-dimensional collagen gels, PDGF and EGF stimulate the migration of adult but not fetal fibroblasts (Ellis et al., 1997), while the action of TGF-β is density- and isoform-dependent (Ellis and Schor, 1998). In particular,

TGF-β1 inhibits fetal cell migration when cultures are confluent, while having no effect on subconfluent ones. In contrast, TGF-β1 is inhibitory for adult fibroblasts in subconfluent cultures but not in confluent ones. Interestingly, TGF-β2 exerts the same action with the TGF-β1 isoform. On the other hand, TGF-β3 inhibits migration in subconfluent cultures in cells from both developmental stages. However, in confluency it is inhibitory for fetal fibroblasts and stimulatory for adult ones (Ellis and Schor, 1998). All of the above strongly indicates that the migration of fetal and adult cells depends largely on intrinsic features as well as a complex network of exogenous growth factors, ECM components, and cell–cell interactions.

2.4.4 Extracellular matrix synthesis and remodeling

Scar formation, as already mentioned, is the hallmark of adult wound healing. Since scar tissue is rich in collagen, predominantly type I, but with a less organized pattern, synthesis and remodeling of collagen is expected to mirror the different repair mechanisms between fetuses and adults. Fetal skin has an increased type III/type I collagen ratio, compared to the adult. Interestingly, this change is reflected at the cellular level, as fetal fibroblasts produce more type III collagen (Merkel et al., 1988; Gosiewska et al., 2001). Furthermore, fetal fibroblasts express increased activity of prolyl hydroxylase, an intracellular enzyme responsible for a rate-limiting step in collagen production (Duncan et al., 1992). Finally, in the fetus, collagen synthesis starts immediately after wound formation, while it is delayed in the adult (Hantash et al., 2008).

In line with the increased expression of type I collagen in the adult skin, it has been shown that in response to TGF-β, a major stimulant of collagen synthesis and accumulation (Narayanan et al., 1989), mid-gestational mouse fibroblasts expressed less type I procollagen (Carter et al., 2009). In cells from late gestational stages type I procollagen production is increased, while that of type III procollagen is decreased (Carter et al., 2009). Other growth factors also display differential effects of fetal versus adult fibroblasts. The latter were found to be stimulated for collagen production by IGF-I but not by IGF-II, while IGF-II but not IGF-I induced collagen synthesis in fetal fibroblasts (Rolfe et al., 2007a).

One major difference between fetal and adult skin is that the former contains more HA (Mast et al., 1991; Buchanan et al., 2009). This change probably gives unique structural and functional features in the fetal tissue, such as increased fibroblast migration in the wounded area, as a relation between high HA synthesis and increased migratory activity in fetal fibroblasts has been proposed (Chen et al., 1989). In addition, fetal fibroblasts expressed more HA receptors than adult fibroblasts (Alaish et al., 1994). Concerning wound formation, an increased deposition of HA in fetal tissues compared to adult ones was observed 6 days after wounding (DePalma et al., 1989).

Differences between fetal and adult fibroblasts in HA production are also directed by cell density. Fetal fibroblasts produced a high level of HA in subconfluent and confluent cultures, while the production by adult fibroblasts was decreased significantly when cells reached confluence (Chen et al., 1989). Fetal and adult fibroblasts also respond differently to growth factors toward HA synthesis. TGF-β1 induces HA synthesis for both cell types in confluent cultures, while in subconfluent ones it inhibits adult but not fetal cells (Ellis and Schor, 1996). In addition, TNF-α regulates differentially the three enzymes synthesizing HA, ie, HAS-1, HAS-2, and HAS-3; in fetal cells it increases HAS-1 mRNA levels, while in adult fibroblasts only HAS-3 gene expression is upregulated (Kennedy et al., 2000).

Integrins are important mediators of cell–ECM communication (Juliano and Haskill, 1993) and are central in the repair process (Eckes et al., 2010). Furthermore, extensive cross-talk exists between integrins and growth factors (Nikitovic et al., 2012), exemplified especially in the effects of TGF-β on myofibroblasts and fibrosis (Hinz and McCulloch, 2012). In this setting, the observation that in human fetal skin, the fibroblast α1 and α3 integrin subunit expression was lower but the α2 subunit was higher as compared with adult, was suggested to influence contraction and scar formation during wound repair (Moulin and Plamondon, 2002). Moreover, TGF-β treatment of fetal fibroblasts decreased α1, α2, and β1 integrin expression without affecting α3 integrin, in contrast to adult fibroblasts, where TGF-β increased α3 and β1 integrin subunits without affecting α1 and α2 levels (Moulin et al., 2001). On the other hand, in excisional wounds in the α2β1-null mouse, epithelialization, wound contraction, and myofibroblast differentiation were similar to those in the wild-type mouse, while the most important difference was the strong enhancement of neovascularization of granulation tissue caused by α2β1 ablation (Zweers et al., 2007).

ECM-degrading enzymes play a significant role in all phases of tissue repair. Prominent among them are the MMPs, a 23-member (in humans) family of zinc-dependent endopeptidases, which are synthesized in latent forms and activated by limited proteolysis (Gill and Parks, 2008; Birkedal-Hansen et al., 1993). Their action is also regulated by their specific inhibitors, TIMPs (Gill and Parks, 2008; Baker et al., 2002). In addition, their production is influenced by the environment; the secretion of MMPs is increased greatly by fibroblasts grown in three-dimensional collagen gels (Zervolea et al., 2000).

As scar remodeling is an important task in adult tissue repair, largely regulated by MMPs, the differential expression of MMPs was investigated in human and animal tissues. In nonwounded human and mouse tissues, it was found that the expression of several MMPs and TIMPs increases from early to late gestational stages and even more in adult tissues (Chen et al., 2007; Dang et al., 2003). It was hypothesized that the lower TIMP expression may be crucial for scarless repair, facilitating the turnover of ECM components as well as other important functions of tissue formation, such as cell migration (Chen et al., 2007; Dang et al., 2003). Fetal human fibroblasts secrete much more of the active MMP-9 form compared with their adult counterparts, while that of MMP-2 remain largely similar (Cullen et al., 1997). In the same study, a differential effect of TGF-β1 and PDGF-AB on the expression of these MMPs was reported. TGF-β1 induces the secretion of MMP-2 and MMP-9 only in fetal and not in adult cells. On the other hand, PDGF-AB reduces MMP-2 secretion in fetal cells, while it stimulates MMP-9 secretion in both cell types (Cullen et al., 1997).

2.4.5 Contraction

Wound contraction is a major process in adult wound healing, and it is mediated by fibroblasts that transdifferentiate into myofibroblasts. Myofibroblasts are characterized by an increased expression of α-smooth muscle actin expression, a contractile cytoskeletal protein (Gabbiani et al., 1978). On the other hand, one of the unique features of fetal skin healing is the absence of contracture. In order to understand if these differences are intrinsic to the fetal tissue, in vitro studies have been performed by several laboratories by using as a model the ability of cells from both developmental stages to contract cell-populated three-dimensional gels of polymerized collagen (Grinnell, 2003). A wide diversity of results was obtained from various studies, since a variety of factors seems to affect fibroblast contractibility. For example, it was initially reported that human fetal and adult skin fibroblasts possess equal contractile capacities (Moulin et al., 2001), while the same group using freshly isolated cultures reported that fetal fibroblasts have a considerably reduced contractile capacity, compared to adult cells (Moulin and Plamondon, 2002). Furthermore, human fetal dermal fibroblasts were reported to contract free-floating collagen gels more efficiently than adult ones (Parekh et al., 2007), while the same group observed the opposite effect when utilizing an anchored collagen gel model tethered at two ends (Parekh et al., 2009). In this direction, studies on cells from other species, such as mice or lamb, indicate that skin fibroblasts from early gestational stages display a lower ability to contract collagen gels, compared to cells from late gestational stages or from adult animals (Coleman et al., 1998; Piscatelli et al., 1994).

TGF-β provokes the differentiation of fibroblasts to myofibroblasts by enhancing α-smooth muscle actin (Desmouliere et al., 1993). In murine skin fibroblasts, the differences in the contractile ability between early fetal cells and late fetal and adult cells are linked to the secretion of total and active TGF-β (Coleman et al., 1998). In human fetal lung fibroblasts, the levels of TGF-β receptor interacting protein 1, a selective inhibitor of certain effects of this growth factor (Choy and Derynck, 1998), were found to be higher than in adult cells, thus explaining the diminished collagen contraction ability of fetal versus adult cells (Navarro et al., 2009). Notably, in one study, TGF-β has been reported to inhibit contraction in fetal human skin fibroblasts, being unable to increase α-smooth muscle actin expression, while it stimulates contraction in adult ones (Moulin et al., 2001). On the other hand, other investigators have shown that TGF-β is able to induce collagen gel contraction by fetal cells (Piscatelli et al., 1994; our unpublished observations). Similarly to TGF-β, PDGF is able to potently stimulate contraction in adult skin fibroblasts (Tingström et al., 1992; Clark et al., 1989). We have found that PDGF, when used in high concentrations, can also enhance the contractile capacity of both fetal and adult cells (our unpublished observations), indicating that it is not only the intrinsic features of fibroblasts but also the local concentrations of circulating growth factors that regulate the overall phenomenon of tissue repair in these two developmental stages.

2.5 Growth factors, senescence, and wound healing

Normal cells, like skin fibroblasts, do not proliferate indefinitely, but they have a limited lifespan when cultured in vitro. After their initial isolation from the tissue they proliferate rigorously under appropriate culture conditions; however, after a certain number of cell doublings they enter a state called senescence, characterized by their inability to proliferate (Hayflick, 1965; Kletsas, 2003). This type of senescence, termed replicative senescence, has been shown to be the outcome of the continuous shortening of telomeres (the end of chromosomes) after each cell doubling (Counter et al., 1992). When this shortening comes to a critical point, it activates a classical DNA damage response, including the activation of the ATM-Chk2-p53 axis, which upregulates the cyclin-dependent kinase inhibitor $p21^{WAF1}$. This leads to the hypophosphorylation of the pRB oncosuppressor protein and subsequently to cell cycle arrest (Campisi and d'Adda di Fagagna, 2007). In contrast to normal cells, cancer cells and other immortalized cells express the catalytic domain of the enzyme telomerase (TERT), thus preserving the telomere length and the cells' abilities to proliferate ad infinitum (Counter et al., 1992). Expression of the catalytic domain of telomerase in normal cells can lead to the extension of their lifespan or even to immortalization (Bodnar et al., 1998). Beyond repeated cell doublings, the exposure of cells to external genotoxic stresses, such as ultraviolet and ionizing radiation, oxidative stress, and genotoxic drugs, can lead to premature senescence, referred to as stress-induced premature senescence (SIPS; Toussaint et al., 2000). Normal cells can also undergo SIPS after exposure to several oncogenes after the activation of a DNA damage response, indicating that senescence represents a major anticancer mechanism (Bartkova et al., 2006; Serrano et al., 1997). Importantly, senescent cells can be identified in vivo by the use of specific staining procedures, such as senescence-associated β galactosidase staining (Dimri et al., 1995), Sudan Black B staining (Georgakopoulou et al., 2013), or the expression of specific senescence molecular markers, such as the cyclin-dependent kinase inhibitor $p16^{INK4a}$ (Sindrilaru et al., 2011). At the functional level, senescent cells express a proinflammatory and catabolic phenotype, characterized by the overexpression of MMPs, inflammatory cytokines, several growth factors, and other inflammatory molecules (Campisi and d'Adda di Fagagna, 2007; Gorgoulis et al., 2005; Shelton et al., 1999), suggesting that the accumulation of senescent cells can affect locally tissue homeostasis, thus playing a role in degenerative diseases.

A major characteristic of senescent cells is their inability to respond to growth factors and proliferate. After exposure to growth stimuli, intracellular signaling pathways are activated in senescent cells. However, it has been reported that activated kinases, eg, ERK MAPK, cannot translocate into the nucleus and execute the proliferative program (Lorenzini et al., 2002), although the most important molecular alteration is the overexpression of cyclin-dependent inhibitors $p21^{WAF1}$ and $p16^{INK4a}$ that inhibit growth factor signaling more downstream, thus blocking the transition of the cells from the G1 to the S phase of the cell cycle. On the other hand, inflammatory cytokines (like TNF-α) and growth factors can accelerate the senescence process, making cells less responsive to growth signals (Ågren et al., 2000). The most characteristic example is TGF-β, which was found to provoke premature senescence of human skin fibroblasts, most probably due to the induction of an oxidative stress (Hubackova et al., 2012; Thannickal and Fanburg, 1995).

It has been suggested that the accumulation of senescent fibroblasts may play a role in impaired wound healing. In this vein, a high percentage of senescent cells have been found among fibroblasts from venous ulcers (Mendez et al., 1998). This is probably due to the effect of the wound fluid containing increased concentrations of TNF-α, the latter known to be able to provoke premature senescence (Mendez et al., 1999). In addition, fibroblasts from pressure ulcers also show premature replicative senescence and high levels of plasmin, plasmin activator inhibitor-1, and TGF-β1 that are features of senescent cells (Vande Berg et al., 1998, 2005). Furthermore, fibroblasts isolated from chronic venous leg ulcers express a phenotype reminiscent of that of senescent cells, and they also have a reduced response to mitogenic signals, ie, PDGF, despite unaltered levels of the α and β PDGFR. Furthermore, the percentage of senescent cells increased with ulcer duration, possibly due to exhaustion of the replicative potential by the long-term exposure to inflammatory cytokines (Ågren et al., 1999).

On the other hand, another study reports that fibroblasts from chronic venous leg ulcers show no evidence of senescence (Stephens et al., 2003). Nevertheless, senescent human skin fibroblasts, beyond their inability to proliferate, seem to respond in vitro comparably with early passage ones to anticontractile and antifibrotic activities of factor(s) secreted by neonatal skin fibroblasts or living cell constructs containing fibroblasts and keratinocytes (Pratsinis et al., 2013). Yet, fibroblasts from diabetic patients have an increased oxidative load, leading to premature senescence and a decrease in the ability of PDGF to activate the ERK and Akt pathways and to stimulate proliferation and migration, thus possibly leading to impaired wound healing (Bitar et al., 2013). Uncontrolled macrophage activation is considered to impair wound healing. Interestingly, macrophages with an unrestrained proinflammatory M1 activation state release increased amounts of TNF-α and hydroxyl radicals, thus inducing $p16^{INK4a}$ activation and consequently senescence in resident fibroblasts, leading to impaired healing (Sindrilaru et al., 2011). Premature fibroblast senescence can also be provoked by infections in the wound area. Pyocyanin, a virulence factor released by *Pseudomonas aeruginosa*, an important nosocomial pathogen in burn wounds, seems to be able to induce senescence via an oxidative stress mechanism and the activation of the p38 MAPK pathway, thus inhibiting the repair process (Muller et al., 2009).

The studies mentioned here indicate that the increased number of senescent cells may play a role in problematic wound healing. However, it has been reported that the matricellular protein CCN1 can induce a DNA damage response, p53 activation, and an oxidative stress, leading to $p16^{INK4a}$ activation and premature senescence and expression of antifibrotic genes. Studies with senescence-defective CCN1 mutant mice indicate that this senescence program is crucial for the resolution of fibrosis (Jun and Lau, 2010). This is in agreement with similar studies showing that the execution of senescence in some stage during the wound healing process is needed for the resolution of fibrosis also in the liver (Krizhanovsky et al., 2008). The above indicate a complex role of senescence in normal and impaired wound healing.

Cellular senescence had previously been studied only on adult humans and animal models. However, two publications have shown the existence of a senescent-like state in mice fetuses, probably participating in the tissue remodeling process at this developmental stage (Munoz-Espin et al., 2013; Storer et al., 2013). These senescent cells seem to differ from the adult senescent cells as they have no signs of DNA damage response and p53 activation. However, the role of TGF-β1 was found to be crucial in the development of this type of senescence. In this direction, we have found that TGF-β1 can provoke premature senescence in fetal and adult human skin fibroblasts (our unpublished observations), indicating a more complex role of this growth factor in tissue repair.

2.6 Conclusion

In conclusion, fetal and adult cells, and especially fibroblasts, respond differently, quantitatively and/or qualitatively, to many of the growth factors present in the wound environment. These responses may significantly affect the speed and quality of tissue repair, and they are most probably dictated by intrinsic cell factors, yet they are regulated through a dynamic interaction with ECM components. Although this field has been intensively studied for more than three decades, further research of the fine mechanisms underlying these interactions is needed. Moreover, studies on replicative or stress-induced cell senescence as well as on the role of stem cells in tissue repair (Hu et al., 2014) may be indispensable for developing therapeutic interventions toward the improvement of the quality of wound repair in adults.

References

Aghdam, S.Y., Eming, S.A., Willenborg, S., Neuhaus, B., Niessen, C.M., Partridge, L., Krieg, T., Bruning, J.C., 2012. Vascular endothelial insulin/IGF-I signaling controls skin wound vascularization. Biochem. Biophys. Res. Commun. 421 (2), 197–202.

Ågren, M.S., Eaglstein, W.H., Ferguson, M.W., Harding, K.G., Moore, K., Saarialho-Kere, U.K., Schultz, G.S., 2000. Causes and effects of the chronic inflammation in venous leg ulcers. Acta Derm. Venereol. Suppl. (Stockh.) 210, 3–17.

Ågren, M.S., Steenfos, H.H., Dabelsteen, S., Hansen, J.B., Dabelsteen, E., 1999. Proliferation and mitogenic response to PDGF-BB of fibroblasts isolated from chronic venous leg ulcers is ulcer-age dependent. J. Invest. Dermatol. 112 (4), 463–469.

Alaish, S.M., Yager, D., Diegelmann, R.F., Cohen, I.K., 1994. Biology of fetal wound healing: hyaluronate receptor expression in fetal fibroblasts. J. Pediatr. Surg. 29 (8), 1040–1043.

Ankrapp, D.P., Bevan, D.R., 1993. Insulin-like growth factor-I and human lung fibroblast-derived insulin-like growth factor-I stimulate the proliferation of human lung carcinoma cells in vitro. Cancer Res. 53 (14), 3399–3404.

Antoniades, H.N., Stathakos, D., Scher, C.D., 1975. Isolation of a cationic polypeptide from human serum that stimulates proliferation of 3T3 cells. Proc. Natl. Acad. Sci. U. S. A. 72 (7), 2635–2639.

Armatas, A.A., Pratsinis, H., Mavrogonatou, E., Angelopoulou, M.T., Kouroumalis, A., Karamanos, N.K., Kletsas, D., 2014. The differential proliferative response of fetal and adult human skin fibroblasts to TGF-β is retained when cultured in the presence of fibronectin or collagen. Biochim. Biophys. Acta 1840 (8), 2635–2642.

Armstrong, J.R., Ferguson, M.W., 1995. Ontogeny of the skin and the transition from scar-free to scarring phenotype during wound healing in the pouch young of a marsupial, monodelphis domestica. Dev. Biol. 169 (1), 242–260.

Artuc, M., Steckelings, U.M., Henz, B.M., 2002. Mast cell-fibroblast interactions: human mast cells as source and inducers of fibroblast and epithelial growth factors. J. Invest. Dermatol. 118 (3), 391–395. http://dx.doi.org/10.1046/j.0022-202x.2001.01705.x.

Ashcroft, G.S., Horan, M.A., Ferguson, M.W., 1997. The effects of ageing on wound healing: immunolocalisation of growth factors and their receptors in a murine incisional model. J. Anat. 190 (Pt 3), 351–365.

Ashcroft, G.S., Yang, X., Glick, A.B., Weinstein, M., Letterio, J.L., Mizel, D.E., Anzano, M., Greenwell-Wild, T., Wahl, S.M., Deng, C., Roberts, A.B., 1999. Mice lacking Smad3 show accelerated wound healing and an impaired local inflammatory response. Nat. Cell. Biol. 1 (5), 260–266.

Baker, A.H., Edwards, D.R., Murphy, G., 2002. Metalloproteinase inhibitors: biological actions and therapeutic opportunities. J. Cell. Sci. 115 (Pt 19), 3719–3727.

Barreca, A., De Luca, M., Del Monte, P., Bondanza, S., Damonte, G., Cariola, G., Di Marco, E., Giordano, G., Cancedda, R., Minuto, F., 1992. In vitro paracrine regulation of human keratinocyte growth by fibroblast-derived insulin-like growth factors. J. Cell. Physiol. 151 (2), 262–268. http://dx.doi.org/10.1002/jcp.1041510207.

Barrientos, S., Stojadinovic, O., Golinko, M.S., Brem, H., Tomic-Canic, M., 2008. Growth factors and cytokines in wound healing. Wound Repair Regen. 16 (5), 585–601. http://dx.doi.org/10.1111/j.1524-475X.2008.00410.x pii:WRR410.

Bartkova, J., Rezaei, N., Liontos, M., Karakaidos, P., Kletsas, D., Issaeva, N., Vassiliou, L.V., Kolettas, E., Niforou, K., Zoumpourlis, V.C., Takaoka, M., Nakagawa, H., Tort, F., Fugger, K., Johansson, F., Sehested, M., Andersen, C.L., Dyrskjot, L., Orntoft, T., Lukas, J., Kittas, C., Helleday, T., Halazonetis, T.D., Bartek, J., Gorgoulis, V.G., 2006. Oncogene-induced senescence is part of the tumorigenesis barrier imposed by DNA damage checkpoints. Nature 444 (7119), 633–637.

Baserga, R., Hongo, A., Rubini, M., Prisco, M., Valentinis, B., 1997. The IGF-I receptor in cell growth, transformation and apoptosis. Biochim. Biophys. Acta 1332 (3), F105–126. http://dx.doi.org/10.1016/S0304-419X(97)00007-3.

Beer, H.D., Fassler, R., Werner, S., 2000. Glucocorticoid-regulated gene expression during cutaneous wound repair. Vitam. Horm. 59, 217–239.

Beer, H.D., Longaker, M.T., Werner, S., 1997. Reduced expression of PDGF and PDGF receptors during impaired wound healing. J. Invest. Dermatol. 109 (2), 132–138.

Bergsten, E., Uutela, M., Li, X., Pietras, K., Ostman, A., Heldin, C.H., Alitalo, K., Eriksson, U., 2001. PDGF-D is a specific, protease-activated ligand for the PDGF β-receptor. Nat. Cell. Biol. 3 (5), 512–516.

Betsholtz, C., Westermark, B., 1984. Growth factor-induced proliferation of human fibroblasts in serum-free culture depends on cell density and extracellular calcium concentration. J. Cell. Physiol. 118 (2), 203–210. http://dx.doi.org/10.1002/jcp.1041180213.

Birkedal-Hansen, H., Moore, W.G., Bodden, M.K., Windsor, L.J., Birkedal-Hansen, B., DeCarlo, A., Engler, J.A., 1993. Matrix metalloproteinases: a review. Crit. Rev. Oral. Biol. Med. 4 (2), 197–250.

Bitar, M.S., Abdel-Halim, S.M., Al-Mulla, F., 2013. Caveolin-1/PTRF upregulation constitutes a mechanism for mediating p53-induced cellular senescence: implications for evidence-based therapy of delayed wound healing in diabetes. Am. J. Physiol. Endocrinol. Metab. 305 (8), E951–963.

Bodnar, A.G., Ouellette, M., Frolkis, M., Holt, S.E., Chiu, C.P., Morin, G.B., Harley, C.B., Shay, J.W., Lichtsteiner, S., Wright, W.E., 1998. Extension of life-span by introduction of telomerase into normal human cells. Science 279 (5349), 349–352.

Bonner, J.C., 2004. Regulation of PDGF and its receptors in fibrotic diseases. Cytokine Growth Factor Rev. 15 (4), 255–273. http://dx.doi.org/10.1016/j.cytogfr.2004.03.006, S1359610104000164.

Bradham, D.M., Igarashi, A., Potter, R.L., Grotendorst, G.R., 1991. Connective tissue growth factor: a cysteine-rich mitogen secreted by human vascular endothelial cells is related to the SRC-induced immediate early gene product CEF-10. J. Cell. Biol. 114 (6), 1285–1294.

Broadley, K.N., Aquino, A.M., Woodward, S.C., Buckley-Sturrock, A., Sato, Y., Rifkin, D.B., Davidson, J.M., 1989. Monospecific antibodies implicate basic fibroblast growth factor in normal wound repair. Lab. Invest. 61 (5), 571–575.

Broker, B.J., Chakrabarti, R., Blynman, T., Roesler, J., Wang, M.B., Srivatsan, E.S., 1999. Comparison of growth factor expression in fetal and adult fibroblasts: a preliminary report. Arch. Otolaryngol. Head Neck Surg. 125 (6), 676–680.

Brown, R.L., Ormsby, I., Doetschman, T.C., Greenhalgh, D.G., 1995. Wound healing in the transforming growth factor-β-deficient mouse. Wound Repair Regen. 3 (1), 25–36.

Buchanan, E.P., Longaker, M.T., Lorenz, H.P., 2009. Fetal skin wound healing. Adv. Clin. Chem. 48, 137–161.

Buckley-Sturrock, A., Woodward, S.C., Senior, R.M., Griffin, G.L., Klagsbrun, M., Davidson, J.M., 1989. Differential stimulation of collagenase and chemotactic activity in fibroblasts derived from rat wound repair tissue and human skin by growth factors. J. Cell. Physiol. 138 (1), 70–78. http://dx.doi.org/10.1002/jcp.1041380111.

Burrington, J.D., 1971. Wound healing in the fetal lamb. J. Pediatr. Surg. 6 (5), 523–528.

Campisi, J., d'Adda di Fagagna, F., 2007. Cellular senescence: when bad things happen to good cells. Nat. Rev. Mol. Cell. Biol. 8 (9), 729–740.

Carter, R., Jain, K., Sykes, V., Lanning, D., 2009. Differential expression of procollagen genes between mid- and late-gestational fetal fibroblasts. J. Surg. Res. 156 (1), 90–94.

Cass, D.L., Bullard, K.M., Sylvester, K.G., Yang, E.Y., Longaker, M.T., Adzick, N.S., 1997. Wound size and gestational age modulate scar formation in fetal wound repair. J. Pediatr. Surg. 32 (3), 411–415.

Chen, W., Fu, X., Ge, S., Sun, T., Sheng, Z., 2007. Differential expression of matrix metalloproteinases and tissue-derived inhibitors of metalloproteinase in fetal and adult skins. Int. J. Biochem. Cell. Biol. 39 (5), 997–1005.

Chen, W.Y., Grant, M.E., Schor, A.M., Schor, S.L., 1989. Differences between adult and foetal fibroblasts in the regulation of hyaluronate synthesis: correlation with migratory activity. J. Cell Sci. 94 (Pt 3), 577–584.

Choy, L., Derynck, R., 1998. The type II transforming growth factor (TGF)-β receptor-interacting protein TRIP-1 acts as a modulator of the TGF-β response. J. Biol. Chem. 273 (47), 31455–31462.

Chrissouli, S., Pratsinis, H., Velissariou, V., Anastasiou, A., Kletsas, D., 2010. Human amniotic fluid stimulates the proliferation of human fetal and adult skin fibroblasts: the roles of BFGF and PDGF and of the ERK and AKT signaling pathways. Wound Repair Regen. 18 (6), 643–654.

Clark, J.G., Madtes, D.K., Raghu, G., 1993. Effects of platelet-derived growth factor isoforms on human lung fibroblast proliferation and procollagen gene expression. Exp. Lung Res. 19 (3), 327–344.

Clark, R.A., Folkvord, J.M., Hart, C.E., Murray, M.J., McPherson, J.M., 1989. Platelet isoforms of platelet-derived growth factor stimulate fibroblasts to contract collagen matrices. J. Clin. Invest. 84 (3), 1036–1040.

Clark, R.A.F., 1996. Wound repair: overview and general considerations. In: Clark, R.A.F. (Ed.), The Molecular and Cellular Biology of Wound Repair. Plenum Press, New York and London, pp. 3–50.

Coleman, C., Tuan, T.L., Buckley, S., Anderson, K.D., Warburton, D., 1998. Contractility, transforming growth factor-β, and plasmin in fetal skin fibroblasts: role in scarless wound healing. Pediatr. Res. 43 (3), 403–409.

Colwell, A.S., Krummel, T.M., Longaker, M.T., Lorenz, H.P., 2006. Fetal and adult fibroblasts have similar TGF-β-mediated, Smad-dependent signaling pathways. Plast. Reconstr. Surg. 117 (7), 2277–2283.

Colwell, A.S., Yun, R., Krummel, T.M., Longaker, M.T., Lorenz, H.P., 2007. Keratinocytes modulate fetal and postnatal fibroblast transforming growth factor-β and Smad expression in co-culture. Plast. Reconstr. Surg. 119 (5), 1440–1445.

Coppe, J.P., Kauser, K., Campisi, J., Beausejour, C.M., 2006. Secretion of vascular endothelial growth factor by primary human fibroblasts at senescence. J. Biol. Chem. 281 (40), 29568–29574. http://dx.doi.org/10.1074/jbc.M603307200 pii:M603307200.

Counter, C.M., Avilion, A.A., LeFeuvre, C.E., Stewart, N.G., Greider, C.W., Harley, C.B., Bacchetti, S., 1992. Telomere shortening associated with chromosome instability is arrested in immortal cells which express telomerase activity. EMBO J. 11 (5), 1921–1929.

Cullen, B., Silcock, D., Brown, L.J., Gosiewska, A., Geesin, J.C., 1997. The differential regulation and secretion of proteinases from fetal and neonatal fibroblasts by growth factors. Int. J. Biochem. Cell Biol. 29 (1), 241–250.

Dang, C.M., Beanes, S.R., Lee, H., Zhang, X., Soo, C., Ting, K., 2003. Scarless fetal wounds are associated with an increased matrix metalloproteinase-to-tissue-derived inhibitor of metalloproteinase ratio. Plast. Reconstr. Surg. 111 (7), 2273–2285.

DePalma, R.L., Krummel, T.M., Durham 3rd, L.A., Michna, B.A., Thomas, B.L., Nelson, J.M., Diegelmann, R.F., 1989. Characterization and quantitation of wound matrix in the fetal rabbit. Matrix 9 (3), 224–231.

Derynck, R., Zhang, Y.E., 2003. Smad-dependent and Smad-independent pathways in TGF-β family signalling. Nature 425 (6958), 577–584.

Desmouliere, A., Geinoz, A., Gabbiani, F., Gabbiani, G., 1993. Transforming growth factor-β1 induces α-smooth muscle actin expression in granulation tissue myofibroblasts and in quiescent and growing cultured fibroblasts. J. Cell Biol. 122 (1), 103–111.

Dimri, G.P., Lee, X., Basile, G., Acosta, M., Scott, G., Roskelley, C., Medrano, E.E., Linskens, M., Rubelj, I., Pereira-Smith, O., et al., 1995. A biomarker that identifies senescent human cells in culture and in aging skin in vivo. Proc. Natl. Acad. Sci. U. S. A. 92 (20), 9363–9367.

Duncan, B.W., Qian, J., Liu, X., Bhatnagar, R.S., 1992. Regulation of prolyl hydroxylase activity in fetal and adult fibroblasts. In: Adzick, N.S., Longaker, M.T. (Eds.), Fetal Wound Healing. Elsevier Scientific Press, New York (NY), pp. 303–323.

Durante, C., Agostini, F., Abbruzzese, L., Toffola, R.T., Zanolin, S., Suine, C., Mazzucato, M., 2013. Growth factor release from platelet concentrates: analytic quantification and characterization for clinical applications. Vox Sang. 105 (2), 129–136.

Eckes, B., Nischt, R., Krieg, T., 2010. Cell-matrix interactions in dermal repair and scarring. Fibrogenesis Tissue Repair 3, 4.

Ellis, I., Banyard, J., Schor, S.L., 1997. Differential response of fetal and adult fibroblasts to cytokines: cell migration and hyaluronan synthesis. Development 124 (8), 1593–1600.

Ellis, I.R., Schor, S.L., 1996. Differential effects of TGF-β1 on hyaluronan synthesis by fetal and adult skin fibroblasts: implications for cell migration and wound healing. Exp. Cell Res. 228 (2), 326–333.

Ellis, I.R., Schor, S.L., 1998. Differential motogenic and biosynthetic response of fetal and adult skin fibroblasts to TGF-β isoforms. Cytokine 10 (4), 281–289.

Eming, S.A., Krieg, T., Davidson, J.M., 2007. Inflammation in wound repair: molecular and cellular mechanisms. J. Invest. Dermatol. 127 (3), 514–525.

Eppley, B.L., Woodell, J.E., Higgins, J., 2004. Platelet quantification and growth factor analysis from platelet-rich plasma: implications for wound healing. Plast. Reconstr. Surg. 114 (6), 1502–1508.

Eswarakumar, V.P., Lax, I., Schlessinger, J., 2005. Cellular signaling by fibroblast growth factor receptors. Cytokine Growth Factor Rev. 16 (2), 139–149.

Ferrara, N., Gerber, H.P., LeCouter, J., 2003. The biology of VEGF and its receptors. Nat. Med. 9 (6), 669–676.

Frantz, F.W., Bettinger, D.A., Haynes, J.H., Johnson, D.E., Harvey, K.M., Dalton, H.P., Yager, D.R., Diegelmann, R.F., Cohen, I.K., 1993. Biology of fetal repair: the presence of bacteria in fetal wounds induces an adult-like healing response. J. Pediatr. Surg. 28 (3), 428–433 (discussion 433–424).

Frazier, K., Williams, S., Kothapalli, D., Klapper, H., Grotendorst, G.R., 1996. Stimulation of fibroblast cell growth, matrix production, and granulation tissue formation by connective tissue growth factor. J. Invest. Dermatol. 107 (3), 404–411.

Fredriksson, L., Li, H., Eriksson, U., 2004. The PDGF family: four gene products form five dimeric isoforms. Cytokine Growth Factor Rev. 15 (4), 197–204.

Gabbiani, G., Chaponnier, C., Huttner, I., 1978. Cytoplasmic filaments and gap junctions in epithelial cells and myofibroblasts during wound healing. J. Cell Biol. 76 (3), 561–568.

Gale, N.W., Yancopoulos, G.D., 1999. Growth factors acting via endothelial cell-specific receptor tyrosine kinases: Vegfs, angiopoietins, and ephrins in vascular development. Genes Dev. 13 (9), 1055–1066.

Georgakopoulou, E.A., Tsimaratou, K., Evangelou, K., Fernandez Marcos, P.J., Zoumpourlis, V., Trougakos, I.P., Kletsas, D., Bartek, J., Serrano, M., Gorgoulis, V.G., 2013. Specific lipofuscin staining as a novel biomarker to detect replicative and stress-induced senescence. A method applicable in cryo-preserved and archival tissues. Aging (Albany N. Y.) 5 (1), 37–50.

Giannouli, C.C., Kletsas, D., 2006. TGF-β regulates differentially the proliferation of fetal and adult human skin fibroblasts via the activation of PKA and the autocrine action of FGF-2. Cell. Signal. 18 (9), 1417–1429.

Gill, S.E., Parks, W.C., 2008. Metalloproteinases and their inhibitors: regulators of wound healing. Int. J. Biochem. Cell Biol. 40 (6-7), 1334–1347.

Gorgoulis, V.G., Pratsinis, H., Zacharatos, P., Demoliou, C., Sigala, F., Asimacopoulos, P.J., Papavassiliou, A.G., Kletsas, D., 2005. P53-dependent ICAM-1 overexpression in senescent human cells identified in atherosclerotic lesions. Lab. Invest. 85 (4), 502–511.

Gosiewska, A., Yi, C.F., Brown, L.J., Cullen, B., Silcock, D., Geesin, J.C., 2001. Differential expression and regulation of extracellular matrix-associated genes in fetal and neonatal fibroblasts. Wound Repair Regen. 9 (3), 213–222.

Greaves, N.S., Ashcroft, K.J., Baguneid, M., Bayat, A., 2013. Current understanding of molecular and cellular mechanisms in fibroplasia and angiogenesis during acute wound healing. J. Dermatol. Sci. 72 (3), 206–217.

Grinnell, F., 2003. Fibroblast biology in three-dimensional collagen matrices. Trends Cell Biol. 13 (5), 264–269.

Hanasono, M.M., Kita, M., Mikulec, A.A., Lonergan, D., Koch, R.J., 2003. Autocrine growth factor production by fetal, keloid, and normal dermal fibroblasts. Arch. Facial. Plast. Surg. 5 (1), 26–30. qoa10029.

Hantash, B.M., Zhao, L., Knowles, J.A., Lorenz, H.P., 2008. Adult and fetal wound healing. Front. Biosci. 13, 51–61. 2559.

Hardwicke, J., Schmaljohann, D., Boyce, D., Thomas, D., 2008. Epidermal growth factor therapy and wound healing–past, present and future perspectives. Surgeon 6 (3), 172–177.

Hayflick, L., 1965. The limited in vitro lifetime of human diploid cell strains. Exp. Cell Res. 37, 614–636.

Heldin, C.H., Eriksson, U., Östman, A., 2002. New members of the platelet-derived growth factor family of mitogens. Arch. Biochem. Biophys. 398 (2), 284–290. http://dx.doi.org/10.1006/abbi.2001.2707 pii:S0003986101927079.

Heldin, C.H., Westermark, B., 1999. Mechanism of action and in vivo role of platelet-derived growth factor. Physiol. Rev. 79 (4), 1283–1316.

Hinz, B., McCulloch, C.A., 2012. Integrin function in heart fibrosis: mechanical strain, transforming growth factor-β1 activation, and collagen glycation. In: Karamanos, N. (Ed.), Extracellular Matrix: Pathobiology and Signaling. De Gruyter, Berlin/Boston, pp. 406–431.

Holley, R.W., Baldwin, J.H., Greenfield, S., Armour, R., 1985. A growth regulatory factor that can both inhibit and stimulate growth. Ciba Found. Symp. 116, 241–252.

Howdieshell, T.R., Callaway, D., Webb, W.L., Gaines, M.D., Procter Jr., C.D., Sathyanarayana, P.J.S., Brock, T.L., McNeil, P.L., 2001. Antibody neutralization of vascular endothelial growth factor inhibits wound granulation tissue formation. J. Surg. Res. 96 (2), 173–182.

Hu, M.S., Rennert, R.C., McArdle, A., Chung, M.T., Walmsley, G.G., Longaker, M.T., Lorenz, H.P., 2014. The role of stem cells during scarless skin wound healing. Adv. Wound Care (New Rochelle) 3 (4), 304–314.

Hubackova, S., Krejcikova, K., Bartek, J., Hodny, Z., 2012. IL1- and TGF-β-Nox4 signaling, oxidative stress and DNA damage response are shared features of replicative, oncogene-induced, and drug-induced paracrine 'bystander senescence'. Aging (Albany N. Y.) 4 (12), 932–951.

Ihn, H., 2002. Pathogenesis of fibrosis: role of TGF-beta and CTGF. Curr. Opin. Rheumatol. 14 (6), 681–685.

Ishikawa, O., LeRoy, E.C., Trojanowska, M., 1990. Mitogenic effect of transforming growth factor β1 on human fibroblasts involves the induction of platelet-derived growth factor α receptors. J. Cell Physiol. 145 (1), 181–186. http://dx.doi.org/10.1002/jcp.1041450124.

Itoh, N., Ornitz, D.M., 2011. Fibroblast growth factors: from molecular evolution to roles in development, metabolism and disease. J. Biochem. 149 (2), 121–130.

Jeong, D.U., Lee, C.R., Lee, J.H., Pak, J., Kang, L.W., Jeong, B.C., Lee, S.H., 2014. Clinical applications of platelet-rich plasma in patellar tendinopathy. Biomed. Res. Int. 2014, 249498.

Johnson, N.R., Wang, Y., 2013. Controlled delivery of heparin-binding EGF-like growth factor yields fast and comprehensive wound healing. J. Control. Release 166 (2), 124–129.

Jones, J.I., Clemmons, D.R., 1995. Insulin-like growth factors and their binding proteins: biological actions. Endocr. Rev. 16 (1), 3–34.

Juliano, R.L., Haskill, S., 1993. Signal transduction from the extracellular matrix. J. Cell Biol. 120 (3), 577–585.

Jun, J.I., Lau, L.F., 2010. The matricellular protein CCN1 induces fibroblast senescence and restricts fibrosis in cutaneous wound healing. Nat. Cell Biol. 12 (7), 676–685.

Kardassis, D., Murphy, C., Fotsis, T., Moustakas, A., Stournaras, C., 2009. Control of transforming growth factor β signal transduction by small GTPases. FEBS J. 276 (11), 2947–2965.

Kennedy, C.I., Diegelmann, R.F., Haynes, J.H., Yager, D.R., 2000. Proinflammatory cytokines differentially regulate hyaluronan synthase isoforms in fetal and adult fibroblasts. J. Pediatr. Surg. 35 (6), 874–879.

Khorramizadeh, M.R., Tredget, E.E., Telasky, C., Shen, Q., Ghahary, A., 1999. Aging differentially modulates the expression of collagen and collagenase in dermal fibroblasts. Mol. Cell Biochem. 194 (1–2), 99–108.

Kletsas, D., 2003. Aging of fibroblasts. In: Kaul, S.C., Wadhwa, R. (Eds.), Aging of Cells in and Outside the Body. Kluwer Academic Publishers, pp. 27–46.

Ko, J., Jun, H., Chung, H., Yoon, C., Kim, T., Kwon, M., Lee, S., Jung, S., Kim, M., Park, J.H., 2011. Comparison of EGF with VEGF non-viral gene therapy for cutaneous wound healing of streptozotocin diabetic mice. Diabetes Metab. J. 35 (3), 226–235.

Koch, R.M., Roche, N.S., Parks, W.T., Ashcroft, G.S., Letterio, J.J., Roberts, A.B., 2000. Incisional wound healing in transforming growth factor-β1 null mice. Wound Repair Regen. 8 (3), 179–191.

Kondo, H., Matsuda, R., Yonezawa, Y., 1993. Autonomous migration of human fetal skin fibroblasts into a denuded area in a cell monolayer is mediated by basic fibroblast growth factor and collagen. In Vitro Cell. Dev. Biol. Anim. 29A (12), 929–935.

Kondo, H., Yonezawa, Y., 1995. Fetal-adult phenotype transition, in terms of the serum dependency and growth factor requirements, of human skin fibroblast migration. Exp. Cell Res. 220 (2), 501–504.

Krizhanovsky, V., Xue, W., Zender, L., Yon, M., Hernando, E., Lowe, S.W., 2008. Implications of cellular senescence in tissue damage response, tumor suppression, and stem cell biology. Cold Spring Harb. Symp. Quant. Biol. 73, 513–522.

Kurobe, M., Furukawa, S., Hayashi, K., 1985. Synthesis and secretion of an epidermal growth factor (EGF) by human fibroblast cells in culture. Biochem. Biophys. Res. Commun. 131 (3), 1080–1085. http://dx.doi.org/10.1016/0006-291X(85)90201-3.

Laestander, C., Engstrom, W., 2014. Role of fibroblast growth factors in elicitation of cell responses. Cell Prolif. 47 (1), 3–11.

LaRochelle, W.J., Jeffers, M., McDonald, W.F., Chillakuru, R.A., Giese, N.A., Lokker, N.A., Sullivan, C., Boldog, F.L., Yang, M., Vernet, C., Burgess, C.E., Fernandes, E., Deegler, L.L., Rittman, B., Shimkets, J., Shimkets, R.A., Rothberg, J.M., Lichenstein, H.S., 2001. PDGF-D, a new protease-activated growth factor. Nat. Cell Biol. 3 (5), 517–521.

Lee, N.J., Wang, S.J., Durairaj, K.K., Srivatsan, E.S., Wang, M.B., 2000. Increased expression of transforming growth factor-β1, acidic fibroblast growth factor, and basic fibroblast growth factor in fetal versus adult fibroblast cell lines. Laryngoscope 110 (4), 616–619. http://dx.doi.org/10.1097/00005537-200004000-00015.

Levine, A.C., Ren, M., Huber, G.K., Kirschenbaum, A., 1992. The effect of androgen, estrogen, and growth factors on the proliferation of cultured fibroblasts derived from human fetal and adult prostates. Endocrinology 130 (4), 2413–2419.

Levitzki, A., 2004. PDGF receptor kinase inhibitors for the treatment of PDGF driven diseases. Cytokine Growth Factor Rev 15 (4), 229–235.

Li, X., Ponten, A., Aase, K., Karlsson, L., Abramsson, A., Uutela, M., Backstrom, G., Hellstrom, M., Bostrom, H., Li, H., Soriano, P., Betsholtz, C., Heldin, C.H., Alitalo, K., Ostman, A., Eriksson, U., 2000. PDGF-C is a new protease-activated ligand for the PDGF α-receptor. Nat. Cell Biol. 2 (5), 302–309.

Lonergan, D.M., Mikulec, A.A., Hanasono, M.M., Kita, M., Koch, R.J., 2003. Growth factor profile of irradiated human dermal fibroblasts using a serum-free method. Plast. Reconstr. Surg. 111 (6), 1960–1968. http://dx.doi.org/10.1097/01.PRS.0000055065.41599.75.

Longaker, M.T., Adzick, N.S., Hall, J.L., Stair, S.E., Crombleholme, T.M., Duncan, B.W., Bradley, S.M., Harrison, M.R., Stern, R., 1990. Studies in fetal wound healing, VII. Fetal wound healing may be modulated by hyaluronic acid stimulating activity in amniotic fluid. J. Pediatr. Surg. 25 (4), 430–433.

Longaker, M.T., Whitby, D.J., Ferguson, M.W., Lorenz, H.P., Harrison, M.R., Adzick, N.S., 1994. Adult skin wounds in the fetal environment heal with scar formation. Ann. Surg. 219 (1), 65–72.

Lorenz, H.P., Longaker, M.T., Perkocha, L.A., Jennings, R.W., Harrison, M.R., Adzick, N.S., 1992. Scarless wound repair: a human fetal skin model. Development 114 (1), 253–259.

Lorenzini, A., Tresini, M., Mawal-Dewan, M., Frisoni, L., Zhang, H., Allen, R.G., Sell, C., Cristofalo, V.J., 2002. Role of the Raf/MEK/ERK and the PI3K/Akt(PKB) pathways in fibroblast senescence. Exp. Gerontol. 37 (10–11), 1149–1156.

Luetteke, N.C., Qiu, T.H., Peiffer, R.L., Oliver, P., Smithies, O., Lee, D.C., 1993. TGFα deficiency results in hair follicle and eye abnormalities in targeted and waved-1 mice. Cell 73 (2), 263–278.

Mann, G.B., Fowler, K.J., Gabriel, A., Nice, E.C., Williams, R.L., Dunn, A.R., 1993. Mice with a null mutation of the TGF-α gene have abnormal skin architecture, wavy hair, and curly whiskers and often develop corneal inflammation. Cell 73 (2), 249–261.

Massague, J., 1990. The transforming growth factor-β family. Annu. Rev. Cell Biol. 6, 597–641.

Mast, B.A., Flood, L.C., Haynes, J.H., DePalma, R.L., Cohen, I.K., Diegelmann, R.F., Krummel, T.M., 1991. Hyaluronic acid is a major component of the matrix of fetal rabbit skin and wounds: implications for healing by regeneration. Matrix 11 (1), 63–68.

Mauviel, A., 2005. Transforming growth factor-β: a key mediator of fibrosis. Methods Mol. Med. 117, 69–80.

McCallion, R.L., Ferguson, M.W.J., 1996. Fetal wound healing and the development of anti-scarring therapies for adult wound healing. In: Clark, R.A.F. (Ed.), The Molecular and Cellular Biology of Wound Repair. Plenum Press, New York and London, pp. 561–600.

Mendez, M.V., Raffetto, J.D., Phillips, T., Menzoian, J.O., Park, H.Y., 1999. The proliferative capacity of neonatal skin fibroblasts is reduced after exposure to venous ulcer wound fluid: a potential mechanism for senescence in venous ulcers. J. Vasc. Surg. 30 (4), 734–743.

Mendez, M.V., Stanley, A., Park, H.Y., Shon, K., Phillips, T., Menzoian, J.O., 1998. Fibroblasts cultured from venous ulcers display cellular characteristics of senescence. J. Vasc. Surg. 28 (5), 876–883.

Merkel, J.R., DiPaolo, B.R., Hallock, G.G., Rice, D.C., 1988. Type I and type III collagen content of healing wounds in fetal and adult rats. Proc. Soc. Exp. Biol. Med. 187 (4), 493–497.

Micallef, L., Vedrenne, N., Billet, F., Coulomb, B., Darby, I.A., Desmouliere, A., 2012. The myofibroblast, multiple origins for major roles in normal and pathological tissue repair. Fibrogenesis Tissue Repair 5 (Suppl. 1), S5. Proceedings of Fibroproliferative disorders: from biochemical analysis to targeted therapies Petro E. Petrides and David Brenner.

Moulin, V., Plamondon, M., 2002. Differential expression of collagen integrin receptor on fetal versus adult skin fibroblasts: implication in wound contraction during healing. Br. J. Dermatol. 147 (5), 886–892.

Moulin, V., Tam, B.Y., Castilloux, G., Auger, F.A., O'Connor-McCourt, M.D., Philip, A., Germain, L., 2001. Fetal and adult human skin fibroblasts display intrinsic differences in contractile capacity. J. Cell Physiol. 188 (2), 211–222.

Moustakas, A., Heldin, C.H., 2009. The regulation of TGF-β signal transduction. Development 136 (22), 3699–3714.

Muller, M., Li, Z., Maitz, P.K., 2009. Pseudomonas pyocyanin inhibits wound repair by inducing premature cellular senescence: role for p38 mitogen-activated protein kinase. Burns 35 (4), 500–508.

Munoz-Espin, D., Canamero, M., Maraver, A., Gomez-Lopez, G., Contreras, J., Murillo-Cuesta, S., Rodriguez-Baeza, A., Varela-Nieto, I., Ruberte, J., Collado, M., Serrano, M., 2013. Programmed cell senescence during mammalian embryonic development. Cell 155 (5), 1104–1118.

Munz, B., Smola, H., Engelhardt, F., Bleuel, K., Brauchle, M., Lein, I., Evans, L.W., Huylebroeck, D., Balling, R., Werner, S., 1999. Overexpression of activin A in the skin of transgenic mice reveals new activities of activin in epidermal morphogenesis, dermal fibrosis and wound repair. EMBO J. 18 (19), 5205–5215.

Nanney, L.B., King, L.E., 1996. Epidermal growth factor and transforming growth factor-α. In: Clark, R.A.F. (Ed.), The Molecular and Cellular Biology of Wound Repair. Plenum Press, New York and London, pp. 171–194.

Narayanan, A.S., Page, R.C., Swanson, J., 1989. Collagen synthesis by human fibroblasts. Regulation by transforming growth factor-β in the presence of other inflammatory mediators. Biochem. J. 260 (2), 463–469.

Nath, R.K., LaRegina, M., Markham, H., Ksander, G.A., Weeks, P.M., 1994. The expression of transforming growth factor type β in fetal and adult rabbit skin wounds. J. Pediatr. Surg. 29 (3), 416–421.

Navarro, A., Rezaiekhaligh, M., Keightley, J.A., Mabry, S.M., Perez, R.E., Ekekezie, I.I., 2009. Higher trip-1 level explains diminished collagen contraction ability of fetal versus adult fibroblasts. Am. J. Physiol. Lung Cell. Mol. Physiol. 296 (6), L928–L935.

Nikitovic, D., Pratsinis, H., Berdiaki, A., Gialeli, C., Kletsas, D., Tzanakakis George, N., 2012. Growth factor signaling and extracellular matrix. In: Karamanos, N. (Ed.), Extracellular Matrix: Pathobiology and Signaling. De Gruyter, Berlin/Boston, pp. 741–762.

Nissen, N.N., Polverini, P.J., Koch, A.E., Volin, M.V., Gamelli, R.L., DiPietro, L.A., 1998. Vascular endothelial growth factor mediates angiogenic activity during the proliferative phase of wound healing. Am. J. Pathol. 152 (6), 1445–1452.

Okada-Ban, M., Thiery, J.P., Jouanneau, J., 2000. Fibroblast growth factor-2. Int. J. Biochem. Cell Biol. 32 (3), 263–267.

Oladipupo, S.S., Smith, C., Santeford, A., Park, C., Sene, A., Wiley, L.A., Osei-Owusu, P., Hsu, J., Zapata, N., Liu, F., Nakamura, R., Lavine, K.J., Blumer, K.J., Choi, K., Apte, R.S., Ornitz, D.M., 2014. Endothelial cell FGF signaling is required for injury response but not for vascular homeostasis. Proc. Natl. Acad. Sci. U. S. A. 111 (37), 13379–13384.

Ortega, S., Ittmann, M., Tsang, S.H., Ehrlich, M., Basilico, C., 1998. Neuronal defects and delayed wound healing in mice lacking fibroblast growth factor 2. Proc. Natl. Acad. Sci. U. S. A. 95 (10), 5672–5677.

Ostman, A., Andersson, M., Hellman, U., Heldin, C.H., 1991. Identification of three amino acids in the platelet-derived growth factor (PDGF) b-chain that are important for binding to the PDGF β-receptor. J. Biol. Chem. 266 (16), 10073–10077.

Parekh, A., Sandulache, V.C., Lieb, A.S., Dohar, J.E., Hebda, P.A., 2007. Differential regulation of free-floating collagen gel contraction by human fetal and adult dermal fibroblasts in response to prostaglandin e2 mediated by an ep2/camp-dependent mechanism. Wound Repair Regen. 15 (3), 390–398.

Parekh, A., Sandulache, V.C., Singh, T., Cetin, S., Sacks, M.S., Dohar, J.E., Hebda, P.A., 2009. Prostaglandin e2 differentially regulates contraction and structural reorganization of anchored collagen gels by human adult and fetal dermal fibroblasts. Wound Repair Regen. 17 (1), 88–98.

Park, S.H., 2005. Fine tuning and cross-talking of TGF-β signal by inhibitory Smads. J. Biochem. Mol. Biol. 38 (1), 9–16.

Paulsson, Y., Beckmann, M.P., Westermark, B., Heldin, C.H., 1988. Density-dependent inhibition of cell growth by transforming growth factor-β1 in normal human fibroblasts. Growth Factors 1 (1), 19–27.

Peled, Z.M., Rhee, S.J., Hsu, M., Chang, J., Krummel, T.M., Longaker, M.T., 2001. The ontogeny of scarless healing II: EGF and PDGF-b gene expression in fetal rat skin and fibroblasts as a function of gestational age. Ann. Plast. Surg. 47 (4), 417–424.

Peng, C., Chen, B., Kao, H.K., Murphy, G., Orgill, D.P., Guo, L., 2011. Lack of FGF-7 further delays cutaneous wound healing in diabetic mice. Plast. Reconstr. Surg. 128 (6), 673e–684e.

Pierce, G.F., Tarpley, J.E., Tseng, J., Bready, J., Chang, D., Kenney, W.C., Rudolph, R., Robson, M.C., Vande Berg, J., Reid, P., et al., 1995. Detection of platelet-derived growth factor (PDGF)-AA in actively healing human wounds treated with recombinant PDGF-BB and absence of PDGF in chronic nonhealing wounds. J. Clin. Invest. 96 (3), 1336–1350.

Piscatelli, S.J., Michaels, B.M., Gregory, P., Jennings, R.W., Longaker, M.T., Harrison, M.R., Siebert, J.W., 1994. Fetal fibroblast contraction of collagen matrices in vitro: the effects of epidermal growth factor and transforming growth factor-β. Ann. Plast. Surg. 33 (1), 38–45.

Powers, C.J., McLeskey, S.W., Wellstein, A., 2000. Fibroblast growth factors, their receptors and signaling. Endocr. Relat. Cancer 7 (3), 165–197.

Pratsinis, H., Armatas, A., Dimozi, A., Lefaki, M., Vassiliu, P., Kletsas, D., 2013. Paracrine anti-fibrotic effects of neonatal cells and living cell constructs on young and senescent human dermal fibroblasts. Wound Repair Regen. 21 (6), 842–851.

Pratsinis, H., Giannouli, C.C., Zervolea, I., Psarras, S., Stathakos, D., Kletsas, D., 2004. Differential proliferative response of fetal and adult human skin fibroblasts to transforming growth factor-β. Wound Repair Regen. 12 (3), 374–383.

Pratsinis, H., Kletsas, D., Stathakos, D., 1997. Autocrine growth regulation in fetal and adult human fibroblasts. Biochem. Biophys. Res. Commun. 237 (2), 348–353. http://dx.doi.org/10.1006/bbrc.1997.7136 pii:S0006-291X(97)97136-9.

Rojas-Valencia, L., Montiel, F., Montano, M., Selman, M., Pardo, A., 1995. Expression of a 2.8-kb PDGF-B/c-sis transcript and synthesis of PDGF-like protein by human lung fibroblasts. Chest 108 (1), 240–245.

Rolfe, K.J., Cambrey, A.D., Richardson, J., Irvine, L.M., Grobbelaar, A.O., Linge, C., 2007a. Dermal fibroblasts derived from fetal and postnatal humans exhibit distinct responses to insulin like growth factors. BMC Dev. Biol. 7, 124. http://dx.doi.org/10.1186/1471-213X-7-124 pii:1471-213X-7-124.

Rolfe, K.J., Irvine, L.M., Grobbelaar, A.O., Linge, C., 2007b. Differential gene expression in response to transforming growth factor-β1 by fetal and postnatal dermal fibroblasts. Wound Repair Regen. 15 (6), 897–906. http://dx.doi.org/10.1111/j.1524-475X.2007.00314.x pii:WRR314.

Rolfe, K.J., Richardson, J., Vigor, C., Irvine, L.M., Grobbelaar, A.O., Linge, C., 2007c. A role for TGF-β1-induced cellular responses during wound healing of the non-scarring early human fetus? J. Invest. Dermatol. 127 (11), 2656–2667. http://dx.doi.org/10.1038/sj.jid.5700951 pii:5700951.

Rozman, P., Bolta, Z., 2007. Use of platelet growth factors in treating wounds and soft-tissue injuries. Acta Dermatovenerol. Alp. Pannonica. Adriat. 16 (4), 156–165.

Ryan, P.D., Goss, P.E., 2008. The emerging role of the insulin-like growth factor pathway as a therapeutic target in cancer. Oncologist 13 (1), 16–24. http://dx.doi.org/10.1634/theoncologist.2007-0199 pii:13/1/16.

Schiller, M., Javelaud, D., Mauviel, A., 2004. TGF-β-induced Smad signaling and gene regulation: consequences for extracellular matrix remodeling and wound healing. J. Dermatol. Sci. 35 (2), 83–92.

Schor, S.L., 1994. Cytokine control of cell motility: modulation and mediation by the extracellular matrix. Prog. Growth Factor Res. 5 (2), 223–248.

Schor, S.L., Schor, A.M., Grey, A.M., Chen, J., Rushton, G., Grant, M.E., Ellis, I., 1989. Mechanism of action of the migration stimulating factor produced by fetal and cancer patient fibroblasts: effect on hyaluronic and synthesis. In Vitro Cell Dev. Biol. 25 (8), 737–746.

Schor, S.L., Schor, A.M., Grey, A.M., Rushton, G., 1988. Foetal and cancer patient fibroblasts produce an autocrine migration-stimulating factor not made by normal adult cells. J. Cell. Sci. 90 (Pt 3), 391–399.

Schor, S.L., Schor, A.M., Rushton, G., Smith, L., 1985. Adult, foetal and transformed fibroblasts display different migratory phenotypes on collagen gels: evidence for an isoformic transition during foetal development. J. Cell Sci. 73, 221–234.
Schultz, G., Rotatori, D.S., Clark, W., 1991. EGF and TGF-α in wound healing and repair. J. Cell Biochem. 45 (4), 346–352. http://dx.doi.org/10.1002/jcb.240450407.
Secker, G.A., Shortt, A.J., Sampson, E., Schwarz, Q.P., Schultz, G.S., Daniels, J.T., 2008. TGF-β stimulated re-epithelialisation is regulated by CTGF and Ras/MEK/ERK signalling. Exp. Cell Res. 314 (1), 131–142.
Serrano, M., Lin, A.W., McCurrach, M.E., Beach, D., Lowe, S.W., 1997. Oncogenic ras provokes premature cell senescence associated with accumulation of p53 and $p16^{INK4a}$. Cell 88 (5), 593–602.
Shah, M., Foreman, D.M., Ferguson, M.W., 1995. Neutralisation of TGF-β1 and TGF-β2 or exogenous addition of TGF-β3 to cutaneous rat wounds reduces scarring. J. Cell Sci. 108 (Pt 3), 985–1002.
Shelton, D.N., Chang, E., Whittier, P.S., Choi, D., Funk, W.D., 1999. Microarray analysis of replicative senescence. Curr. Biol. 9 (17), 939–945.
Shi, Y., Massague, J., 2003. Mechanisms of TGF-β signaling from cell membrane to the nucleus. Cell 113 (6), 685–700.
Sindrilaru, A., Peters, T., Wieschalka, S., Baican, C., Baican, A., Peter, H., Hainzl, A., Schatz, S., Qi, Y., Schlecht, A., Weiss, J.M., Wlaschek, M., Sunderkotter, C., Scharffetter-Kochanek, K., 2011. An unrestrained proinflammatory m1 macrophage population induced by iron impairs wound healing in humans and mice. J. Clin. Invest. 121 (3), 985–997. http://dx.doi.org/10.1172/JCI44490 pii:44490.
Soma, Y., Grotendorst, G.R., 1989. TGF-β stimulates primary human skin fibroblast DNA synthesis via an autocrine production of PDGF-related peptides. J. Cell Physiol. 140 (2), 246–253. http://dx.doi.org/10.1002/jcp.1041400209.
Stathakos, D., Kletsas, D., Psarras, S., 1991. Growth factors from human platelets: a composite group of cell-regulatory peptides. Review of clinical pharmacology and pharmacokinetics. Int. Ed. 5 (1), 12–38.
Stephens, P., Cook, H., Hilton, J., Jones, C.J., Haughton, M.F., Wyllie, F.S., Skinner, J.W., Harding, K.G., Kipling, D., Thomas, D.W., 2003. An analysis of replicative senescence in dermal fibroblasts derived from chronic leg wounds predicts that telomerase therapy would fail to reverse their disease-specific cellular and proteolytic phenotype. Exp. Cell Res. 283 (1), 22–35.
Storer, M., Mas, A., Robert-Moreno, A., Pecoraro, M., Ortells, M.C., Di Giacomo, V., Yosef, R., Pilpel, N., Krizhanovsky, V., Sharpe, J., Keyes, W.M., 2013. Senescence is a developmental mechanism that contributes to embryonic growth and patterning. Cell 155 (5), 1119–1130.
Tallquist, M., Kazlauskas, A., 2004. PDGF signaling in cells and mice. Cytokine Growth Factor Rev. 15 (4), 205–213.
Thannickal, V.J., Fanburg, B.L., 1995. Activation of an H_2O_2-generating NADH oxidase in human lung fibroblasts by transforming growth factor β1. J. Biol. Chem. 270 (51), 30334–30338.
Tian, M., Schiemann, W.P., 2009. The TGF-β paradox in human cancer: an update. Future Oncol. 5 (2), 259–271.
Tingström, A., Heldin, C.H., Rubin, K., 1992. Regulation of fibroblast-mediated collagen gel contraction by platelet-derived growth factor, interleukin-1α and transforming growth factor-β1. J. Cell Sci. 102 (Pt 2), 315–322.
Tolino, M.A., Block, E.R., Klarlund, J.K., 2011. Brief treatment with heparin-binding EGF-like growth factor, but not with EGF, is sufficient to accelerate epithelial wound healing. Biochim. Biophys. Acta 1810 (9), 875–878.

Toussaint, O., Medrano, E.E., von Zglinicki, T., 2000. Cellular and molecular mechanisms of stress-induced premature senescence (sips) of human diploid fibroblasts and melanocytes. Exp. Gerontol. 35 (8), 927–945.

Tsou, R., Fathke, C., Wilson, L., Wallace, K., Gibran, N., Isik, F., 2002. Retroviral delivery of dominant-negative vascular endothelial growth factor receptor type 2 to murine wounds inhibits wound angiogenesis. Wound Repair Regen. 10 (4), 222–229.

Tugues, S., Koch, S., Gualandi, L., Li, X., Claesson-Welsh, L., 2011. Vascular endothelial growth factors and receptors: anti-angiogenic therapy in the treatment of cancer. Mol. Aspects Med. 32 (2), 88–111.

Vande Berg, J.S., Rose, M.A., Haywood-Reid, P.L., Rudolph, R., Payne, W.G., Robson, M.C., 2005. Cultured pressure ulcer fibroblasts show replicative senescence with elevated production of plasmin, plasminogen activator inhibitor-1, and transforming growth factor-β1. Wound Repair Regen. 13 (1), 76–83.

Vande Berg, J.S., Rudolph, R., Hollan, C., Haywood-Reid, P.L., 1998. Fibroblast senescence in pressure ulcers. Wound Repair Regen. 6 (1), 38–49.

Walraven, M., Gouverneur, M., Middelkoop, E., Beelen, R.H., Ulrich, M.M., 2014. Altered TGF-β signaling in fetal fibroblasts: What is known about the underlying mechanisms? Wound Repair Regen. 22 (1), 3–13.

Werner, S., 1998. Keratinocyte growth factor: a unique player in epithelial repair processes. Cytokine Growth Factor Rev. 9 (2), 153–165.

Werner, S., Grose, R., 2003. Regulation of wound healing by growth factors and cytokines. Physiol. Rev. 83 (3), 835–870. http://dx.doi.org/10.1152/physrev.00031.2002 pii:83/3/835.

Werner, S., Smola, H., Liao, X., Longaker, M.T., Krieg, T., Hofschneider, P.H., Williams, L.T., 1994. The function of KGF in morphogenesis of epithelium and reepithelialization of wounds. Science 266 (5186), 819–822.

Whitby, D.J., Ferguson, M.W., 1991. Immunohistochemical localization of growth factors in fetal wound healing. Dev. Biol. 147 (1), 207–215. S0012-1606(05)80018-1.

Wilson, K.J., Gilmore, J.L., Foley, J., Lemmon, M.A., Riese 2nd, D.J., 2009. Functional selectivity of EGF family peptide growth factors: implications for cancer. Pharmacol. Ther. 122 (1), 1–8.

Wu, W.S., Wang, F.S., Yang, K.D., Huang, C.C., Kuo, Y.R., 2006. Dexamethasone induction of keloid regression through effective suppression of VEGF expression and keloid fibroblast proliferation. J Invest Dermatol 126 (6), 1264–1271.

Yamakage, A., Kikuchi, K., Smith, E.A., LeRoy, E.C., Trojanowska, M., 1992. Selective upregulation of platelet-derived growth factor α receptors by transforming growth factor β in scleroderma fibroblasts. J. Exp. Med. 175 (5), 1227–1234.

Zachary, I., 1998. Vascular endothelial growth factor. Int. J. Biochem. Cell Biol. 30 (11), 1169–1174. S1357-2725(98)00082-X.

Zeng, G., McCue, H.M., Mastrangelo, L., Millis, A.J., 1996. Endogenous TGF-β activity is modified during cellular aging: effects on metalloproteinase and TIMP-1 expression. Exp. Cell Res. 228 (2), 271–276. http://dx.doi.org/10.1006/excr.1996.0326 pii:S0014-4827(96)90326-2.

Zervolea, I., Kletsas, D., Stathakos, D., 2000. Autocrine regulation of proliferation and extracellular matrix homeostasis in human fibroblasts. Biochem. Biophys. Res. Commun. 276 (2), 785–790.

Zuo, G.W., Kohls, C.D., He, B.C., Chen, L., Zhang, W., Shi, Q., Zhang, B.Q., Kang, Q., Luo, J., Luo, X., Wagner, E.R., Kim, S.H., Restegar, F., Haydon, R.C., Deng, Z.L., Luu, H.H., He, T.C., Luo, Q., 2010. The CCN proteins: important signaling mediators in stem cell differentiation and tumorigenesis. Histol. Histopathol. 25 (6), 795–806.

Zweers, M.C., Davidson, J.M., Pozzi, A., Hallinger, R., Janz, K., Quondamatteo, F., Leutgeb, B., Krieg, T., Eckes, B., 2007. Integrin α2β1 is required for regulation of murine wound angiogenesis but is dispensable for reepithelialization. J. Invest. Dermatol. 127 (2), 467–478.

Targeting the myofibroblast to improve wound healing

B. Hinz
University of Toronto, Toronto, ON, Canada

3.1 Introduction

3.1.1 The stages of the wound healing process: timing is everything

Understanding the molecular and cellular regulation of wound healing, or tissue repair in general, is essential to increase the chances of patient survival, eg, after trauma or severe burns. Improving the poor healing of chronic wounds or reducing the scarring of excessively healing wounds will improve the patient's quality of life. Consequently, wound healing research is an important topic for government funding (Richmond et al., 2013) and a market for new drugs (Meier and Nanney, 2006). The wound healing process consists of four main phases: (1) the inflammatory phase, starting immediately after tissue injury by initiation of the coagulation cascade that induces inflammatory cell accumulation and cytokine production (Shaw and Martin, 2009; Werner and Grose, 2003); (2) infiltrating macrophages promote fibroblast migration, proliferation and differentiation, collagen production, and neovascularization in the wound bed, leading to the formation of granulation tissue (Eming et al., 2007; Martin and Leibovich, 2005); (3) tissue continuity is then restored by reducing the size of the wound and the development of a permanent scar by the combination of granulation tissue contraction and epithelialization, starting at the borders of the wound (Tomasek et al., 2002); and (4) during the final maturation phase, the number of cells and vessels is decreased by apoptosis (Desmoulière et al., 2005). "No cell is an island", and dysregulation of any of the overlapping wound healing phases will affect the fine equilibrium required for normal healing. The cytokine cross-talk between inflammatory cells recruited from the circulation and the local populations of fibroblastic, epithelial, and vascular cells is crucial for the proper timing of wound healing (Werner and Grose, 2003). The specific local microenvironment generated by the interplay of all these different cell types provides essential cues for their differentiation state, activity, and survival. Miscommunications ultimately result in either insufficient wound healing, such as observed in chronic wounds, or excessive healing, leading to hypertrophic scarring.

3.1.2 Inflammatory cells prepare the stage for fibroblasts

One example of how intercellular cross-talk influences the outcome of wound healing is the important relation between fibroblastic cells in the connective tissue and

Wound Healing Biomaterials - Volume 1. http://dx.doi.org/10.1016/B978-1-78242-455-0.00003-3

inflammatory cells, in particular macrophages (Brancato and Albina, 2011; Eming et al., 2007; Martin and Leibovich, 2005; Mosser and Edwards, 2008; Wynn et al., 2013). The mechanisms and roles of macrophages in dysregulated tissue repair have been the subjects of excellent reviews (Borthwick et al., 2013; Duffield et al., 2013; Wynn et al., 2013; Wynn and Ramalingam, 2012). The inflammatory cell research community may forgive that the role of inflammatory cells in this myofibroblast-centric chapter is reduced to serving as mere fibroblast providers. Macrophages perform a variety of functions that promote fibroblast activities during normal wound healing, including the following: (1) stimulation of extracellular matrix (ECM) deposition and processing by producing arginase-1, transforming growth factor (TGF)-β1, and interleukin (IL)-13; (2) fibroblast survival and proliferation; and (3) activation into myofibroblasts by secreting TGF-β1 and IL-13 (Amini-Nik et al., 2014; Borthwick et al., 2013; Campbell et al., 2013; Hesse et al., 2001; Lanone et al., 2002; Lee et al., 2001; Mirza et al., 2009; Peters et al., 2005).

Because various animal models lacking inflammatory cells exhibited normal tissue repair, it was suggested that inflammatory cells are not pivotal to healing sterile wounds (Martin et al., 2003; Mori et al., 2008). However, other groups have reported disturbed wound healing in mice that were depleted for macrophages or specific macrophage functions (Amini-Nik et al., 2014; Goren et al., 2009; Mirza et al., 2009; Peters et al., 2005). One frequent observation from inflammation-incompetent animal models is reduced (or no) scar formation, similar to what is seen in embryonic and fetal wounds that heal without inflammation and scarring (Colwell et al., 2003; Larson et al., 2010; Schwartzfarb and Kirsner, 2012). Conceivably, the different strategies to experimentally deplete macrophages and different timings contribute to the apparent conflicting findings. In early stages of fibrosis, conditional depletion of macrophages has been shown to reduce liver fibrosis, whereas depletion in the resolution phase had profibrotic effects (Duffield et al., 2005). Genetic macrophage depletion strategies are based on using different "macrophage-specific" gene promoters that possibly target specific subsets of macrophages. However, macrophages are increasingly understood as a heterogeneous population of different activation (polarization) types that dominate different phases of normal wound healing (Lucas et al., 2010). The classical dichotomy of proinflammatory M1 versus profibrotic M2 subsets of macrophages is likely oversimplified, and macrophages appear to exist in a more continuous spectrum of polarization/activation states (Davies et al., 2013; Friedman et al., 2013; Mantovani et al., 2013; Martinez and Gordon, 2014; Mosser and Edwards, 2008; Wynn et al., 2013).

While the role of macrophages in sterile wound healing may still be a matter of debate, it is clear that macrophage dysregulation fosters the transition of normal into either chronic (Goren et al., 2007; Hong et al., 2014) or fibrotic wound repair (Stramer et al., 2007; Wynn et al., 2013). Hence the modulation of the macrophage phenotype has been suggested as one approach to achieving scarless healing (Cash et al., 2014). However, safety issues will have to be considered since the depletion of cutaneous macrophages and dendritic cells was shown to promote the growth of basal cell carcinoma in mice (Konig et al., 2014).

3.2 From normal to abnormal wound healing: the myofibroblast

3.2.1 Definition of the myofibroblast

The development of granulation tissue in phase 2, contraction in phase 3, and resolution in phase 4 are key in promoting normal wound healing. It has been known since the work of Carrel in the early 20th century that the forces of wound contraction are generated within the granulation tissue itself. Gabbiani et al. (1971) later identified specialized fibroblasts as the active component in granulation tissue contraction, which have been named "myofibroblasts" to account for their ultrastructural similarity to smooth muscle cells.

The differentiation of fibroblasts into myofibroblasts can be described as a two-step process (Fig. 3.1). First, to repopulate damaged tissues, quiescent fibroblasts acquire a migratory phenotype by developing contractile bundles. These in vivo stress fibers are initially composed of cytoplasmic actins and generate comparably small traction forces (Hinz et al., 2001b). We have earlier proposed the term "proto-myofibroblast" to discriminate such activated fibroblasts from quiescent fibroblasts that are devoid of any contractile apparatus in most intact tissues (Tomasek et al., 2002). This first phenotypic change occurs in response to changes in the composition, organization, and mechanical property of the ECM (Hinz, 2010) and to cytokines locally released by inflammatory and resident cells (Werner and Grose, 2003). Second, with increasing stress in the ECM, partly resulting from their own remodeling activity, proto-myofibroblasts further develop into "differentiated myofibroblasts" by neo-expressing α-smooth muscle actin (α-SMA), the most widely used myofibroblast marker. The expression of α-SMA is precisely controlled by the joint action of growth factors like TGF-β1, specialized ECM proteins like the fibronectin (FN) splice variant extradomain (ED)-A FN, and the mechanical microenvironment (Klingberg et al., 2013). The incorporation of α-SMA into stress fibers significantly augments the contractile activity of fibroblastic cells and hallmarks the contraction phase of connective tissue remodeling (Hinz et al., 2001a). The forces generated by stress fibers are transmitted to the ECM at specialized focal adhesions containing transmembrane integrins (Dugina et al., 2001; Hinz et al., 2003; Singer et al., 1985; Fig. 3.1). Importantly, the mechanical conditions generated by the myofibroblasts, ie, strained and stiffened ECM, feedback and result in persistent profibrotic myofibroblast activity (Duscher et al., 2014; Hinz, 2013b).

During physiological wound healing, the contractile activity of myofibroblasts is terminated when the tissue is repaired; α-SMA expression then decreases, and myofibroblasts disappear by apoptosis (Desmoulière et al., 1995). In excessive wound healing, however, myofibroblast activity persists and leads to tissue deformation, which is particularly evident in hypertrophic scars (Gauglitz et al., 2011), the fibrotic phase of scleroderma (Castelino and Varga, 2014), and the palmar fibromatosis of Dupuytren's disease (Hinz and Gabbiani, 2011; Verhoekx et al., 2013; Vi et al., 2009). Myofibroblast-generated contractures are also characteristic for fibrosis affecting vital organs such as the liver (Forbes and Rosenthal, 2014; Liedtke et al., 2013), heart (Davis and Molkentin, 2014;

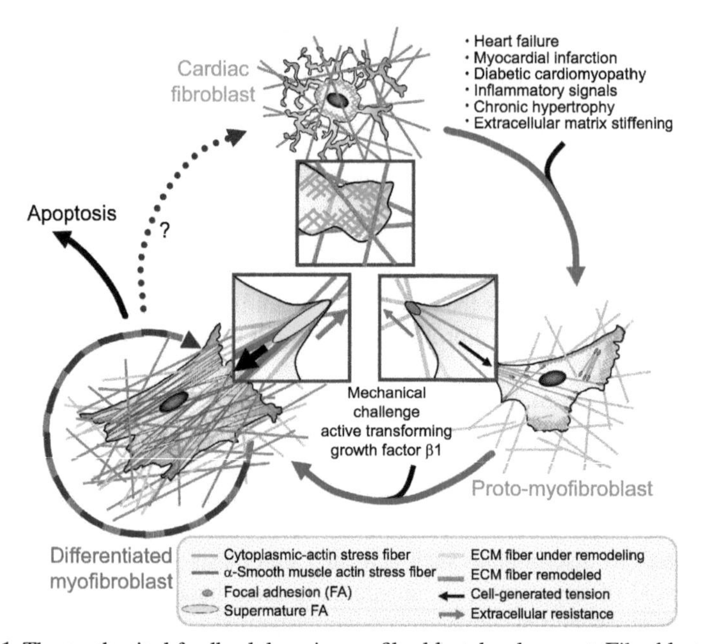

Figure 3.1 The mechanical feedback loop in myofibroblast development. Fibroblast-to-myofibroblast differentiation is controlled by a careful balance of intracellular tension and extracellular matrix (ECM) stiffness. In response to injury, chronic overload, or ECM stiffening, fibroblasts become activated to spread and migrate. Fibroblast remodeling activity leads to a gradual increase in global ECM stiffness that counteracts cell traction forces. The resulting formation of small focal adhesions and stress fibers that contain only cytoplasmic actins characterize the proto-myofibroblast. Continuing ECM fiber alignment creates larger surfaces for adhesion formation; larger adhesions permit the development of stronger stress fibers and the generation of higher contractile forces. In the presence of active transforming growth factor-β1, proto-myofibroblasts begin to express α-smooth muscle actin (α-SMA) that is gradually incorporated into the preexisting stress fibers. The force generated by α-SMA-containing stress fiber is significantly higher than cytoplasmic actin stress fibers, leading to a further maturation of adhesion ECM contraction, thereby establishing a mechanical feed-forward loop. Myofibroblasts may exit this cycle when the original structure of the ECM is reconstituted and again takes over the mechanical load. The failure of stress-released myofibroblasts to undergo apoptosis results in their persisting activity and ultimately leads to fibrosis.
Modified and reproduced with permission from: Hinz, B., 2007. Formation and function of the myofibroblast during tissue repair. J. Invest. Dermatol. 127, 526–537.

Turner and Porter, 2013; Weber et al., 2013), lung (Noble et al., 2012; Sivakumar et al., 2012), and kidney (Campanholle et al., 2013; Duffield, 2014). In addition, myofibroblasts play a role in the stroma reaction against tumors, which promotes cancer progression by creating a stimulating microenvironment for epithelial tumor cells (Ohlund et al., 2014).

3.2.2 Myofibroblast markers

Considering their importance in turning beneficial repair into detrimental scarring, there is a clear need to identify and specifically target the myofibroblast. One of the most frequent questions that is being asked on this topic is, "Is there a unique myofibroblast marker?". You may imagine my frustration (and that of the asker) that the answer is always, "Not that we know of". It is even questionable whether such a marker exists at all, but myofibroblasts can be characterized by a specific set of cytoskeletal proteins (Hinz et al., 2012). Neoexpression of α-SMA in stress fibers is the most commonly used molecular marker for myofibroblasts (Darby et al., 1990; Fig. 3.2).

Inflammatory and epithelial cells are α-SMA negative, although there are always exceptions to the rule that will not be discussed here. Additional markers are required to discriminate between myofibroblasts, vascular smooth muscle cells, and pericytes that are all α-SMA positive (Arnoldi et al., 2012). Smooth muscle markers that are not expressed by myofibroblasts include desmin, smooth muscle myosin heavy chain, h-caldesmon, and smoothelin (Hinz et al., 2007). However, during organ injury, fibrosis, and in cell culture, smooth muscle cells can acquire the myofibroblast phenotype and lose these late differentiation markers. Both vascular smooth muscle and endothelial cells express vimentin that is also typical for fibroblasts and myofibroblasts (Arnoldi et al., 2012). Hence costaining for desmin (smooth muscle) and endothelial cell-specific proteins such as CD31 or VE-cadherin will in most cases be sufficient to discriminate between these main cell types populating a granulation tissue (Fig. 3.3). A great number of other "myofibroblast markers" have been reported over the years but turned out to be less specific than desired.

3.2.3 Myofibroblast origins

Traditionally, local quiescent fibroblasts are described as the main cell population giving rise to myofibroblasts in cutaneous repair (Gabbiani, 2003). That is, using

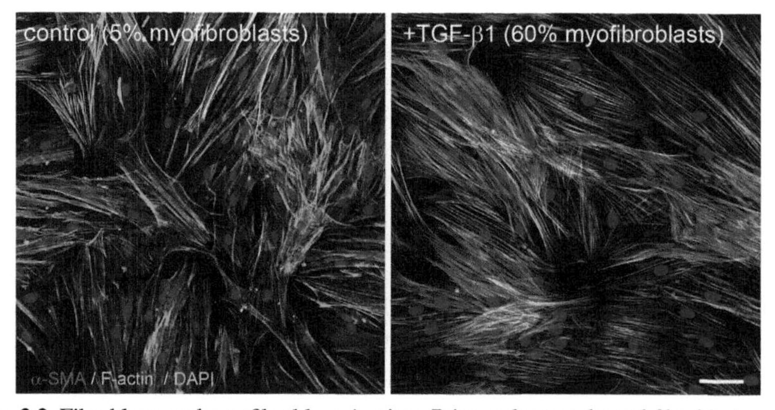

Figure 3.2 Fibroblasts and myofibroblasts in vitro. Primary human dermal fibroblasts were cultured for 4 days on conventional culture plastic dishes in the presence and absence of profibrotic transforming growth factor (TGF)-β1, added to minimum essential medium, supplemented with 5% fetal calf serum. Cells were then stained for F-actin-rich stress fibers (Phalloidin-*green*), α-smooth muscle actin (α-SMA; *red, yellow in overlay*), and nuclei (DAPI-*blue*). Note that all fibroblastic cells form stress fibers, and ~5% spontaneously express α-SMA on stiff culture plastic. Treatment with TGF-β1 (10 ng/ml) increases the fraction of α-SMA positive cells to ~60%. Scale = 50 μm.

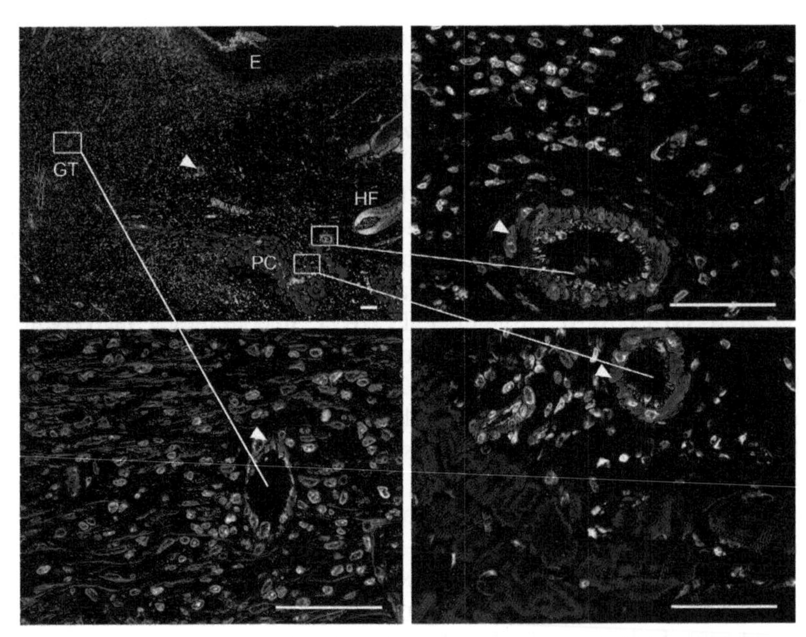

Figure 3.3 Granulation tissue myofibroblasts. Sections of 9-day-old full-thickness rat skin wounds are stained against α-smooth muscle actin (α-SMA-*red*), desmin *(blue)*, CD31 *(green)*, and cell nuclei *(white)*. Myofibroblasts residing under the hyperproliferative epithelium stain exclusively for α-SMA, whereas most smooth muscle cells of small vessels coexpress α-SMA and desmin (*arrowheads*). Endothelial cells lining the vessel lumen only stain for CD31. (*E*, epidermis; *GT*, granulation tissue; *HF*, hair follicle; *PC*, panniculus carnosus). Scale bars = 150 μm.

"fibroblast" in a very general sense and ignoring the often poor definition of this heterogeneous cell population for lack of suitable markers (Driskell et al., 2013). With improved diagnostics, lineage tracing studies, and genetic tools, a number of other myofibroblast precursors have been identified in the skin and adjacent layers, including fibroblasts, smooth muscle cells and pericytes, resident mesenchymal progenitor cells, adipose tissue cells, and bone marrow–derived circulating fibrocytes and mesenchymal stem cells, to list the most prominent candidates (Hinz et al., 2007, 2012). The relative contribution of each of these potential precursors to the myofibroblast population is a matter of ongoing and often heated debate. It emerges that the myofibroblast is in an activation state that can be acquired by multiple cell types with the specific conditions of the insult playing a major role in recruiting different precursors (Hinz et al., 2012).

3.3 Treatment of chronic wounds: stimulating myofibroblast development

Without question, myofibroblast research is mainly fueled by the desire to inhibit their excessive activity in fibrotic conditions, as will be discussed further. However, let us not forget their beneficial contribution to the normal wound healing process and the consequences when this contribution is missing. An estimated 2% of patients in Western countries suffer from chronic wounds (Clark et al., 2007), causing mortality and morbidity in all age groups and representing an important cost factor for the global healthcare system (Banwell, 1999; Bowering, 2001). Predisposed to the development of impaired healing are diabetic patients, the elderly, and vasculopathic patients (Falanga, 2005). The highly inflammatory and hypoxic environment of chronic wounds impedes the formation of functional and contracting granulation tissue, and the lack of myofibroblasts is partly responsible for the poor wound closure observed in chronic nonhealing wounds (Alizadeh et al., 2007; Falanga, 2005). However, other studies reported higher percentages of myofibroblasts in fibroblast cultures initiated from chronic wounds versus acute wounds (Schwarz et al., 2013). Notably, although some differences between tissue fibroblasts are preserved in culture for a limited number of days/passages, the overwhelming mechanical (stiff plastic) and chemical (serum) conditions in cell culture rarely allow for direct in vivo conclusions by characterizing the myofibroblast phenotype only in vitro (Hinz, 2010). In immunohistochemical assessments, a higher number of α-SMA positive myofibroblasts was found in nonhealing venous leg ulcers as compared to acute wound controls (Trøstrup et al., 2011). It is conceivable that the specific conditions in different chronic wounds determine whether myofibroblast development is suppressed or stimulated. For instance, the venous leg ulcer is a type of chronic wound that typically is not hypoxic. This is relevant because hypoxia has been shown to regulate myofibroblast development (Alizadeh et al., 2007).

Many chronic wounds are associated with local tissue ischemia where transcutaneous measured tissue oxygen levels drop from normal 30–50 mm Hg to pathological levels of 5–20 mm Hg. In cutaneous wound healing, hypoxia specifically interferes with three phases of the healing process: (1) in the inflammatory phase, leukocyte activity is impaired because leukocytes require a high amount of oxygen to produce oxidants by NADPH oxidases (NOX); (2) in the granulation phase, fibroblast proliferation ceases at oxygen levels below 15 mm Hg; moreover, collagen production and deposition are reduced because high amounts of oxygen are required for the function of prolyl hydroxylase, a crucial enzyme in the collagen maturation cascade (Scheid et al., 2000); and (3) in the third phase of the wound healing process, granulation tissue contraction is significantly delayed and associated with reduced myofibroblast differentiation (Alizadeh et al., 2007), and epithelialization is delayed by hypoxia (Tandara and Mustoe, 2004). We have previously demonstrated that in vitro hypoxia inhibits dermal myofibroblast differentiation in a process that involves the loss of intracellular force development; importantly, this suppressing effect of hypoxia on dermal myofibroblasts is overcome by applying extracellular strain (Modarressi et al., 2010).

Current strategies to improve chronic wound healing involve the application of grafts that provide paracrine signals, a scaffold for local cells and/or cells that are delivered together with the graft (Clark et al., 2007; Rennert et al., 2013). Whatever the graft and whatever the cell delivered to improve the healing of chronic wounds, simultaneously supporting the formation of myofibroblasts in the granulation tissue will likely enhance the success of the therapy. Previous attempts to promote wound closure by using growth factors to stimulate fibroblast proliferation, differentiation, and motility as well as angiogenesis have been disappointing (Steed, 2006; Wieman, 1998). In contrast, clinically successful strategies to support chronic wound healing include physical approaches such as negative pressure wound therapy (Gregor et al., 2008; Thompson and Marks, 2007). It becomes increasingly accepted that the positive clinical results found for this form of microdeformational wound therapy are through mechanical forces at the cellular level (Greene et al., 2006; Saxena et al., 2004; Scherer et al., 2008; Wiegand and White, 2013). With the knowledge that the application of mechanical tension stimulates myofibroblast differentiation from various precursors (Hinz, 2010), understanding and exploiting the mechanisms of cell mechanotransduction in the wound environment has gained increasing importance and is being considered for clinical strategies (Gurtner et al., 2011; Wong et al., 2011).

3.4 Targeting the myofibroblast as an antiscarring strategy

Antiscarring therapies possibly target the myofibroblast at different levels: (1) myofibroblast recruitment from different precursor cells, (2) myofibroblast contractile and ECM-secreting function, (3) maintenance of the myofibroblast phenotype, and (4) myofibroblast survival and apoptosis. Although skin scarring can be dramatic and disfiguring, it is rarely lethal. Hence even if skin wound healing is often used as an initial model to study tissue, other organ systems are more in the focus of the antiscarring strategy, most notably lung, heart, and kidney. It also has to be noted that rodent models are not particularly suitable to study excessive skin repair since they do not develop hypertrophic scars, although some tricks can be applied to create scar-like conditions (Davidson et al., 2013; Gordillo et al., 2013).

3.4.1 Blocking direct myofibroblast-promoting factors: transforming growth factor-β1

Antifibrosis strategies have been envisioned and designed to prevent the development of tissue contractures by targeting cytokines that directly control myofibroblast differentiation. A plethora of factors have been described to modulate myofibroblast differentiation (Barrientos et al., 2008; Borthwick et al., 2013; Hinz, 2007). However, most of these factors seem to ultimately act by regulating TGF-β signaling. TGF-β1 is considered to be the major growth factor directly promoting myofibroblast development by inducing the expression of α-SMA (Desmoulière et al., 1993), ECM proteins (Werner and Grose, 2003), and a number of cytoskeletal proteins that construct the myofibroblast contractile apparatus (Malmstrom et al., 2004). Concomitantly, TGF-β1 plays fundamental roles in skin fibrosis and wound healing. Tissue levels of active TGF-β1 correlate with the degree of skin fibrosis (Whitfield et al., 2003). Transgenic mice, expressing a constitutively active form of the TGF-β1 receptor, spontaneously develop skin fibrosis (Sonnylal et al., 2007).

Active TGF-β1 assembles a complex of TGF-β1-receptor types I and II in the myofibroblast membrane. The kinase activity of the complex phosphorylates Smad 2/3, which translocates to the nucleus and activates transcription factors that regulate fibrogenic genes; alternatively, TGF-β1 can act through the noncanonical JNK pathway (Massague, 2012). TGF-β1 signaling can be inhibited by blocking either active TGF-β1 or components of the TGF-β1 receptor complex (Massague, 2012). Blocking TGF-β1 with neutralizing antibodies prevents the development of fibrosis in mouse skin (Varga and Pasche, 2008). The humanized monoclonal antibody GC-1008 directed against human TGF-β1, TGF-β2, and TGF-β3 has entered phase 1 clinical trials to treat skin fibrosis in scleroderma patients (Table 3.1). The humanized antibody LY2382770 is tested in phase 2 trials in patients with chronic kidney disease (Table 3.1). The potential of different TGF-β1 receptor type I kinase inhibitors is tested in animal models of lung fibrosis (Bonafoux and Lee, 2009; Bonniaud et al., 2005; Scotton et al., 2013). An obvious risk of these strategies is the fact that the pleiotropic TGF-β1 has many beneficial effects on the wound healing outcome, such as controlling the homeostasis of epidermal, vascular, and inflammatory cells (Henderson and Sheppard, 2013; Hinz, 2013a). Consistently, TGF-β1 knock-out mice have defective vasculogenesis and spontaneously develop skin and lung inflammations (Shull et al., 1992). Hence it is a major challenge to therapeutically inhibit detrimental actions of TGF-β1 without impeding its beneficial effects.

To do so, one needs to consider the biology of TGF-β1 activation. Fibroblasts and other cells secrete TGF-β1 noncovalently associated with its latency-associated propeptide (LAP). This small latent complex covalently binds to the latent TGF-β1 binding protein (LTBP-1), an integral component of the ECM that stores and presents latent TGF-β1 for subsequent activation (Jenkins, 2008; Robertson and Rifkin, 2013; Worthington et al., 2011; Zilberberg et al., 2012; Fig. 3.4). Both LAP and LTBP-1 are upregulated during skin fibrosis, and defects in latent TGF-β1 binding to the ECM results in a variety of diseases that involve TGF-β1 dysregulation (Baldwin et al., 2013; Doyle et al., 2012; Ramirez and Rifkin, 2009; Whitfield et al., 2003). Normal dermis contains high levels of LTBP-1, potentially acting as a TGF-β1 repository (Karonen et al., 1997; Raghunath et al., 1998). The levels of total TGF-β1 are elevated during all stages of wound healing; however, active TGF-β1 only peaks transiently during early inflammation and persistently during the later phase of wound contraction by myofibroblasts (Brunner and Blakytny, 2004; Yang et al., 1999). The activation of latent TGF-β1 by its dissociation from LAP is promoted by different mechanisms, depending on the cell type and physiological context (Henderson and Sheppard, 2013; Minagawa et al., 2014; Sweetwyne and Murphy-Ullrich, 2012; Tatler and Jenkins, 2012; Travis and Sheppard, 2014; Worthington et al., 2011).

Transmembrane cell-ECM receptors, integrins, are crucial in TGF-β1 activation by two main modes of action. The first depends on proteases that are guided by integrin αvβ8 to digest the latent TGF-β1 complex (Minagawa et al., 2014). The second mechanism is independent of proteolysis and seems to dominate in myofibroblasts and fibrosis. The integrins αvβ3, αvβ5, αvβ6, and possibly αvβ1 activate latent TGF-β1 by transmitting cell contractile forces to an arginine-glycine-aspartic acid (RGD) binding site in the LAP portion of the latent TGF-β1 complex (Giacomini et al., 2012; Henderson et al., 2013; Tatler et al., 2011; Wipff and Hinz, 2008; Wipff et al., 2007). Cell pulling on LAP induces a conformational change that liberates the active TGF-β1 only if LAP is physically connected to the ECM through the LTBP-1 (Annes et al.,

Table 3.1 Nonexhaustive list of preclinical and clinical trials with potential relevance for improved wound healing[a]

Effector/drug name[b]	Class and target	Disease	Clinical trial identity[c]	Producer
GC-1008 (fresolimumab)	Antibody directed against human TGF-β	IPF	NCT00125385	Genzyme (Cambridge, MA)
		Scleroderma	NCT01284322	
LY2382770	Humanized antibody directed against TGF-β	Diabetic kidney disease	NCT01113801	Eli Lilly (Indianapolis, IN)
STX-100	Humanized antibody directed against αvβ6 integrin	IPF	NCT01371305	Biogen Idec (Cambridge, MA)
Lebrikizumab	Antibody directed against IL-13	IPF	NCT01872689	Hoffmann-La Roche (Basel, Switzerland)
		Asthma	NCT02104674	
Tralokinumab	Antibody directed against IL-13	IPF	NCT01629667	Medimmune LLC (Gaithersburg, MD)
QAX576	Antibody directed against IL-13	IPF	NCT01266135	Novartis (Basel, Switzerland)
		Keloids	NCT00987545	
SAR156597	Antibody directed against IL-4/IL-13	IPF	NCT01529853	Sanofi (Paris, France)
FG-3019	Humanized antibody directed against CCN2	IPF	NCT01262001, NCT01890265	FibroGen (San Francisco, CA)
		Liver fibrosis	NCT01217632	
GS-6624	Antibody directed against LOXL2	IPF	NCT01362231	Gilead (Foster City, CA)
		Liver fibrosis	NCT01672853	
Imatinib mesylate (Gleevec)	PDGF receptor kinase small molecule inhibitor	IPF	NCT00131274	Novartis
		Scleroderma	NCT00506831	
BIBF-1120 (Nintedanib)	PDGF receptor/tyrosine kinase inhibitor	IPF	NCT01335464	Boehringer Ingelheim (Ingelheim, Germany)
Juvista	Recombinant human TGF-β3	Hypertrophic scars	NCT00742443	Renovo (Manchester, UK)
GSK2126458	Small molecule inhibitor of PI3K	IPF	NCT01725139	GlaxoSmithKline (Brentford, UK)
BMS-986020	Small molecule antagonist of the LPA1 receptor	IPF	NCT01766817	Bristol-Myers Squibb (New York, NY)
Pirfenidone	Small molecule inhibitor of TGF-β and p38 kinase signaling	IPF	NCT00287729, NCT00287716, NCT01366209, NCT01504334	InterMune (Brisbane, CA)

IPF, Idiopathic pulmonary fibrosis.
[a]Sources are the text and excellent reviews (Friedman et al., 2013; Wynn and Ramalingam, 2012).
[b]Arranged in order of first appearance in the text.
[c]ClinicalTrials.gov

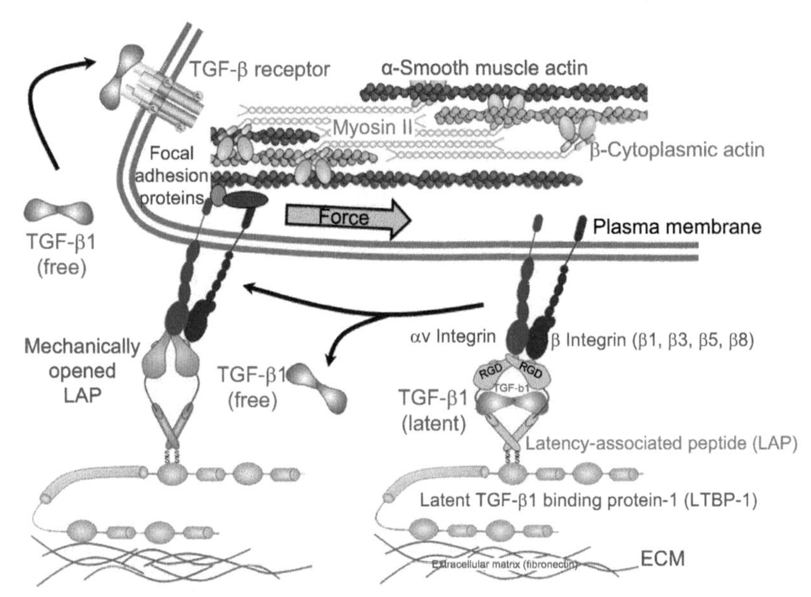

Figure 3.4 Mechanical activation of latent transforming growth factor (TGF)-β1. Latent TGF-β1 (TGF-β1 and the latency-associated propeptide (LAP)) is stored in the ECM together with the latent TGF-β1 binding protein LTBP-1. Upon actin/myosin-promoted myofibroblast contraction, interaction of integrins with arginine-glycine-aspartic acid (RDG) binding sites in the LAP activates TGF-β1 by inducing a putative conformation change in LAP. Reproduced with permission from: Hinz, B., 2013a. It has to be the αv: myofibroblast integrins activate latent TGF-β1. Nat. Med. 19, 1567–1568.

2004; Buscemi et al., 2011; Shi et al., 2011). The mechanical resistance of the ECM at the time of cell pulling on LAP directly influences the efficacy of TGF-β1 activation (Klingberg et al., 2014; Wipff et al., 2007), which establishes a mechanical checkpoint in the progression of tissue repair. In other words, latent TGF-β1 stored in the ECM only becomes available to myofibroblasts and their precursors if the ECM has reached an organization degree, ie, stiffness, which requires the stronger remodeling forces of myofibroblasts.

The inhibition of TGF-β1-activating integrins has promising potential to therapeutically regulate wound healing and fibrosis (Henderson and Sheppard, 2013). The lungs of β6 integrin knock-out mice are protected from bleomycin-induced fibrosis (Munger et al., 1999), and controlled delivery of β6 integrin-blocking antibodies reduces lung fibrosis without negative side effects (Horan et al., 2008). The humanized anti-αvβ6 antibody STX-100 (Table 3.1) has entered phase 2 clinical trials in patients with idiopathic pulmonary fibrosis (IPF). However, integrin αvβ6 is only expressed on epithelial cells, and αvβ6 integrin knock-out mice are not protected against carbon tetrachloride–induced liver fibrosis (Wang et al., 2007). Pericyte-specific deletion of the αv integrin subunit in mice and the use of an αv integrin small molecule inhibitor have been shown to suppress experimentally induced fibrosis in the liver, lung, and kidney (Henderson et al., 2013). Joint blocking of αvβ3 and αvβ5 integrins inhibits cardiac myofibroblast differentiation by reducing TGF-β1 activation (Sarrazy et al., 2014). Similar actions of these integrins have been reported for dermal and scleroderma fibroblasts (Asano, 2010). Together, these studies suggest that the activation of TGF-β1 and fibroblasts in different organs may depend on different αv integrins, which would allow for the development of specific therapeutic designs. Pan-αv integrin inhibitors have not yet been clinically tested.

3.4.2 Blocking direct myofibroblast-promoting factors: interleukin-13 and interleukin-4

The Th2-type cytokines IL-13 and IL-4 exert profibrotic effects, possibly independently from TGF-β1, by acting through the IL-4Rα/Sta6 pathway (Kaviratne et al., 2004; Lafyatis and Farina, 2012; Wynn and Ramalingam, 2012). Both cytokines induce fibroblast-to-myofibroblast differentiation (Hashimoto et al., 2001). Inhibition of IL-4, IL-13, or their receptors was shown to be effective in reducing fibrosis in different experimental settings (Wynn and Ramalingam, 2012), including skin sclerosis (Aliprantis et al., 2007; Le Moine et al., 1999; Ong et al., 1998). The anti-IL-13 antibody lebrikizumab (Table 3.1) has delivered the first positive clinical data after application in asthma patients (Corren et al., 2011). Anti-IL-13 antibodies tralokinumab and QAX576 as well as the anti-IL-4/IL-13 antibody SAR156597 are all in clinical trials for the application in fibrotic diseases (Table 3.1).

3.4.3 Preventing myofibroblast formation by inhibiting transforming growth factor-β1 cooperative factors

Alternatively, or in addition to, interfering directly with TGF-β1 signaling, one may envision targeting effectors of TGF-β1 that may not be able to directly induce myofibroblast differentiation but that have been described as important cofactors. Here, I only provide a selection of these factors and apologize if I miss your favorite one. Notably, many of the myofibroblast-supporting factors are closely associated with the ECM and have been reviewed, such as hyaluronic acid, tenascin C, small leucine-rich proteoglycans, periostin, osteopontin, and collagens (Klingberg et al., 2013).

Angiotensin II: Angiotensin II has been shown to induce fibroblast-to-myofibroblast differentiation in lung, kidney, and heart fibroblasts, presumably by upregulating TGF-β1 production (Mezzano et al., 2001; Rosenkranz, 2004; Sun et al., 1997; Uhal et al., 2007). In these organs, clinically approved inhibitors of angiotensin II signaling, such as losartan and resveratrol, are antifibrotic (Olson et al., 2005; Ramirez and Rifkin, 2012). Losartan has been shown to suppress fibrogenesis of cultured keloid fibroblasts (Ikeda et al., 2013), but none of the drugs are used for skin wound applications.

CCN2: CCN2 (originally named connective tissue growth factor) is locally upregulated in a variety of fibrotic conditions, also reflected in elevated serum levels of subjects with fibrosis (Dendooven et al., 2011; Leask et al., 2009; Phanish et al., 2010). Mutations in the CCN2 gene are associated with human diseases that relate to fibrosis (Jun and Lau, 2011). CCN2 has been shown to potentiate the profibrotic action of TGF-β1 and at the same time to induce fibroblast proliferation (Leask, 2010).

CCN2 acts synergistically with TGF-β1 but does not seem to substitute for its action (Brigstock, 2010; Wang et al., 2011). For instance, the expression of α-SMA remains unchanged after the administration of recombinant CCN2 to cultures of corneal fibroblasts (Folger et al., 2001). Knock-down of CCN2 with different approaches protected experimental animals from developing severe fibrosis and abolished/reduced myofibroblast activation (Brigstock, 2010; Jun and Lau, 2011). The main source for tissue CCN2 in these conditions seems to be fibroblasts and myofibroblasts since the conditional knock-out of CCN2 under control of the proα2(I) collagen gene promoter protected mice against experimentally induced skin fibrosis (Liu et al., 2011). FG-3019, a humanized antibody directed against CCN2, has been tested in many early clinical trials in fibrosis and has advanced to phase 2 trials in IPF patients and hepatic fibrosis (Table 3.1).

ED-A FN: ED-A FN is necessary for myofibroblast differentiation (Serini et al., 1998). ED-A FN is a splice variant of cellular fibronectin that is transiently expressed during embryogenesis and in conditions of tissue injury and repair. In contrast to plasma FN that is produced by hepatocytes and released into the circulation, cellular fibronectins (ED-A and ED-B) are produced by cells in the tissue (Klingberg et al., 2013). The repair of skin and lung after injury is greatly impaired in the absence of ED-A FN, as shown using transgenic mouse models (Muro et al., 2003, 2008). The mechanism(s) by which ED-A FN differentially exerts its effects compared to plasma FN are not fully understood. It is possible that ED-A incorporation in the FN molecule results in conformational shifts that expose neoepitopes in the ED-A or adjacent Type III domains for cell signaling (White et al., 2008). ED-A FN signals via integrins and binding of α4β7 integrin to the ED-A domain promotes TGF-β1-induced myofibroblast differentiation (Kohan et al., 2010). Toll-like receptor 4 (TLR4) has been shown to bind to ED-A FN, and the interaction of both factors is important to promote fibrogenesis in experimentally induced skin fibrosis; both ED-A FN and TLR4 are upregulated in skin biopsies from patients with scleroderma (Bhattacharyya et al., 2014).

Galectin-3: Another cooperative factor of TGF-β1 seems to be the cell surface protein galectin-3. Galectins are a family of carbohydrate-binding proteins that all share their affinity for β-galactoside-containing glycans, which are expressed on the surface of various cells and part of ECM glycoproteins (Yang et al., 2008). Of this family of 15 identified members, gelectin-3 has been shown to be mandatory for TGF-β1-induced myofibroblast differentiation of hepatic stellate cells in culture and is upregulated in experimental liver, kidney, heart, and lung fibrosis (Henderson et al., 2006; Li et al., 2014; Mackinnon et al., 2012). Knock-out of galectin-3 significantly reduces hepatic fibrosis at unchanged levels of TGF-β1 and inhibits myofibroblast activation of cultured stellate cells in the presence of TGF-β1; the addition of recombinant galectin-3 rescues myofibroblast differentiation (Henderson et al., 2006). Galectin-3 is upregulated in the granulation tissue during the healing of rat wounds (Juniantito et al., 2011), and other galectins play a role in dermal myofibroblast differentiation (Dvorankova et al., 2011; Lin et al., 2014).

Lysyl oxidases (LOXs) and LOX-like (LOXL) enzymes: The appreciation that tissue stiffness and mechanical stress drive fibrosis (Hinz, 2013b) has motivated new antifibrotic strategies aiming at interfering with enzymes that cross-link ECM. The copper-dependent LOXs catalyze the formation of covalent cross-links in fibrillar collagen and other ECM proteins by generating reactive aldehyde groups from peptidyl lysines (Kagan, 2000). The expression of LOXs is high during tissue repair and fibrosis (Lopez et al., 2010) and is induced by TGF-β1 in fibroblast cultures (Goto et al., 2005; Roy et al., 1996; Voloshenyuk et al., 2011). LOXs are also instrumental in mediating fibroblast-to-myofibroblast activation in a variety of fibrotic conditions, including skin fibrosis (Chanoki et al., 1995; Counts et al., 1981; Di Donato et al., 1997; Kagan, 1994). This effect is at least partly mediated by increased ECM stiffening, as shown in experimental liver fibrosis (Perepelyuk et al., 2013) and colorectal cancer progression (Baker et al., 2013). In addition to LOXs, LOXL2 has been identified to form fibrosis-specific and stable collagen cross-links (Barker et al., 2011). The antibody GS-6624 (Table 3.1) directed against LOXL2 is in clinical development for the treatment of fibrotic disease after having shown effectiveness in different animal models of fibrosis and cancer (Barry-Hamilton et al., 2010). One of the main effects of blocking LOXL2 is reduced myofibroblast differentiation and collagen secretion (Barry-Hamilton et al., 2010). Other ECM cross-linking enzymes, such as prolyl (Myllyharju, 2008) and lysyl hydroxylases (van der Slot et al., 2005; Wu et al., 2006), are also being reported and studied for their antifibrotic action.

NOX4: The NOX family of NADPH oxidases comprises transmembrane proteins that regulate intracellular redox signaling by reducing extracellular oxygen to superoxide and thereby generating other downstream reactive oxygen species (ROS). NOX control of wound repair is an emerging field, and while high-level ROS generation is involved in direct bacterial killing, low-level ROS generation appears to regulate wound healing (Sen, 2009). NOX1, NOX2, and NOX4 isoforms play roles in regulating the redox environment in wounds (Sen, 2003; Sen et al., 2002). NOX4 has been demonstrated to positively regulate lung fibrosis, at least partly by controlling myofibroblast differentiation (Carnesecchi et al., 2011; Hecker et al., 2009). NOX4 also mediates TGF-β1-induced conversion of cardiac fibroblasts into myofibroblasts by regulating Smad 2/3 activation (Cucoranu et al., 2005). In mice rendered diabetic by streptozotocin, topical galectin-1 increased the closure of excisional full-thickness skin wounds by upregulating NOX4-mediated signaling (Lin et al., 2014). Conversely, overactivity of NOX4 is associated with skin fibrosis and systemic sclerosis (Bohm et al., 2014; Piera-Velazquez and Jimenez, 2015). The different NOX enzymes are being evaluated as novel drug targets in fibrosis and tissue repair, including skin fibrosis (Babalola et al., 2014; Lambeth et al., 2008; Thannickal, 2012).

Platelet-derived growth factor (PDGF): Inhibition of PDGF signaling has long been identified as a potential antifibrotic strategy (Bonner, 2004; Heldin, 2013; Iwayama and Olson, 2013; Kong et al., 2013). Blocking the PDGF receptors α and β with specific antibodies was shown to inhibit myofibroblast formation in experimentally induced lung (Wollin et al., 2014) and kidney fibrosis (Chen et al., 2011; LeBleu and Kalluri, 2011). The PDGF receptor kinase inhibitor imatinib mesylate exhibits antiscarring effects on wound healing and in scleroderma (Gordon and Spiera, 2011; Mori et al., 2008; Rajkumar et al., 2006). Imatinib mesylate (Table 3.1) did not improve survival and lung function in IPF patients (Daniels et al., 2010) but continues to be evaluated as a possible therapeutic for kidney fibrosis (Kay and High, 2008).

The more general tyrosine kinase receptor inhibitor BIBF-1120 (Nintedanib), affecting PDGF, fibroblast growth factor, and vascular endothelial growth factor receptor signaling, is in phase 2 (Ogura et al., 2014; Richeldi et al., 2011) and in phase 3 (Table 3.1) trials in patients with IPF.

Wnt: Activation of Wnt/β-catenin signaling has been reported in kidney, liver, skin (Wei et al., 2011), and cardiac fibrosis (Baarsma et al., 2011). Persistent activation of Wnt/β-catenin signaling in dermal fibroblasts leads to the upregulation of fibrogenesis and expression of various ECM proteins (Hamburg et al., 2014). Pharmacological inhibitors of Wnt effectively reduced cutaneous fibrosis using different experimental models (Beyer et al., 2013; Dees and Distler, 2013). Other blocking studies demonstrated that bleomycin-induced pulmonary fibrosis is highly dependent on the Wnt-1 pathway (Aumiller et al., 2013; Konigshoff et al., 2009). Thus Wnt signaling has emerged as a potential therapeutic target for IPF and several other chronic fibrotic disorders. However, because of the fundamental character of this pathway, it will be a future challenge to develop fibrosis-specific strategies and inhibitors targeting Wnt/β-catenin (Kahn, 2014).

3.4.4 Inhibiting myofibroblast formation by using myofibroblast-suppressing cytokines

IL-1: Attractive with respect to the therapeutic treatment of fibrosis are growth factors that antagonize TGF-β1 and thereby reduce myofibroblast differentiation. The inflammatory mediator IL-1 efficiently inhibits TGF-β1-induced α-SMA expression in cultured dermal fibroblasts when added as a recombinant protein (Shephard et al., 2004); the overexpression of the intracellular IL-1 receptor antagonist (IL-1ra) has similar effects (Kanangat et al., 2006). Coculture experiments with fibroblasts suggest that the time-controlled secretion of IL-1 by keratinocytes may be one mechanism during cutaneous wound healing to suppress fibroblast-to-myofibroblast transition in early wounds (Shephard et al., 2004).

Interferon (IFN)-γ, a cytokine produced by T cells, also reduces α-SMA expression in cultured fibroblasts. This effect appears to be mediated by activation of the repressor protein YB-1, which translocates to the nucleus and interferes with Smad3-mediated transcription of TGF-β1-induced genes (Dooley et al., 2006; Higashi et al., 2003). Full-thickness burn wounds in IFN-γ-deficient mice healed faster than in wild-type mice due to less inflammation and increased collagen deposition, wound contraction, and epithelialization (Shen et al., 2012). It was thus suggested that IFN-γ has the potential to reduce scarring after burn wounds in humans (Shen et al., 2012). However, phase 2 and 3 clinical trials with IFN-γ as an antifibrotic agent in IPF patients did not show therapeutic effects (King et al., 2009; Raghu et al., 2004).

TGF-β3 has been shown to exert myofibroblast-suppressing effects in cell culture experiments and preclinical studies (Serini and Gabbiani, 1996; Shah et al., 1992, 1995), which motivated the development of recombinant human TGF-β3 (rhTGF-β3) into the scar-prevention drug Juvista (Table 3.1). Although Juvista showed promising effects in phase 1/2 clinical trials, it did not meet the endpoints in phase 3 clinical trials (Ågren and Danielsen, 2014).

3.4.5 Inhibiting myofibroblast function reduces tissue deformation

The major activity of differentiated myofibroblasts is enhanced contractile activity, leading to tissue contractures and dysfunction. Surprisingly, few strategies and studies have been published that specifically target this detrimental cell function (Hinz, 2010; Van De Water et al., 2013). A peptide designed to mimic the α-SMA-specific N-terminus was shown to disassemble α-SMA from contractile stress fibers while leaving the β-cytoplasmic actin core of the fibers unaltered (Chaponnier et al., 1995; Clement et al., 2005; Hinz et al., 2002). The loss of stress fiber α-SMA resulted in acutely reduced myofibroblast contraction and subsequently reduced collagen production (Hinz et al., 2002), consistent with earlier reports that this actin isoform promotes "extra" contractile activity in fibroblastic cells (Hinz et al., 2001a). However, α-SMA knock-out mice do not exhibit an obvious wound healing phenotype, which has been attributed to possible functional compensation by γ-smooth muscle and α-skeletal actins (Tomasek et al., 2013). More general approaches to interfere with the integrity and contraction of stress fibers involve the inhibition of the Rho/Rho-associated kinase pathway (Huang et al., 2011; Zhou et al., 2013) and megakaryoblastic leukemia factor-1 (MKL-1). The inhibition of Rho-associated kinase with fasudil reduced contraction and myofibroblast differentiation in rat skin wounds (Bond et al., 2011). MKL-1, also known as myocardin-related transcription factor (MRTF), is central in mediating the polymerization state of actin and the transcriptional activity of muscle-cell genes and is critical for myofibroblast differentiation (Crider et al., 2011; Scharenberg et al., 2014). The inhibition of MRTF with novel compounds has been shown to be effective in reducing experimentally induced skin fibrosis in rodents (Haak et al., 2014) and differentiation of human colonic myofibroblasts (Johnson et al., 2014).

3.4.6 Inducing myofibroblast regression by impeding their resistance to apoptosis

Another promising strategy to reduce fibrosis and hypertrophic scarring is to induce myofibroblast regression by stimulating apoptosis. Several factors have been shown to contribute to the development of an apoptosis-resistant cell phenotype, including cytokines, cell–cell contacts and cell–matrix adhesions. For the myofibroblast and other cell types, two major intracellular pathways have been identified to act prosurvival (or antiapoptotic): focal adhesion kinase (FAK) signaling in cell-ECM adhesions and phosphatidylinositol 3-kinase (PI3K)-AKT signaling. In addition to promoting mechanical myofibroblast differentiation (Wong et al., 2012), FAK activation protects myofibroblasts from entering into apoptosis in response to the loss of cell adhesion, a phenomenon called anoikis (Horowitz et al., 2007). The PI3K-AKT pathway has been shown to be activated by TGF-β1, independently from canonical Smad downstream signaling events but involving a soluble factor and p38 MAP kinase (Horowitz et al., 2004, 2007). Activation of this pathway prevents myofibroblast apoptosis in response to serum deprivation (Horowitz et al., 2004).

Both FAK and PI3K-AKT signaling pathways are inhibited by the protein kinase inhibitor AG1879 (Thannickal et al., 2003), which reduces myofibroblast occurrence in lung fibrosis (Vittal et al., 2005). The small molecule inhibitor GSK2126458 of PI3K is currently in planning for a phase 1 trial in patients with IPF (Table 3.1). Lysophosphatic acid (LPA) has also been shown to promote fibroblast recruitment to fibrotic lesions and resistance to apoptosis by acting through the LPA1 receptor (Sakai et al., 2013; Tager et al., 2008). The small molecule antagonist of the LPA1 receptor, BMS-986020 (Table 3.1), was shown to protect mice from induced skin fibrosis (Castelino et al., 2011) and is being evaluated as an antifibrosis therapy (Rancoule et al., 2011).

3.5 Conclusions

After having been discovered in the early 1970s by Professor Gabbiani, the myofibroblast has not lost any of its attractiveness. Quite the opposite, myofibroblast research seems more intensive than ever, and excellent reviews provide a concise overview of antifibrotic drugs in preclinical and clinical trials (Friedman et al., 2013; Wynn and Ramalingam, 2012). At least three main factors have stimulated a boost and/or revival in the interest of pharmaceutical companies to develop novel antifibrosis treatments that target this cell phenotype. First, fundamental research has delivered a substantial number of scientifically informed new strategies and molecular targets. Second, existing drugs that have been developed and clinically tested for use in antiinflammation or anticancer treatments are now being repurposed, such as fresolimumab, FG-3019, Gleevec, Nintedanib, or BMS-986020 (Table 3.1). Third, pirfenidone, a small molecule inhibitor of TGF-β and p38 kinase signaling, is the first antifibrosis drug approved in Europe and Japan for use in pulmonary fibrosis. After evaluation in the CAPACITY phase 3 trials (Noble et al., 2011), pirfenidone has been tested in the ASCEND phase 3 trials in the United States (Table 3.1) in IPF patients (King et al., 2014). The mechanism of action of pirfenidone is not entirely clear, but it has antifibrotic and antiinflammatory actions; the latter are possibly mediated by inhibiting the mitogen-activated protein kinase p38 that plays a major role in regulating inflammatory processes (Arthur and Ley, 2013). Pirfenidone has also entered clinical trials to treat patients with hypertrophic scars caused by burn injuries and was shown to improve scar regression when applied topically and repeatedly (Armendariz-Borunda et al., 2012). The endpoints and biomarkers defined and accepted for these trials will facilitate the design of new clinical studies.

Acknowledgments

The research of BH is supported by Canadian Institutes of Health Research (CIHR) (grants #210820, #286920, #286720, and #497202), the Collaborative Health Research Program (CIHR/NSERC) (grants #1004005 and #413783), and the Canada Foundation for Innovation and Ontario Research Fund (CFI/ORF) (grant #26653).

References

Ågren, M.S., Danielsen, P.L., 2014. Antiscarring pharmaceuticals: lost in translation? Wound Repair Regen. 22, 293–294.

Aliprantis, A.O., Wang, J., Fathman, J.W., Lemaire, R., Dorfman, D.M., Lafyatis, R., Glimcher, L.H., 2007. Transcription factor T-bet regulates skin sclerosis through its function in innate immunity and via IL-13. Proc. Natl. Acad. Sci. U.S.A 104, 2827–2830.

Alizadeh, N., Pepper, M.S., Modarressi, A., Alfo, K., Schlaudraff, K., Montandon, D., Gabbiani, G., Bochaton-Piallat, M.L., Pittet, B., 2007. Persistent ischemia impairs myofibroblast development in wound granulation tissue: a new model of delayed wound healing. Wound Repair Regen. 15, 809–816.

Amini-Nik, S., Cambridge, E., Yu, W., Guo, A., Whetstone, H., Nadesan, P., Poon, R., Hinz, B., Alman, B.A., 2014. β-Catenin-regulated myeloid cell adhesion and migration determine wound healing. J. Clin. Invest. 124, 2599–2610.

Annes, J.P., Chen, Y., Munger, J.S., Rifkin, D.B., 2004. Integrin αVβ6-mediated activation of latent TGF-β requires the latent TGF-β binding protein-1. J. Cell Biol. 165, 723–734.

Armendariz-Borunda, J., Lyra-Gonzalez, I., Medina-Preciado, D., Gonzalez-Garcia, I., Martinez-Fong, D., Miranda, R.A., Magana-Castro, R., Pena-Santoyo, P., Garcia-Rocha, S., Bautista, C.A., Godoy, J., Flores-Montana, J., Floresvillar-Mosqueda, J., Armendariz-Vazquez, O., Lucano-Landeros, M.S., Vazquez-Del Mercado, M., Sanchez-Parada, M.G., 2012. A controlled clinical trial with pirfenidone in the treatment of pathological skin scarring caused by burns in pediatric patients. Ann. Plast. Surg. 68, 22–28.

Arnoldi, R., Chaponnier, C., Gabbiani, G., Hinz, B., 2012. Heterogeneity of smooth muscle. In: Hill, J. (Ed.), Muscle: Fundamental Biology and Mechanisms of Disease. Elsevier Inc., pp. 1183–1195.

Arthur, J.S., Ley, S.C., 2013. Mitogen-activated protein kinases in innate immunity. Nat. Rev. Immunol. 13, 679–692.

Asano, Y., 2010. Future treatments in systemic sclerosis. J. Dermatol. 37, 54–70.

Aumiller, V., Balsara, N., Wilhelm, J., Gunther, A., Konigshoff, M., 2013. WNT/β-catenin signaling induces IL-1β expression by alveolar epithelial cells in pulmonary fibrosis. Am. J. Respir. Cell Mol. Biol. 49, 96–104.

Baarsma, H.A., Spanjer, A.I., Haitsma, G., Engelbertink, L.H., Meurs, H., Jonker, M.R., Timens, W., Postma, D.S., Kerstjens, H.A., Gosens, R., 2011. Activation of WNT/β-catenin signaling in pulmonary fibroblasts by TGF-β_1 is increased in chronic obstructive pulmonary disease. PLoS One 6, e25450.

Babalola, O., Mamalis, A., Lev-Tov, H., Jagdeo, J., 2014. NADPH oxidase enzymes in skin fibrosis: molecular targets and therapeutic agents. Arch. Dermatol. Res. 306, 313–330.

Baker, A.M., Bird, D., Lang, G., Cox, T.R., Erler, J.T., 2013. Lysyl oxidase enzymatic function increases stiffness to drive colorectal cancer progression through FAK. Oncogene 32, 1863–1868.

Baldwin, A.K., Simpson, A., Steer, R., Cain, S.A., Kielty, C.M., 2013. Elastic fibres in health and disease. Expert Rev. Mol. Med. 15, e8.

Banwell, P.E., 1999. Topical negative pressure therapy in wound care. J. Wound Care 8, 79–84.

Barker, H.E., Chang, J., Cox, T.R., Lang, G., Bird, D., Nicolau, M., Evans, H.R., Gartland, A., Erler, J.T., 2011. LOXL2-mediated matrix remodeling in metastasis and mammary gland involution. Cancer Res. 71, 1561–1572.

Barrientos, S., Stojadinovic, O., Golinko, M.S., Brem, H., Tomic-Canic, M., 2008. Growth factors and cytokines in wound healing. Wound Repair Regen. 16, 585–601.

Barry-Hamilton, V., Spangler, R., Marshall, D., McCauley, S., Rodriguez, H.M., Oyasu, M., Mikels, A., Vaysberg, M., Ghermazien, H., Wai, C., Garcia, C.A., Velayo, A.C., Jorgensen, B., Biermann, D., Tsai, D., Green, J., Zaffryar-Eilot, S., Holzer, A., Ogg, S., Thai, D., Neufeld, G., Van Vlasselaer, P., Smith, V., 2010. Allosteric inhibition of lysyl oxidase-like-2 impedes the development of a pathologic microenvironment. Nat. Med. 16, 1009–1017.

Beyer, C., Reichert, H., Akan, H., Mallano, T., Schramm, A., Dees, C., Palumbo-Zerr, K., Lin, N.Y., Distler, A., Gelse, K., Varga, J., Distler, O., Schett, G., Distler, J.H., 2013. Blockade of canonical Wnt signalling ameliorates experimental dermal fibrosis. Ann. Rheum. Dis. 72, 1255–1258.

Bhattacharyya, S., Tamaki, Z., Wang, W., Hinchcliff, M., Hoover, P., Getsios, S., White, E.S., Varga, J., 2014. FibronectinEDA promotes chronic cutaneous fibrosis through Toll-like receptor signaling. Sci. Transl. Med. 6, 232ra50.

Bohm, M., Dosoki, H., Kerkhoff, C., 2014. Is Nox4 a key regulator of the activated state of fibroblasts in systemic sclerosis? Exp. Dermatol. 23, 679–681.

Bonafoux, D., Lee, W.C., 2009. Strategies for TGF-β modulation: a review of recent patents. Expert Opin. Ther. Pat. 19, 1759–1769.

Bond, J.E., Kokosis, G., Ren, L., Selim, M.A., Bergeron, A., Levinson, H., 2011. Wound contraction is attenuated by fasudil inhibition of Rho-associated kinase. Plast. Reconstr. Surg. 128, 438e–450e.

Bonner, J.C., 2004. Regulation of PDGF and its receptors in fibrotic diseases. Cytokine Growth Factor Rev. 15, 255–273.

Bonniaud, P., Margetts, P.J., Kolb, M., Schroeder, J.A., Kapoun, A.M., Damm, D., Murphy, A., Chakravarty, S., Dugar, S., Higgins, L., Protter, A.A., Gauldie, J., 2005. Progressive transforming growth factor β1-induced lung fibrosis is blocked by an orally active ALK5 kinase inhibitor. Am. J. Respir. Crit. Care Med. 171, 889–898.

Borthwick, L.A., Wynn, T.A., Fisher, A.J., 2013. Cytokine mediated tissue fibrosis. Biochim. Biophys. Acta 1832, 1049–1060.

Bowering, C.K., 2001. Diabetic foot ulcers. Pathophysiology, assessment, and therapy. Can. Fam. Physician 47, 1007–1016.

Brancato, S.K., Albina, J.E., 2011. Wound macrophages as key regulators of repair: origin, phenotype, and function. Am. J. Pathol. 178, 19–25.

Brigstock, D.R., 2010. Connective tissue growth factor (CCN2, CTGF) and organ fibrosis: lessons from transgenic animals. J. Cell Commun. Signal. 4, 1–4.

Brunner, G., Blakytny, R., 2004. Extracellular regulation of TGF-β activity in wound repair: growth factor latency as a sensor mechanism for injury. Thromb. Haemost. 92, 253–261.

Buscemi, L., Ramonet, D., Klingberg, F., Formey, A., Smith-Clerc, J., Meister, J.J., Hinz, B., 2011. The single-molecule mechanics of the latent TGF-β1 complex. Curr. Biol. 21, 2046–2054.

Campanholle, G., Ligestri, G., Gharib, S.A., Duffield, J.S., 2013. Cellular mechanisms of tissue fibrosis. 3. Novel mechanisms of kidney fibrosis. Am. J. Physiol. Cell Physiol. 304 (7), C591–C603.

Campbell, L., Saville, C.R., Murray, P.J., Cruickshank, S.M., Hardman, M.J., 2013. Local arginase 1 activity is required for cutaneous wound healing. J. Invest. Dermatol. 133, 2461–2470.

Carnesecchi, S., Deffert, C., Donati, Y., Basset, O., Hinz, B., Preynat-Seauve, O., Guichard, C., Arbiser, J.L., Banfi, B., Pache, J.C., Barazzone-Argiroffo, C., Krause, K.H., 2011. A key role for NOX4 in epithelial cell death during development of lung fibrosis. Antioxid. Redox Signal. 15, 607–619.

Cash, J.L., Bass, M.D., Campbell, J., Barnes, M., Kubes, P., Martin, P., 2014. Resolution mediator chemerin15 reprograms the wound microenvironment to promote repair and reduce scarring. Curr. Biol. 24, 1406–1414.

Castelino, F.V., Seiders, J., Bain, G., Brooks, S.F., King, C.D., Swaney, J.S., Lorrain, D.S., Chun, J., Luster, A.D., Tager, A.M., 2011. Amelioration of dermal fibrosis by genetic deletion or pharmacologic antagonism of lysophosphatidic acid receptor 1 in a mouse model of scleroderma. Arthritis Rheum. 63, 1405–1415.

Castelino, F.V., Varga, J., 2014. Emerging cellular and molecular targets in fibrosis: implications for scleroderma pathogenesis and targeted therapy. Curr. Opin. Rheumatol. 26, 607–614.

Chanoki, M., Ishii, M., Kobayashi, H., Fushida, H., Yashiro, N., Hamada, T., Ooshima, A., 1995. Increased expression of lysyl oxidase in skin with scleroderma. Br. J. Dermatol. 133, 710–715.

Chaponnier, C., Goethals, M., Janmey, P.A., Gabbiani, F., Gabbiani, G., Vandekerckhove, J., 1995. The specific NH2-terminal sequence Ac-EEED of α-smooth muscle actin plays a role in polymerization in vitro and in vivo. J. Cell Biol. 130, 887–895.

Chen, Y.T., Chang, F.C., Wu, C.F., Chou, Y.H., Hsu, H.L., Chiang, W.C., Shen, J., Chen, Y.M., Wu, K.D., Tsai, T.J., Duffield, J.S., Lin, S.L., 2011. Platelet-derived growth factor receptor signaling activates pericyte-myofibroblast transition in obstructive and post-ischemic kidney fibrosis. Kidney Int. 80, 1170–1181.

Clark, R.A., Ghosh, K., Tonnesen, M.G., 2007. Tissue engineering for cutaneous wounds. J. Invest. Dermatol. 127, 1018–1029.

Clement, S., Hinz, B., Dugina, V., Gabbiani, G., Chaponnier, C., 2005. The N-terminal Ac-EEED sequence plays a role in α-smooth-muscle actin incorporation into stress fibers. J. Cell Sci. 118, 1395–1404.

Colwell, A.S., Longaker, M.T., Lorenz, H.P., 2003. Fetal wound healing. Front. Biosci. 8, s1240–s1248.

Corren, J., Lemanske, R.F., Hanania, N.A., Korenblat, P.E., Parsey, M.V., Arron, J.R., Harris, J.M., Scheerens, H., Wu, L.C., Su, Z., Mosesova, S., Eisner, M.D., Bohen, S.P., Matthews, J.G., 2011. Lebrikizumab treatment in adults with asthma. N. Engl. J. Med. 365, 1088–1098.

Counts, D.F., Evans, J.N., Dipetrillo, T.A., Sterling Jr., K.M., Kelley, J., 1981. Collagen lysyl oxidase activity in the lung increases during bleomycin-induced lung fibrosis. J. Pharmacol. Exp. Ther. 219, 675–678.

Crider, B.J., Risinger Jr., G.M., Haaksma, C.J., Howard, E.W., Tomasek, J.J., 2011. Myocardin-related transcription factors A and B are key regulators of TGF-β1-induced fibroblast to myofibroblast differentiation. J. Invest. Dermatol. 131, 2378–2385.

Cucoranu, I., Clempus, R., Dikalova, A., Phelan, P.J., Ariyan, S., Dikalov, S., Sorescu, D., 2005. NAD(P)H oxidase 4 mediates transforming growth factor-β1-induced differentiation of cardiac fibroblasts into myofibroblasts. Circ. Res. 97, 900–907.

Daniels, C.E., Lasky, J.A., Limper, A.H., Mieras, K., Gabor, E., Schroeder, D.R., Imatinib, I.P.F.S.I., 2010. Imatinib treatment for idiopathic pulmonary fibrosis: randomized placebo-controlled trial results. Am. J. Respir. Crit. Care Med. 181, 604–610.

Darby, I., Skalli, O., Gabbiani, G., 1990. α-Smooth muscle actin is transiently expressed by myofibroblasts during experimental wound healing. Lab. Invest. 63, 21–29.

Davidson, J.M., Yu, F., Opalenik, S.R., 2013. Splinting strategies to overcome confounding wound contraction in experimental animal models. Adv. Wound Care (New Rochelle) 2, 142–148.

Davies, L.C., Jenkins, S.J., Allen, J.E., Taylor, P.R., 2013. Tissue-resident macrophages. Nat. Immunol. 14, 986–995.

Davis, J., Molkentin, J.D., 2014. Myofibroblasts: trust your heart and let fate decide. J. Mol. Cell. Cardiol. 70, 9–18.
Dees, C., Distler, J.H., 2013. Canonical Wnt signalling as a key regulator of fibrogenesis – implications for targeted therapies? Exp. Dermatol. 22, 710–713.
Dendooven, A., Gerritsen, K.G., Nguyen, T.Q., Kok, R.J., Goldschmeding, R., 2011. Connective tissue growth factor (CTGF/CCN2) ELISA: a novel tool for monitoring fibrosis. Biomarkers 16, 289–301.
Desmoulière, A., Chaponnier, C., Gabbiani, G., 2005. Tissue repair, contraction, and the myofibroblast. Wound Repair Regen. 13, 7–12.
Desmoulière, A., Geinoz, A., Gabbiani, F., Gabbiani, G., 1993. Transforming growth factor-β1 induces α-smooth muscle actin expression in granulation tissue myofibroblasts and in quiescent and growing cultured fibroblasts. J. Cell Biol. 122, 103–111.
Desmoulière, A., Redard, M., Darby, I., Gabbiani, G., 1995. Apoptosis mediates the decrease in cellularity during the transition between granulation tissue and scar. Am. J. Pathol. 146, 56–66.
Di Donato, A., Ghiggeri, G.M., Di Duca, M., Jivotenko, E., Acinni, R., Campolo, J., Ginevri, F., Gusmano, R., 1997. Lysyl oxidase expression and collagen cross-linking during chronic adriamycin nephropathy. Nephron 76, 192–200.
Dooley, S., Said, H.M., Gressner, A.M., Floege, J., En-Nia, A., Mertens, P.R., 2006. Y-box protein-1 is the crucial mediator of antifibrotic interferon-γ effects. J. Biol. Chem. 281, 1784–1795.
Doyle, J.J., Gerber, E.E., Dietz, H.C., 2012. Matrix-dependent perturbation of TGFβ signaling and disease. FEBS Lett. 586, 2003–2015.
Driskell, R.R., Lichtenberger, B.M., Hoste, E., Kretzschmar, K., Simons, B.D., Charalambous, M., Ferron, S.R., Herault, Y., Pavlovic, G., Ferguson-Smith, A.C., Watt, F.M., 2013. Distinct fibroblast lineages determine dermal architecture in skin development and repair. Nature 504, 277–281.
Duffield, J.S., 2014. Cellular and molecular mechanisms in kidney fibrosis. J. Clin. Invest. 124, 2299–2306.
Duffield, J.S., Forbes, S.J., Constandinou, C.M., Clay, S., Partolina, M., Vuthoori, S., Wu, S., Lang, R., Iredale, J.P., 2005. Selective depletion of macrophages reveals distinct, opposing roles during liver injury and repair. J. Clin. Invest. 115, 56–65.
Duffield, J.S., Lupher, M., Thannickal, V.J., Wynn, T.A., 2013. Host responses in tissue repair and fibrosis. Annu. Rev. Pathol. 8, 241–276.
Dugina, V., Fontao, L., Chaponnier, C., Vasiliev, J., Gabbiani, G., 2001. Focal adhesion features during myofibroblastic differentiation are controlled by intracellular and extracellular factors. J. Cell Sci. 114, 3285–3296.
Duscher, D., Maan, Z.N., Wong, V.W., Rennert, R.C., Januszyk, M., Rodrigues, M., Hu, M., Whitmore, A.J., Whittam, A.J., Longaker, M.T., Gurtner, G.C., 2014. Mechanotransduction and fibrosis. J. Biomech. 47, 1997–2005.
Dvorankova, B., Szabo, P., Lacina, L., Gal, P., Uhrova, J., Zima, T., Kaltner, H., Andre, S., Gabius, H.J., Sykova, E., Smetana Jr., K., 2011. Human galectins induce conversion of dermal fibroblasts into myofibroblasts and production of extracellular matrix: potential application in tissue engineering and wound repair. Cells Tissues Organs 194, 469–480.
Eming, S.A., Krieg, T., Davidson, J.M., 2007. Inflammation in wound repair: molecular and cellular mechanisms. J. Invest. Dermatol. 127, 514–525.
Falanga, V., 2005. Wound healing and its impairment in the diabetic foot. Lancet 366, 1736–1743.
Folger, P.A., Zekaria, D., Grotendorst, G., Masur, S.K., 2001. Transforming growth factor-β-stimulated connective tissue growth factor expression during corneal myofibroblast differentiation. Invest. Ophthalmol. Vis. Sci. 42, 2534–2541.

Forbes, S.J., Rosenthal, N., 2014. Preparing the ground for tissue regeneration: from mechanism to therapy. Nat. Med. 20, 857–869.

Friedman, S.L., Sheppard, D., Duffield, J.S., Violette, S., 2013. Therapy for fibrotic diseases: nearing the starting line. Sci. Transl. Med. 5, 167sr161.

Gabbiani, G., 2003. The myofibroblast in wound healing and fibrocontractive diseases. J. Pathol. 200, 500–503.

Gabbiani, G., Ryan, G.B., Majno, G., 1971. Presence of modified fibroblasts in granulation tissue and their possible role in wound contraction. Experientia 27, 549–550.

Gauglitz, G.G., Korting, H.C., Pavicic, T., Ruzicka, T., Jeschke, M.G., 2011. Hypertrophic scarring and keloids: pathomechanisms and current and emerging treatment strategies. Mol. Med. 17, 113–125.

Giacomini, M.M., Travis, M.A., Kudo, M., Sheppard, D., 2012. Epithelial cells utilize cortical actin/myosin to activate latent TGF-β through integrin αvβ6-dependent physical force. Exp. Cell Res. 318, 716–722.

Gordillo, G.M., Bernatchez, S.F., Diegelmann, R., Di Pietro, L.A., Eriksson, E., Hinz, B., Hopf, H.W., Kirsner, R., Liu, P., Parnell, L.K., Sandusky, G.E., Sen, C.K., Tomic-Canic, M., Volk, S.W., Baird, A., 2013. Preclinical models of wound healing: is man the model? Proceedings of the wound healing society symposium. Adv. Wound Care (New Rochelle) 2, 1–4.

Gordon, J., Spiera, R., 2011. Imatinib and the treatment of fibrosis: recent trials and tribulations. Curr. Rheumatol. Rep. 13, 51–58.

Goren, I., Allmann, N., Yogev, N., Schurmann, C., Linke, A., Holdener, M., Waisman, A., Pfeilschifter, J., Frank, S., 2009. A transgenic mouse model of inducible macrophage depletion: effects of diphtheria toxin-driven lysozyme M-specific cell lineage ablation on wound inflammatory, angiogenic, and contractive processes. Am. J. Pathol. 175, 132–147.

Goren, I., Muller, E., Schiefelbein, D., Christen, U., Pfeilschifter, J., Muhl, H., Frank, S., 2007. Systemic anti-TNFα treatment restores diabetes-impaired skin repair in ob/ob mice by inactivation of macrophages. J. Invest. Dermatol. 127, 2259–2267.

Goto, Y., Uchio-Yamada, K., Anan, S., Yamamoto, Y., Ogura, A., Manabe, N., 2005. Transforming growth factor-β1 mediated up-regulation of lysyl oxidase in the kidneys of hereditary nephrotic mouse with chronic renal fibrosis. Virchows Arch. 447, 859–868.

Greene, A.K., Puder, M., Roy, R., Arsenault, D., Kwei, S., Moses, M.A., Orgill, D.P., 2006. Microdeformational wound therapy: effects on angiogenesis and matrix metalloproteinases in chronic wounds of 3 debilitated patients. Ann. Plast. Surg. 56, 418–422.

Gregor, S., Maegele, M., Sauerland, S., Krahn, J.F., Peinemann, F., Lange, S., 2008. Negative pressure wound therapy: a vacuum of evidence? Arch. Surg. 143, 189–196.

Gurtner, G.C., Dauskardt, R.H., Wong, V.W., Bhatt, K.A., Wu, K., Vial, I.N., Padois, K., Korman, J.M., Longaker, M.T., 2011. Improving cutaneous scar formation by controlling the mechanical environment: large animal and phase I studies. Ann. Surg. 254, 217–225.

Haak, A.J., Tsou, P.S., Amin, M.A., Ruth, J.H., Campbell, P., Fox, D.A., Khanna, D., Larsen, S.D., Neubig, R.R., 2014. Targeting the myofibroblast genetic switch: inhibitors of myocardin-related transcription factor/serum response factor-regulated gene transcription prevent fibrosis in a murine model of skin injury. J. Pharmacol. Exp. Ther. 349, 480–486.

Hamburg, E., DiNuoscio, G.J., Mullin, N.K., Lafayatis, R., Atit, R.P., 2014. Sustained β-catenin activity in dermal fibroblasts promotes fibrosis by up-regulating expression of extracellular matrix protein-coding genes. J. Pathol. 235 (5), 686–697.

Hashimoto, S., Gon, Y., Takeshita, I., Maruoka, S., Horie, T., 2001. IL-4 and IL-13 induce myofibroblastic phenotype of human lung fibroblasts through c-Jun NH2-terminal kinase-dependent pathway. J. Allergy Clin. Immunol. 107, 1001–1008.

Hecker, L., Vittal, R., Jones, T., Jagirdar, R., Luckhardt, T.R., Horowitz, J.C., Pennathur, S., Martinez, F.J., Thannickal, V.J., 2009. NADPH oxidase-4 mediates myofibroblast activation and fibrogenic responses to lung injury. Nat. Med. 15, 1077–1081.

Heldin, C.H., 2013. Targeting the PDGF signaling pathway in tumor treatment. Cell. Commun. Signal. 11, 97.

Henderson, N.C., Arnold, T.D., Katamura, Y., Giacomini, M.M., Rodriguez, J.D., McCarty, J.H., Pellicoro, A., Raschperger, E., Betsholtz, C., Ruminski, P.G., Griggs, D.W., Prinsen, M.J., Maher, J.J., Iredale, J.P., Lacy-Hulbert, A., Adams, R.H., Sheppard, D., 2013. Targeting of αv integrin identifies a core molecular pathway that regulates fibrosis in several organs. Nat. Med. 19, 1617–1624.

Henderson, N.C., Mackinnon, A.C., Farnworth, S.L., Poirier, F., Russo, F.P., Iredale, J.P., Haslett, C., Simpson, K.J., Sethi, T., 2006. Galectin-3 regulates myofibroblast activation and hepatic fibrosis. Proc. Natl. Acad. Sci. U.S.A 103, 5060–5065.

Henderson, N.C., Sheppard, D., 2013. Integrin-mediated regulation of TGFβ in fibrosis. Biochim. Biophys. Acta 1832, 891–896.

Hesse, M., Modolell, M., La Flamme, A.C., Schito, M., Fuentes, J.M., Cheever, A.W., Pearce, E.J., Wynn, T.A., 2001. Differential regulation of nitric oxide synthase-2 and arginase-1 by type 1/type 2 cytokines in vivo: granulomatous pathology is shaped by the pattern of L-arginine metabolism. J. Immunol. 167, 6533–6544.

Higashi, K., Inagaki, Y., Fujimori, K., Nakao, A., Kaneko, H., Nakatsuka, I., 2003. Interferon-γ interferes with transforming growth factor-β signaling through direct interaction of YB-1 with Smad3. J. Biol. Chem. 278, 43470–43479.

Hinz, B., 2007. Formation and function of the myofibroblast during tissue repair. J. Invest. Dermatol. 127, 526–537.

Hinz, B., 2010. The myofibroblast: paradigm for a mechanically active cell. J. Biomech. 43, 146–155.

Hinz, B., 2013a. It has to be the αv: myofibroblast integrins activate latent TGF-β1. Nat. Med. 19, 1567–1568.

Hinz, B., 2013b. Matrix mechanics and regulation of the fibroblast phenotype. Periodontology 2000 63, 14–28.

Hinz, B., Celetta, G., Tomasek, J.J., Gabbiani, G., Chaponnier, C., 2001a. α-Smooth muscle actin expression upregulates fibroblast contractile activity. Mol. Biol. Cell 12, 2730–2741.

Hinz, B., Dugina, V., Ballestrem, C., Wehrle-Haller, B., Chaponnier, C., 2003. α-Smooth muscle actin is crucial for focal adhesion maturation in myofibroblasts. Mol. Biol. Cell 14, 2508–2519.

Hinz, B., Gabbiani, G., 2011. The role of myofibroblasts in Dupuytren's disease: fundamental aspects of contraction and therapeutic perspectives. In: Eaton, M.H.S.C. (Ed.), Morbus Dupuytren and Related Hyperproliferative Disorders: Principles, Research, and Clinical Perspectives. Springer Verlag.

Hinz, B., Gabbiani, G., Chaponnier, C., 2002. The NH2-terminal peptide of α-smooth muscle actin inhibits force generation by the myofibroblast in vitro and in vivo. J. Cell Biol. 157, 657–663.

Hinz, B., Mastrangelo, D., Iselin, C.E., Chaponnier, C., Gabbiani, G., 2001b. Mechanical tension controls granulation tissue contractile activity and myofibroblast differentiation. Am. J. Pathol. 159, 1009–1020.

Hinz, B., Phan, S.H., Thannickal, V.J., Galli, A., Bochaton-Piallat, M.L., Gabbiani, G., 2007. The myofibroblast: one function, multiple origins. Am. J. Pathol. 170, 1807–1816.

Hinz, B., Phan, S.H., Thannickal, V.J., Prunotto, M., Desmouliere, A., Varga, J., De Wever, O., Mareel, M., Gabbiani, G., 2012. Recent developments in myofibroblast biology: paradigms for connective tissue remodeling. Am. J. Pathol. 180, 1340–1355.

Hong, S., Tian, H., Lu, Y., Laborde, J.M., Muhale, F.A., Wang, Q., Alapure, B.V., Serhan, C.N., Bazan, N.G., 2014. Neuroprotectin/protectin D1: endogenous biosynthesis and actions on diabetic macrophages in promoting wound healing and innervation impaired by diabetes. Am. J. Physiol. Cell Physiol. 307 (11), C1058. http://dx.doi.org/10.1152/ajpcell.00270.2014.

Horan, G.S., Wood, S., Ona, V., Li, D.J., Lukashev, M.E., Weinreb, P.H., Simon, K.J., Hahm, K., Allaire, N.E., Rinaldi, N.J., Goyal, J., Feghali-Bostwick, C.A., Matteson, E.L., O'Hara, C., Lafyatis, R., Davis, G.S., Huang, X., Sheppard, D., Violette, S.M., 2008. Partial inhibition of integrin $\alpha_v\beta_6$ prevents pulmonary fibrosis without exacerbating inflammation. Am. J. Respir. Crit. Care Med. 177, 56–65.

Horowitz, J.C., Lee, D.Y., Waghray, M., Keshamouni, V.G., Thomas, P.E., Zhang, H., Cui, Z., Thannickal, V.J., 2004. Activation of the pro-survival phosphatidylinositol 3-kinase/AKT pathway by transforming growth factor-β1 in mesenchymal cells is mediated by p38 MAPK-dependent induction of an autocrine growth factor. J. Biol. Chem. 279, 1359–1367.

Horowitz, J.C., Rogers, D.S., Sharma, V., Vittal, R., White, E.S., Cui, Z., Thannickal, V.J., 2007. Combinatorial activation of FAK and AKT by transforming growth factor-β1 confers an anoikis-resistant phenotype to myofibroblasts. Cell Signal. 19, 761–771.

Huang, X., Gai, Y., Yang, N., Lu, B., Samuel, C.S., Thannickal, V.J., Zhou, Y., 2011. Relaxin regulates myofibroblast contractility and protects against lung fibrosis. Am. J. Pathol. 179, 2751–2765.

Ikeda, K., Torigoe, T., Matsumoto, Y., Fujita, T., Sato, N., Yotsuyanagi, T., 2013. Resveratrol inhibits fibrogenesis and induces apoptosis in keloid fibroblasts. Wound Repair Regen. 21, 616–623.

Iwayama, T., Olson, L.E., 2013. Involvement of PDGF in fibrosis and scleroderma: recent insights from animal models and potential therapeutic opportunities. Curr. Rheumatol. Rep. 15, 304.

Jenkins, G., 2008. The role of proteases in transforming growth factor-β activation. Int. J. Biochem. Cell Biol. 40, 1068–1078.

Johnson, L.A., Rodansky, E.S., Haak, A.J., Larsen, S.D., Neubig, R.R., Higgins, P.D., 2014. Novel Rho/MRTF/SRF inhibitors block matrix-stiffness and TGF-β-induced fibrogenesis in human colonic myofibroblasts. Inflamm. Bowel Dis. 20, 154–165.

Jun, J.I., Lau, L.F., 2011. Taking aim at the extracellular matrix: CCN proteins as emerging therapeutic targets. Nat. Rev. Drug Discov. 10, 945–963.

Juniantito, V., Izawa, T., Yamamoto, E., Murai, F., Kuwamura, M., Yamate, J., 2011. Heterogeneity of macrophage populations and expression of galectin-3 in cutaneous wound healing in rats. J. Comp. Pathol. 145, 378–389.

Kagan, H.M., 1994. Lysyl oxidase: mechanism, regulation and relationship to liver fibrosis. Pathol. Res. Pract. 190, 910–919.

Kagan, H.M., 2000. Intra- and extracellular enzymes of collagen biosynthesis as biological and chemical targets in the control of fibrosis. Acta Trop. 77, 147–152.

Kahn, M., 2014. Can we safely target the WNT pathway? Nat. Rev. Drug Discov. 13, 513–532.

Kanangat, S., Postlethwaite, A.E., Higgins, G.C., Hasty, K.A., 2006. Novel functions of intracellular IL-1ra in human dermal fibroblasts: implications in the pathogenesis of fibrosis. J. Invest. Dermatol. 126, 756–765.

Karonen, T., Jeskanen, L., Keski-Oja, J., 1997. Transforming growth factor β1 and its latent form binding protein-1 associate with elastic fibres in human dermis: accumulation in actinic damage and absence in anetoderma. Br. J. Dermatol. 137, 51–58.

Kaviratne, M., Hesse, M., Leusink, M., Cheever, A.W., Davies, S.J., McKerrow, J.H., Wakefield, L.M., Letterio, J.J., Wynn, T.A., 2004. IL-13 activates a mechanism of tissue fibrosis that is completely TGF-β independent. J. Immunol. 173, 4020–4029.

Kay, J., High, W.A., 2008. Imatinib mesylate treatment of nephrogenic systemic fibrosis. Arthritis Rheum. 58, 2543–2548.

King Jr., T.E., Albera, C., Bradford, W.Z., Costabel, U., Hormel, P., Lancaster, L., Noble, P.W., Sahn, S.A., Szwarcberg, J., Thomeer, M., Valeyre, D., du Bois, R.M., Group, I.S., 2009. Effect of interferon γ-1b on survival in patients with idiopathic pulmonary fibrosis (INSPIRE): a multicentre, randomised, placebo-controlled trial. Lancet 374, 222–228.

King Jr., T.E., Bradford, W.Z., Castro-Bernardini, S., Fagan, E.A., Glaspole, I., Glassberg, M.K., Gorina, E., Hopkins, P.M., Kardatzke, D., Lancaster, L., Lederer, D.J., Nathan, S.D., Pereira, C.A., Sahn, S.A., Sussman, R., Swigris, J.J., Noble, P.W., Group, A.S., 2014. A phase 3 trial of pirfenidone in patients with idiopathic pulmonary fibrosis. N. Engl. J. Med. 370, 2083–2092.

Klingberg, F., Chow, M.L., Koehler, A., Boo, S., Buscemi, L., Quinn, T.M., Costell, M., Alman, B.A., Genot, E., Hinz, B., 2014. Prestress in the extracellular matrix sensitizes latent TGF-β1 for activation. J. Cell Biol. 207, 283–297.

Klingberg, F., Hinz, B., White, E.S., 2013. The myofibroblast matrix: implications for tissue repair and fibrosis. J. Pathol. 229, 298–309.

Kohan, M., Muro, A.F., White, E.S., Berkman, N., 2010. EDA-containing cellular fibronectin induces fibroblast differentiation through binding to α4β7 integrin receptor and MAPK/Erk 1/2-dependent signaling. FASEB J. 24, 4503–4512.

Kong, P., Christia, P., Frangogiannis, N.G., 2013. The pathogenesis of cardiac fibrosis. Cell. Mol. Life Sci. 71 (4), 549–574.

Konig, S., Nitzki, F., Uhmann, A., Dittmann, K., Theiss-Suennemann, J., Herrmann, M., Reichardt, H.M., Schwendener, R., Pukrop, T., Schulz-Schaeffer, W., Hahn, H., 2014. Depletion of cutaneous macrophages and dendritic cells promotes growth of basal cell carcinoma in mice. PLoS One 9, e93555.

Konigshoff, M., Kramer, M., Balsara, N., Wilhelm, J., Amarie, O.V., Jahn, A., Rose, F., Fink, L., Seeger, W., Schaefer, L., Gunther, A., Eickelberg, O., 2009. WNT1-inducible signaling protein-1 mediates pulmonary fibrosis in mice and is upregulated in humans with idiopathic pulmonary fibrosis. J. Clin. Invest. 119, 772–787.

Lafyatis, R., Farina, A., 2012. New insights into the mechanisms of innate immune receptor signalling in fibrosis. Open Rheumatol. J. 6, 72–79.

Lambeth, J.D., Krause, K.H., Clark, R.A., 2008. NOX enzymes as novel targets for drug development. Semin. Immunopathol. 30, 339–363.

Lanone, S., Zheng, T., Zhu, Z., Liu, W., Lee, C.G., Ma, B., Chen, Q., Homer, R.J., Wang, J., Rabach, L.A., Rabach, M.E., Shipley, J.M., Shapiro, S.D., Senior, R.M., Elias, J.A., 2002. Overlapping and enzyme-specific contributions of matrix metalloproteinases-9 and -12 in IL-13-induced inflammation and remodeling. J. Clin. Invest. 110, 463–474.

Larson, B.J., Longaker, M.T., Lorenz, H.P., 2010. Scarless fetal wound healing: a basic science review. Plast. Reconstr. Surg. 126, 1172–1180.

Le Moine, A., Flamand, V., Demoor, F.X., Noel, J.C., Surquin, M., Kiss, R., Nahori, M.A., Pretolani, M., Goldman, M., Abramowicz, D., 1999. Critical roles for IL-4, IL-5, and eosinophils in chronic skin allograft rejection. J. Clin. Invest. 103, 1659–1667.

Leask, A., 2010. Potential therapeutic targets for cardiac fibrosis: TGFβ, angiotensin, endothelin, CCN2, and PDGF, partners in fibroblast activation. Circ. Res. 106, 1675–1680.

Leask, A., Parapuram, S.K., Shi-Wen, X., Abraham, D.J., 2009. Connective tissue growth factor (CTGF, CCN2) gene regulation: a potent clinical bio-marker of fibroproliferative disease? J. Cell. Commun. Signal. 3, 89–94.

LeBleu, V.S., Kalluri, R., 2011. Blockade of PDGF receptor signaling reduces myofibroblast number and attenuates renal fibrosis. Kidney Int. 80, 1119–1121.

Lee, C.G., Homer, R.J., Zhu, Z., Lanone, S., Wang, X., Koteliansky, V., Shipley, J.M., Gotwals, P., Noble, P., Chen, Q., Senior, R.M., Elias, J.A., 2001. Interleukin-13 induces tissue fibrosis by selectively stimulating and activating transforming growth factor β_1. J. Exp. Med. 194, 809–821.

Li, L.C., Li, J., Gao, J., 2014. Functions of galectin-3 and its role in fibrotic diseases. J. Pharmacol. Exp. Ther. 351, 336–343.

Liedtke, C., Luedde, T., Sauerbruch, T., Scholten, D., Streetz, K., Tacke, F., Tolba, R., Trautwein, C., Trebicka, J., Weiskirchen, R., 2013. Experimental liver fibrosis research: update on animal models, legal issues and translational aspects. Fibrogenesis Tissue Repair 6, 19.

Lin, Y.T., Chen, J.S., Wu, M.H., Hsieh, I.S., Liang, C.H., Hsu, C.L., Hong, T.M., Chen, Y.L., 2014. Galectin-1 accelerates wound healing by regulating the neuropilin-1/Smad3/NOX4 pathway and ROS production in myofibroblasts. J. Invest. Dermatol. 135 (1), 258–268.

Liu, S., Shi-wen, X., Abraham, D.J., Leask, A., 2011. CCN2 is required for bleomycin-induced skin fibrosis in mice. Arthritis Rheum. 63, 239–246.

Lopez, B., Gonzalez, A., Hermida, N., Valencia, F., de Teresa, E., Diez, J., 2010. Role of lysyl oxidase in myocardial fibrosis: from basic science to clinical aspects. Am. J. Physiol. Heart Circ. Physiol. 299, H1–H9.

Lucas, T., Waisman, A., Ranjan, R., Roes, J., Krieg, T., Muller, W., Roers, A., Eming, S.A., 2010. Differential roles of macrophages in diverse phases of skin repair. J. Immunol. 184, 3964–3977.

Mackinnon, A.C., Gibbons, M.A., Farnworth, S.L., Leffler, H., Nilsson, U.J., Delaine, T., Simpson, A.J., Forbes, S.J., Hirani, N., Gauldie, J., Sethi, T., 2012. Regulation of transforming growth factor-β1-driven lung fibrosis by galectin-3. Am. J. Respir. Crit. Care Med. 185, 537–546.

Malmstrom, J., Lindberg, H., Lindberg, C., Bratt, C., Wieslander, E., Delander, E.L., Sarnstrand, B., Burns, J.S., Mose-Larsen, P., Fey, S., Marko-Varga, G., 2004. Transforming growth factor-β1 specifically induce proteins involved in the myofibroblast contractile apparatus. Mol. Cell. Proteomics 3, 466–477.

Mantovani, A., Biswas, S.K., Galdiero, M.R., Sica, A., Locati, M., 2013. Macrophage plasticity and polarization in tissue repair and remodelling. J. Pathol. 229, 176–185.

Martin, P., D'Souza, D., Martin, J., Grose, R., Cooper, L., Maki, R., McKercher, S.R., 2003. Wound healing in the PU.1 null mouse–tissue repair is not dependent on inflammatory cells. Curr. Biol. 13, 1122–1128.

Martin, P., Leibovich, S.J., 2005. Inflammatory cells during wound repair: the good, the bad and the ugly. Trends Cell Biol. 15, 599–607.

Martinez, F.O., Gordon, S., 2014. The M1 and M2 paradigm of macrophage activation: time for reassessment. F1000Prime Rep. 6, 13.

Massague, J., 2012. TGFβ signalling in context. Nat. Rev. Mol. Cell Biol. 13, 616–630.

Meier, K., Nanney, L.B., 2006. Emerging new drugs for scar reduction. Expert Opin. Emerg. Drugs 11, 39–47.

Mezzano, S.A., Ruiz-Ortega, M., Egido, J., 2001. Angiotensin II and renal fibrosis. Hypertension 38, 635–638.

Minagawa, S., Lou, J., Seed, R.I., Cormier, A., Wu, S., Cheng, Y., Murray, L., Tsui, P., Connor, J., Herbst, R., Govaerts, C., Barker, T., Cambier, S., Yanagisawa, H., Goodsell, A., Hashimoto, M., Brand, O.J., Cheng, R., Ma, R., McKnelly, K.J., Wen, W., Hill, A., Jablons, D., Wolters, P., Kitamura, H., Araya, J., Barczak, A.J., Erle, D.J., Reichardt, L.F., Marks, J.D., Baron, J.L., Nishimura, S.L., 2014. Selective targeting of TGF-β activation to treat fibroinflammatory airway disease. Sci. Transl. Med. 6, 241ra79.

Mirza, R., DiPietro, L.A., Koh, T.J., 2009. Selective and specific macrophage ablation is detrimental to wound healing in mice. Am. J. Pathol. 175, 2454–2462.

Modarressi, A., Pietramaggiori, G., Godbout, C., Vigato, E., Pittet, B., Hinz, B., 2010. Hypoxia impairs skin myofibroblast differentiation and function. J. Invest. Dermatol. 130, 2818–2827.

Mori, R., Shaw, T.J., Martin, P., 2008. Molecular mechanisms linking wound inflammation and fibrosis: knockdown of osteopontin leads to rapid repair and reduced scarring. J. Exp. Med. 205, 43–51.

Mosser, D.M., Edwards, J.P., 2008. Exploring the full spectrum of macrophage activation. Nat. Rev. Immunol. 8, 958–969.

Munger, J.S., Huang, X., Kawakatsu, H., Griffiths, M.J., Dalton, S.L., Wu, J., Pittet, J.F., Kaminski, N., Garat, C., Matthay, M.A., Rifkin, D.B., Sheppard, D., 1999. The integrin αvβ6 binds and activates latent TGF β1: a mechanism for regulating pulmonary inflammation and fibrosis. Cell 96, 319–328.

Muro, A.F., Chauhan, A.K., Gajovic, S., Iaconcig, A., Porro, F., Stanta, G., Baralle, F.E., 2003. Regulated splicing of the fibronectin EDA exon is essential for proper skin wound healing and normal lifespan. J. Cell Biol. 162, 149–160.

Muro, A.F., Moretti, F.A., Moore, B.B., Yan, M., Atrasz, R.G., Wilke, C.A., Flaherty, K.R., Martinez, F.J., Tsui, J.L., Sheppard, D., Baralle, F.E., Toews, G.B., White, E.S., 2008. An essential role for fibronectin extra type III domain A in pulmonary fibrosis. Am. J. Respir. Crit. Care Med. 177, 638–645.

Myllyharju, J., 2008. Prolyl 4-hydroxylases, key enzymes in the synthesis of collagens and regulation of the response to hypoxia, and their roles as treatment targets. Ann. Med. 40, 402–417.

Noble, P.W., Albera, C., Bradford, W.Z., Costabel, U., Glassberg, M.K., Kardatzke, D., King Jr., T.E., Lancaster, L., Sahn, S.A., Szwarcberg, J., Valeyre, D., du Bois, R.M., Group, C.S., 2011. Pirfenidone in patients with idiopathic pulmonary fibrosis (CAPACITY): two randomised trials. Lancet 377, 1760–1769.

Noble, P.W., Barkauskas, C.E., Jiang, D., 2012. Pulmonary fibrosis: patterns and perpetrators. J. Clin. Invest. 122, 2756–2762.

Ogura, T., Taniguchi, H., Azuma, A., Inoue, Y., Kondoh, Y., Hasegawa, Y., Bando, M., Abe, S., Mochizuki, Y., Chida, K., Kluglich, M., Fujimoto, T., Okazaki, K., Tadayasu, Y., Sakamoto, W., Sugiyama, Y., 2014. Safety and pharmacokinetics of nintedanib and pirfenidone in idiopathic pulmonary fibrosis. Eur. Respir. J. 45 (5), 1382–1392.

Ohlund, D., Elyada, E., Tuveson, D., 2014. Fibroblast heterogeneity in the cancer wound. J. Exp. Med. 211, 1503–1523.

Olson, E.R., Naugle, J.E., Zhang, X., Bomser, J.A., Meszaros, J.G., 2005. Inhibition of cardiac fibroblast proliferation and myofibroblast differentiation by resveratrol. Am. J. Physiol. Heart Circ. Physiol. 288, H1131–1138.

Ong, C., Wong, C., Roberts, C.R., Teh, H.S., Jirik, F.R., 1998. Anti-IL-4 treatment prevents dermal collagen deposition in the tight-skin mouse model of scleroderma. Eur. J. Immunol. 28, 2619–2629.

Perepelyuk, M., Terajima, M., Wang, A.Y., Georges, P.C., Janmey, P.A., Yamauchi, M., Wells, R.G., 2013. Hepatic stellate cells and portal fibroblasts are the major cellular sources of collagens and lysyl oxidases in normal liver and early after injury. Am. J. Physiol. Gastrointest. Liver Physiol. 304, G605–614.

Peters, T., Sindrilaru, A., Hinz, B., Hinrichs, R., Menke, A., Al-Azzeh, E.A., Holzwarth, K., Oreshkova, T., Wang, H., Kess, D., Walzog, B., Sulyok, S., Sunderkotter, C., Friedrich, W., Wlaschek, M., Krieg, T., Scharffetter-Kochanek, K., 2005. Wound-healing defect of $CD18^{-/-}$ mice due to a decrease in TGF-β1 and myofibroblast differentiation. Embo. J. 24, 3400–3410.

Phanish, M.K., Winn, S.K., Dockrell, M.E., 2010. Connective tissue growth factor-(CTGF, CCN2) – a marker, mediator and therapeutic target for renal fibrosis. Nephron Exp. Nephrol. 114, e83–92.

Piera-Velazquez, S., Jimenez, S.A., 2015. Role of cellular senescence and NOX4-mediated oxidative stress in systemic sclerosis pathogenesis. Curr. Rheumatol. Rep. 17, 473.

Raghu, G., Brown, K.K., Bradford, W.Z., Starko, K., Noble, P.W., Schwartz, D.A., King Jr., T.E., 2004. A placebo-controlled trial of interferon γ-1b in patients with idiopathic pulmonary fibrosis. N. Engl. J. Med. 350, 125–133.

Raghunath, M., Unsold, C., Kubitscheck, U., Bruckner-Tuderman, L., Peters, R., Meuli, M., 1998. The cutaneous microfibrillar apparatus contains latent transforming growth factor-β binding protein-1 (LTBP-1) and is a repository for latent TGF-β1. J. Invest. Dermatol. 111, 559–564.

Rajkumar, V.S., Shiwen, X., Bostrom, M., Leoni, P., Muddle, J., Ivarsson, M., Gerdin, B., Denton, C.P., Bou-Gharios, G., Black, C.M., Abraham, D.J., 2006. Platelet-derived growth factor-β receptor activation is essential for fibroblast and pericyte recruitment during cutaneous wound healing. Am. J. Pathol. 169, 2254–2265.

Ramirez, F., Rifkin, D.B., 2009. Extracellular microfibrils: contextual platforms for TGFβ and BMP signaling. Curr. Opin. Cell Biol. 21, 616–622.

Ramirez, F., Rifkin, D.B., 2012. Is losartan the drug for all seasons? Curr. Opin. Pharmacol. 12, 223–224.

Rancoule, C., Pradere, J.P., Gonzalez, J., Klein, J., Valet, P., Bascands, J.L., Schanstra, J.P., Saulnier-Blache, J.S., 2011. Lysophosphatidic acid-1-receptor targeting agents for fibrosis. Expert Opin. Invest. Drugs 20, 657–667.

Rennert, R.C., Rodrigues, M., Wong, V.W., Duscher, D., Hu, M., Maan, Z., Sorkin, M., Gurtner, G.C., Longaker, M.T., 2013. Biological therapies for the treatment of cutaneous wounds: phase III and launched therapies. Expert. Opin. Biol. Ther. 13, 1523–1541.

Richeldi, L., Costabel, U., Selman, M., Kim, D.S., Hansell, D.M., Nicholson, A.G., Brown, K.K., Flaherty, K.R., Noble, P.W., Raghu, G., Brun, M., Gupta, A., Juhel, N., Kluglich, M., du Bois, R.M., 2011. Efficacy of a tyrosine kinase inhibitor in idiopathic pulmonary fibrosis. N. Engl. J. Med. 365, 1079–1087.

Richmond, N.A., Lamel, S.A., Davidson, J.M., Martins-Green, M., Sen, C.K., Tomic-Canic, M., Vivas, A.C., Braun, L.R., Kirsner, R.S., 2013. US-National Institutes of Health-funded research for cutaneous wounds in 2012. Wound Repair Regen. 21 (6), 789–792.

Robertson, I.B., Rifkin, D.B., 2013. Unchaining the beast; insights from structural and evolutionary studies on TGFβ secretion, sequestration, and activation. Cytokine Growth Factor Rev. 24, 355–372.

Rosenkranz, S., 2004. TGF-β1 and angiotensin networking in cardiac remodeling. Cardiovasc. Res. 63, 423–432.

Roy, R., Polgar, P., Wang, Y., Goldstein, R.H., Taylor, L., Kagan, H.M., 1996. Regulation of lysyl oxidase and cyclooxygenase expression in human lung fibroblasts: interactions among TGF-β, IL-1β, and prostaglandin E. J. Cell Biochem. 62, 411–417.

Sakai, N., Chun, J., Duffield, J.S., Wada, T., Luster, A.D., Tager, A.M., 2013. LPA1-induced cytoskeleton reorganization drives fibrosis through CTGF-dependent fibroblast proliferation. FASEB J. 27 (5), 1830–1846.

Sarrazy, V., Koehler, A., Chow, M.L., Zimina, E., Li, C.X., Kato, H., Caldarone, C.A., Hinz, B., 2014. Integrins αvβ5 and αvβ3 promote latent TGF-β1 activation by human cardiac fibroblast contraction. Cardiovasc. Res. 102, 407–417.

Saxena, V., Hwang, C.W., Huang, S., Eichbaum, Q., Ingber, D., Orgill, D.P., 2004. Vacuum-assisted closure: microdeformations of wounds and cell proliferation. Plast. Reconstr. Surg. 114, 1086–1096 discussion 1097–1088.

Scharenberg, M.A., Pippenger, B.E., Sack, R., Zingg, D., Ferralli, J., Schenk, S., Martin, I., Chiquet-Ehrismann, R., 2014. TGF-β-induced differentiation into myofibroblasts involves specific regulation of two MKL1 isoforms. J. Cell Sci. 127, 1079–1091.

Scheid, A., Wenger, R.H., Christina, H., Camenisch, I., Ferenc, A., Stauffer, U.G., Gassmann, M., Meuli, M., 2000. Hypoxia-regulated gene expression in fetal wound regeneration and adult wound repair. Pediatr. Surg. Int. 16, 232–236.

Scherer, S.S., Pietramaggiori, G., Mathews, J.C., Prsa, M.J., Huang, S., Orgill, D.P., 2008. The mechanism of action of the vacuum-assisted closure device. Plast. Reconstr. Surg. 122, 786–797.

Schwartzfarb, E., Kirsner, R.S., 2012. Understanding scarring: scarless fetal wound healing as a model. J. Invest. Dermatol. 132, 260.

Schwarz, F., Jennewein, M., Bubel, M., Holstein, J.H., Pohlemann, T., Oberringer, M., 2013. Soft tissue fibroblasts from well healing and chronic human wounds show different rates of myofibroblasts in vitro. Mol. Biol. Rep. 40, 1721–1733.

Scotton, C.J., Hayes, B., Alexander, R., Datta, A., Forty, E.J., Mercer, P.F., Blanchard, A., Chambers, R.C., 2013. Ex vivo micro-computed tomography analysis of bleomycin-induced lung fibrosis for preclinical drug evaluation. Eur. Respir. J. 42, 1633–1645.

Sen, C.K., 2003. The general case for redox control of wound repair. Wound Repair Regen. 11, 431–438.

Sen, C.K., 2009. Wound healing essentials: let there be oxygen. Wound Repair Regen. 17, 1–18.

Sen, C.K., Khanna, S., Gordillo, G., Bagchi, D., Bagchi, M., Roy, S., 2002. Oxygen, oxidants, and antioxidants in wound healing: an emerging paradigm. Ann. N. Y. Acad. Sci. 957, 239–249.

Serini, G., Bochaton-Piallat, M.L., Ropraz, P., Geinoz, A., Borsi, L., Zardi, L., Gabbiani, G., 1998. The fibronectin domain ED-A is crucial for myofibroblastic phenotype induction by transforming growth factor-β1. J. Cell Biol. 142, 873–881.

Serini, G., Gabbiani, G., 1996. Modulation of α-smooth muscle actin expression in fibroblasts by transforming growth factor-β isoforms: an in vivo and in vitro study. Wound Rep Reg. 4, 278–287.

Shah, M., Foreman, D.M., Ferguson, M.W., 1992. Control of scarring in adult wounds by neutralising antibody to transforming growth factor β. Lancet 339, 213–214.

Shah, M., Foreman, D.M., Ferguson, M.W., 1995. Neutralisation of TGF-β1 and TGF-β2 or exogenous addition of TGF-β3 to cutaneous rat wounds reduces scarring. J. Cell Sci. 108 (Pt 3), 985–1002.

Shaw, T.J., Martin, P., 2009. Wound repair at a glance. J. Cell Sci. 122, 3209–3213.

Shen, H., Yao, P., Lee, E., Greenhalgh, D., Soulika, A.M., 2012. Interferon-γ inhibits healing post scald burn injury. Wound Repair Regen. 20, 580–591.

Shephard, P., Martin, G., Smola-Hess, S., Brunner, G., Krieg, T., Smola, H., 2004. Myofibroblast differentiation is induced in keratinocyte-fibroblast co-cultures and is antagonistically regulated by endogenous transforming growth factor-β and interleukin-1. Am. J. Pathol. 164, 2055–2066.

Shi, M., Zhu, J., Wang, R., Chen, X., Mi, L., Walz, T., Springer, T.A., 2011. Latent TGF-β structure and activation. Nature 474, 343–349.

Shull, M.M., Ormsby, I., Kier, A.B., Pawlowski, S., Diebold, R.J., Yin, M., Allen, R., Sidman, C., Proetzel, G., Calvin, D., Annunziata, N., Doetschman, T., 1992. Targeted disruption of the mouse transforming growth factor-β1 gene results in multifocal inflammatory disease. Nature 359, 693–699.

Singer, I.I., Kazazis, D.M., Kawka, D.W., 1985. Localization of the fibronexus at the surface of granulation tissue myofibroblasts using double-label immunogold electron microscopy on ultrathin frozen sections. Eur. J. Cell Biol. 38, 94–101.

Sivakumar, P., Ntolios, P., Jenkins, G., Laurent, G., 2012. Into the matrix: targeting fibroblasts in pulmonary fibrosis. Curr. Opin. Pulm. Med. 18 (5), 462–469.

Sonnylal, S., Denton, C.P., Zheng, B., Keene, D.R., He, R., Adams, H.P., Vanpelt, C.S., Geng, Y.J., Deng, J.M., Behringer, R.R., de Crombrugghe, B., 2007. Postnatal induction of transforming growth factor β signaling in fibroblasts of mice recapitulates clinical, histologic, and biochemical features of scleroderma. Arthritis Rheum. 56, 334–344.

Steed, D.L., 2006. Clinical evaluation of recombinant human platelet-derived growth factor for the treatment of lower extremity ulcers. Plast. Reconstr. Surg. 117, 143S–149S discussion 150S–151S.

Stramer, B.M., Mori, R., Martin, P., 2007. The inflammation-fibrosis link? A Jekyll and Hyde role for blood cells during wound repair. J. Invest. Dermatol. 127, 1009–1017.

Sun, Y., Ramires, F.J., Zhou, G., Ganjam, V.K., Weber, K.T., 1997. Fibrous tissue and angiotensin II. J. Mol. Cell. Cardiol. 29, 2001–2012.

Sweetwyne, M.T., Murphy-Ullrich, J.E., 2012. Thrombospondin1 in tissue repair and fibrosis: TGF-β-dependent and independent mechanisms. Matrix Biol. 31 (3), 178–186.

Tager, A.M., LaCamera, P., Shea, B.S., Campanella, G.S., Selman, M., Zhao, Z., Polosukhin, V., Wain, J., Karimi-Shah, B.A., Kim, N.D., Hart, W.K., Pardo, A., Blackwell, T.S., Xu, Y., Chun, J., Luster, A.D., 2008. The lysophosphatidic acid receptor LPA1 links pulmonary fibrosis to lung injury by mediating fibroblast recruitment and vascular leak. Nat. Med. 14, 45–54.

Tandara, A.A., Mustoe, T.A., 2004. Oxygen in wound healing – more than a nutrient. World J. Surg. 28, 294–300.

Tatler, A.L., Jenkins, G., 2012. TGF-β activation and lung fibrosis. Proc. Am. Thorac. Soc. 9, 130–136.

Tatler, A.L., John, A.E., Jolly, L., Habgood, A., Porte, J., Brightling, C., Knox, A.J., Pang, L., Sheppard, D., Huang, X., Jenkins, G., 2011. Integrin αvβ5-mediated TGF-β activation by airway smooth muscle cells in asthma. J. Immunol. 187, 6094–6107.

Thannickal, V.J., 2012. Mechanisms of pulmonary fibrosis: role of activated myofibroblasts and NADPH oxidase. Fibrogenesis Tissue Repair 5, S23.

Thannickal, V.J., Lee, D.Y., White, E.S., Cui, Z., Larios, J.M., Chacon, R., Horowitz, J.C., Day, R.M., Thomas, P.E., 2003. Myofibroblast differentiation by transforming growth factor-β1 is dependent on cell adhesion and integrin signaling via focal adhesion kinase. J. Biol. Chem. 278, 12384–12389.

Thompson, J.T., Marks, M.W., 2007. Negative pressure wound therapy. Clin. Plast. Surg. 34, 673–684.

Tomasek, J.J., Gabbiani, G., Hinz, B., Chaponnier, C., Brown, R.A., 2002. Myofibroblasts and mechano-regulation of connective tissue remodelling. Nat. Rev. Mol. Cell Biol. 3, 349–363.

Tomasek, J.J., Haaksma, C.J., Schwartz, R.J., Howard, E.W., 2013. Whole animal knockout of smooth muscle α-actin does not alter excisional wound healing or the fibroblast-to-myofibroblast transition. Wound Repair Regen. 21, 166–176.

Travis, M.A., Sheppard, D., 2014. TGF-β activation and function in immunity. Annu. Rev. Immunol. 32, 51–82.

Trøstrup, H., Lundquist, R., Christensen, L.H., Jorgensen, L.N., Karlsmark, T., Haab, B.B., Ågren, M.S., 2011. S100A8/A9 deficiency in nonhealing venous leg ulcers uncovered by multiplexed antibody microarray profiling. Br. J. Dermatol. 165, 292–301.

Turner, N.A., Porter, K.E., 2013. Function and fate of myofibroblasts after myocardial infarction. Fibrogenesis Tissue Repair 6, 5.

Uhal, B.D., Kim, J.K., Li, X., Molina-Molina, M., 2007. Angiotensin-TGF-β1 crosstalk in human idiopathic pulmonary fibrosis: autocrine mechanisms in myofibroblasts and macrophages. Curr. Pharm. Des. 13, 1247–1256.
van der Slot, A.J., van Dura, E.A., de Wit, E.C., De Groot, J., Huizinga, T.W., Bank, R.A., Zuurmond, A.M., 2005. Elevated formation of pyridinoline cross-links by profibrotic cytokines is associated with enhanced lysyl hydroxylase 2b levels. Biochim. Biophys. Acta 1741, 95–102.
Van De Water, L., Varney, S., Tomasek, J.J., 2013. Mechanoregulation of the myofibroblast in wound contraction, scarring, and fibrosis: opportunities for new therapeutic intervention. Adv. Wound Care (New Rochelle) 2, 122–141.
Varga, J., Pasche, B., 2008. Antitransforming growth factor-β therapy in fibrosis: recent progress and implications for systemic sclerosis. Curr. Opin. Rheumatol. 20, 720–728.
Verhoekx, J.S., Verjee, L.S., Izadi, D., Chan, J.K., Nicolaidou, V., Davidson, D., Midwood, K.S., Nanchahal, J., 2013. Isometric contraction of Dupuytren's myofibroblasts is inhibited by blocking intercellular junctions. J. Invest. Dermatol. 133, 2664–2671.
Vi, L., Feng, L., Zhu, R.D., Wu, Y., Satish, L., Gan, B.S., O'Gorman, D.B., 2009. Periostin differentially induces proliferation, contraction and apoptosis of primary Dupuytren's disease and adjacent palmar fascia cells. Exp. Cell Res. 315, 3574–3586.
Vittal, R., Horowitz, J.C., Moore, B.B., Zhang, H., Martinez, F.J., Toews, G.B., Standiford, T.J., Thannickal, V.J., 2005. Modulation of prosurvival signaling in fibroblasts by a protein kinase inhibitor protects against fibrotic tissue injury. Am. J. Pathol. 166, 367–375.
Voloshenyuk, T.G., Landesman, E.S., Khoutorova, E., Hart, A.D., Gardner, J.D., 2011. Induction of cardiac fibroblast lysyl oxidase by TGF-β1 requires PI3K/Akt, Smad3, and MAPK signaling. Cytokine 55, 90–97.
Wang, B., Dolinski, B.M., Kikuchi, N., Leone, D.R., Peters, M.G., Weinreb, P.H., Violette, S.M., Bissell, D.M., 2007. Role of αvβ6 integrin in acute biliary fibrosis. Hepatology 46, 1404–1412.
Wang, Q., Usinger, W., Nichols, B., Gray, J., Xu, L., Seeley, T.W., Brenner, M., Guo, G., Zhang, W., Oliver, N., Lin, A., Yeowell, D., 2011. Cooperative interaction of CTGF and TGF-β in animal models of fibrotic disease. Fibrogenesis Tissue Repair 4, 4.
Weber, K.T., Sun, Y., Bhattacharya, S.K., Ahokas, R.A., Gerling, I.C., 2013. Myofibroblast-mediated mechanisms of pathological remodelling of the heart. Nat. Rev. Cardiol. 10, 15–26.
Wei, J., Melichian, D., Komura, K., Hinchcliff, M., Lam, A.P., Lafyatis, R., Gottardi, C.J., MacDougald, O.A., Varga, J., 2011. Canonical Wnt signaling induces skin fibrosis and subcutaneous lipoatrophy: a novel mouse model for scleroderma? Arthritis Rheum. 63, 1707–1717.
Werner, S., Grose, R., 2003. Regulation of wound healing by growth factors and cytokines. Physiol. Rev. 83, 835–870.
White, E.S., Baralle, F.E., Muro, A.F., 2008. New insights into form and function of fibronectin splice variants. J. Pathol. 216, 1–14.
Whitfield, M.L., Finlay, D.R., Murray, J.I., Troyanskaya, O.G., Chi, J.T., Pergamenschikov, A., McCalmont, T.H., Brown, P.O., Botstein, D., Connolly, M.K., 2003. Systemic and cell type-specific gene expression patterns in scleroderma skin. Proc. Natl. Acad. Sci. U.S.A 100, 12319–12324.
Wiegand, C., White, R., 2013. Microdeformation in wound healing. Wound Repair Regen. 21, 793–799.
Wieman, T.J., 1998. Clinical efficacy of becaplermin (rhPDGF-BB) gel. Becaplermin gel studies group. Am. J. Surg. 176, 74S–79S.

Wipff, P.J., Hinz, B., 2008. Integrins and the activation of latent transforming growth factor β1-an intimate relationship. Eur. J. Cell Biol. 87, 601–615.
Wipff, P.J., Rifkin, D.B., Meister, J.J., Hinz, B., 2007. Myofibroblast contraction activates latent TGF-β1 from the extracellular matrix. J. Cell Biol. 179, 1311–1323.
Wollin, L., Maillet, I., Quesniaux, V., Holweg, A., Ryffel, B., 2014. Antifibrotic and anti-inflammatory activity of the tyrosine kinase inhibitor nintedanib in experimental models of lung fibrosis. J. Pharmacol. Exp. Ther. 349, 209–220.
Wong, V.W., Akaishi, S., Longaker, M.T., Gurtner, G.C., 2011. Pushing back: wound mechanotransduction in repair and regeneration. J. Invest. Dermatol. 131, 2186–2196.
Wong, V.W., Rustad, K.C., Akaishi, S., Sorkin, M., Glotzbach, J.P., Januszyk, M., Nelson, E.R., Levi, K., Paterno, J., Vial, I.N., Kuang, A.A., Longaker, M.T., Gurtner, G.C., 2012. Focal adhesion kinase links mechanical force to skin fibrosis via inflammatory signaling. Nat. Med. 18, 148–152.
Worthington, J.J., Klementowicz, J.E., Travis, M.A., 2011. TGFβ: a sleeping giant awoken by integrins. Trends Biochem. Sci. 36, 47–54.
Wu, J., Reinhardt, D.P., Batmunkh, C., Lindenmaier, W., Far, R.K., Notbohm, H., Hunzelmann, N., Brinckmann, J., 2006. Functional diversity of lysyl hydroxylase 2 in collagen synthesis of human dermal fibroblasts. Exp. Cell Res. 312, 3485–3494.
Wynn, T.A., Chawla, A., Pollard, J.W., 2013. Macrophage biology in development, homeostasis and disease. Nature 496, 445–455.
Wynn, T.A., Ramalingam, T.R., 2012. Mechanisms of fibrosis: therapeutic translation for fibrotic disease. Nat. Med. 18, 1028–1040.
Yang, L., Qiu, C.X., Ludlow, A., Ferguson, M.W., Brunner, G., 1999. Active transforming growth factor-β in wound repair: determination using a new assay. Am. J. Pathol. 154, 105–111.
Yang, R.Y., Rabinovich, G.A., Liu, F.T., 2008. Galectins: structure, function and therapeutic potential. Expert Rev. Mol. Med. 10, e17.
Zhou, Y., Huang, X., Hecker, L., Kurundkar, D., Kurundkar, A., Liu, H., Jin, T.H., Desai, L., Bernard, K., Thannickal, V.J., 2013. Inhibition of mechanosensitive signaling in myofibroblasts ameliorates experimental pulmonary fibrosis. J. Clin. Invest. 123, 1096–1108.
Zilberberg, L., Todorovic, V., Dabovic, B., Horiguchi, M., Courousse, T., Sakai, L.Y., Rifkin, D.B., 2012. Specificity of latent TGF-β binding protein (LTBP) incorporation into matrix: role of fibrillins and fibronectin. J. Cell. Physiol. 227, 3828–3836.

Manipulating the healing response 4

B. Azzimonti[1], *M. Sabbatini*[2], *L. Rimondini*[1], *M. Cannas*[1]
[1]Department of Health Sciences, University of Piemonte Orientale, UPO, Alessandria, Novara Vercelli, Italy; [2]Department of Science and Innovation Technology (DISIT), University of Piemonte Orientale, UPO, Alessandria, Novara Vercelli, Italy

4.1 Skin self-renewal

The skin protects us from dehydration and the entry of pathogens. It also regulates thermic dispersion and connects the organism with the environment via specific cutaneous receptors. Similarly to other body sites, skin is also subjected to self-renewal to maintain tissue homeostasis and integrity. This process is of primary importance since the skin is exposed to many physical, chemical, and biological insults that initiate the repair machinery (Blanpain and Fuchs, 2009).

The skin is subdivided into the epidermis, the dermis, and the hypodermis. The resident cells of the dermis and hypodermis originate from bone marrow stem cells (SCs): the hematopoietic and the mesenchymal SCs. Hematopoietic SCs are precursors to neutrophils, macrophages, and lymphocytes, among others. These cells are essential in physiological conditions such as surveillance and host defense during wound healing. Endothelial progenitor cells are also hematopoietic in origin and contribute to neovascularization. Mesenchymal SCs are able to differentiate into fibroblasts, adipocytes, and osteoblasts (Cha and Falanga, 2007; Rea et al., 2009). In this context, the intercellular communications, maintained by humoral cytokine signaling and the physiological integrity of physical communications, are of primary importance.

Epidermal SCs reside in the basal layer. Two cellular populations have been identified in mouse epidermis by the specific markers keratin 14 and involucrin. They differ in gene expression profiles and proliferative and tissue-repair capacities. This finding confirms that the epidermis contains long-living quiescent SCs as well as committed progenitors to self-promote a general strategy for tissue renewal (De Rosa and De Luca, 2012). Epidermal SCs are also found in the base of sebaceous glands and the bulge area of hair follicles (Blanpain and Fuchs, 2009). Epidermal SCs are organized in epidermal proliferative units able to self-renew for extended periods. Experimental evidence indicates that the SCs of the hair follicle bulge contribute to epithelialization during wound healing and not to epidermal homeostasis (Ito et al., 2005; Ito and Cotsarelis, 2008).

In nonhealing or chronic wounds the skin cells show an altered behavior. For example, keratinocytes from the margin of chronic wounds possess a particular gene expression profile with different and inefficient healing capacities (Brem et al., 2007). These

Wound Healing Biomaterials - Volume 1. http://dx.doi.org/10.1016/B978-1-78242-455-0.00004-5

keratinocytes are unable to close the wound, which remains accessible to pathogens that perpetuate the host immune response. In this context delineation of the molecular signals, leading to the recruitment of bulge SCs into the wound, may be important in the treatment of these problematic wounds.

4.2 Normal skin wound healing

The ultimate aim of the wound healing process is to restore skin integrity and homeostasis. Normal wound healing is regulated by a plethora of paracrine soluble growth factors as well as by specific cell–cell and cell–matrix interactions and is conventionally subdivided into four overlapping stages: hemostasis, inflammation, proliferation, and remodeling (Table 4.1).

4.2.1 Hemostasis

The main goal of the hemostatic phase is to stop bleeding. Hemostasis starts immediately after injury by narrowing of the damaged vessels promoted by serotonin, thromboxane A2, and epinephrine. Endothelial cells cease secreting heparin-like molecules and thrombomodulin, which initiates the aggregation of platelets, the first modulators of the healing process. The platelets then become activated and start producing von Willebrand factor (Clemetson, 2012; Szántó et al., 2012). These processes promote coagulation by catalytic transformation of fibrinogen into fibrin. The formed blood clot also contains fibronectin, and it protects the structural integrity of vessels and provides a provisional matrix in the wound bed. Platelets release growth factors, such as platelet-derived growth factor (PDGF), transforming growth factor (TGF)-β, TGF-α, fibroblast growth factor (FGF), insulin-like growth factor (IGF)-1, and vascular

Table 4.1 Wound healing process divided into the four stages with dominant cell and extracellular matrix types

	Hemostasis	Inflammation	Proliferation	Remodeling
Dominant cell type	• Platelets	• Neutrophils • Macrophages (M1) • Lymphocytes (B, T, γδ T) • Fibroblasts • Endothelial cells • Keratinocytes	• Lymphocytes (B, T, γδ T) • Macrophages (M2) • Fibroblasts • Endothelial cells • Keratinocytes	• Lymphocytes (γδ T) • Macrophage (M2) • Fibroblasts • Myofibro-blasts • Keratinocytes
Extracellular matrix evolution	• Blood clot (fibrin-fibronectin matrix)	• Provisional matrix	• Granulation tissue	• Scar tissue

endothelial growth factor (VEGF), which stimulate the migration and proliferation of local tissue cells (Ågren et al., 2014; Blair and Flaumenhaft, 2009; Guo and DiPietro, 2010; Reinke and Sorg, 2012; Shah et al., 2012; Yang et al., 2011).

4.2.2 Inflammation

The inflammatory response is characterized by spatially and temporally changing patterns of various leukocyte subsets, whose chronology is essential for optimal repair (Eming et al., 2007). In this stage, the innate immune system is activated by several factors produced by local mast cells, Langerhans cells, and resident dendritic cells (Busti et al., 2005). These soluble molecules then activate the endothelial cells inside the capillaries to produce adhesion molecules that interact with integrins on circulating neutrophils (Kulidjian et al., 1999). The permeability of capillaries increases, which facilitates the immediate influx of polymorphonuclear leukocytes (PMNs) under the influence of chemotactic factors, such as thrombin, fibrin decomposition products, bacterial products (eg, peptidoglycan, lipopolysaccharides), complement components (C5a), histamine, prostaglandin E2, leukotrienes, TGF-β, and PDGF (Brancato and Albina, 2011; Guo and DiPietro, 2010; Hoffman et al., 2006; Laurens et al., 2006; Reinke and Sorg, 2012; Woodfin et al., 2010). Neutrophils protect the injured site by generating reactive oxygen species (ROS) and nitrogen species that kill bacteria and other microorganisms. Moreover, neutrophils contribute to the removal of necrotic tissue by the release of several proteinases such as matrix metalloproteinase (MMP)-8 and MMP-9, elastase, and cathepsin G (Eckes et al., 2010; Taylor et al., 2004; Witte and Barbul, 2002; Yoneda et al., 2003). Degradation products of fibrin and fibronectin of the provisional matrix attract circulating monocytes (Shah et al., 2012; Wu and Chen, 2014). At the wound site, monocytes are transformed into macrophages under the influence of inflammatory mediators, such as TGF-β and thrombin, which dominate in the subsequent days (Delavary et al., 2011; Hoffman et al., 2006; Lucas et al., 2010; Rodero and Khosrotehrani, 2010; Sha et al., 2012; Wu and Chen, 2014). In the early inflammatory phase macrophages actively participate in phagocytosis and the killing of bacteria, removing debris, and releasing a plethora of cytokines such as tumor necrosis factor-α (TNF-α), PDGF, TGF-α, TGF-β, bFGF, heparin-binding EGF-like growth factor, interleukin (IL)-1, and IL-6 (Brancato and Albina, 2011; Diegelmann and Evans, 2004). In a later phase, macrophages also stimulate fibroblasts, endothelial cells, and keratinocytes to modulate collagen accumulation, angiogenesis, and epithelialization (Hoffman et al., 2006; Mosser and Edwards, 2008; Reinke and Sorg, 2012; Willenborg et al., 2012; Wu and Chen, 2014). Experimental data obtained in adult mice have shown that the depletion of macrophages, during the early phase of the skin repair process, reduces scar formation. During the mid-stage, macrophages stabilize vascular structures and mediate the transition from granulation tissue into scar tissue. Macrophage depletion during the late stage of healing does not seem to affect scar tissue formation (Lucas et al., 2010).

Two phenotypes of macrophages have been identified. The short-lived M1 is immediately recruited into the inflamed tissues with proinflammatory activities, eradicating invading microorganisms and promoting type I immune responses. The M1 phenotype expresses high levels of the chemokine receptor CCR2 and low levels of the fractalkine receptor CX3CR1 (Auffray et al., 2009) and overexpress IL-1β, IL-6, iNOS, arginase-1, and VEGF-A. In the later phase, the long-lasting M2 promotes debris scavenging, resolution of inflammation, angiogenesis, and tissue remodeling (Gordon and Martinez, 2010). The M2 phenotype is characterized by a low expression of CCR2 and high expression levels of CX3CR1 (Auffray et al., 2007), an upregulation of scavenger receptors, such as CD163 and CD206, and the antiinflammatory cytokine IL-10 (Willenborg and Eming, 2014).

At the very end of the inflammatory phase, the number of neutrophils declines in a noninfected wound by the process of apoptosis induced by activated macrophages (Meszaros et al., 1999). Lymphocytes are attracted into the wound bed by cytokines, mainly secreted by macrophages to complete the defense against pathogens (Rossi and Zlotnik, 2000). Several studies suggest that decreased T cell infiltration impairs wound healing. CD4 T helper cells have a positive role in wound healing (Park and Barbul, 2004). The direct cell–cell interactions of T lymphocytes with keratinocytes seem to be important for wound healing (Eming et al., 2009). In particular, the γδ T cells, also called dendritic epidermal T cells, have been found to be active in regulating the defense against pathogens and for tissue integrity. Activated γδ T cells support keratinocyte proliferation and survival and mediate the synthesis of hyaluronan by keratinocytes (Jameson et al., 2005). Mice lacking or with defective γδ T cells show delayed wound closure (Jameson and Havran, 2007; Mills et al., 2008). Lymphocytes also influence fibroblast proliferation and collagen biosynthesis (Guo and DiPietro, 2010). Studies in T and B cell-deficient mice indicate a role for lymphocytes in tissue remodeling (Gawronska-Kozak et al., 2006).

B cells were shown to regulate murine wound healing via CD19 signaling (Iwata et al., 2009).

4.2.3 Proliferation

The proliferation phase is characterized by migration and proliferation of PDGF- and IGF-1-activated fibroblasts and endothelial cells that form granulation tissue under the influence of macrophages (Bainbridge, 2013; Midwood et al., 2004; Werner and Grose, 2003). The provisional fibrin/fibronectin matrix is replaced by nonoriented collagen fibers, glycosaminoglycans, proteoglycans, and noncollagenous glycoproteins made by the fibroblasts. The early granulation tissue contains copious amounts of hyaluronic acid and fibronectin. The hyaluronic acid creates a woven structure, which facilitates cell recruitment to the wound area. Fibronectin creates a scaffold facilitating fibrogenesis (Busti et al., 2005). Hyaluronic acid and fibronectin rapidly decrease, while collagen deposition increases accompanied by a gradual decrease in the number of fibroblasts (Hinz, 2007). In the early proliferation phase, type III collagen predominates, rendering tensile strength to the newly formed tissue (Hoffman et al., 2006). Heparan sulfate proteoglycans are the main components of the matrix of early granulation tissue (Gallo et al., 1996; Olczyk et al., 2014).

The creation of new blood vessels (angiogenesis) is a prerequisite for granulation tissue formation (Gurtner et al., 2008; Tonnesen et al., 2000). Angiogenesis starts when endothelial cells migrate to the provisional matrix of the wound, after the secretion of

MMPs that digest basement membranes and release growth factors, which are sequestrated in the extracellular matrix (ECM), to be available to the cells migrated into the wound area (Reinke and Sorg, 2012). Other mediators involved in angiogenesis include bFGF, TGF-β, TNF-α, VEGF, angiogenin, and angiotropin (Busti et al., 2005; Sinno and Prakash, 2013). Following migration, endothelial cells proliferate and subsequently create a tubular network (Reinke and Sorg, 2012; Schultz et al., 2011).

Chondroitin/dermatan sulfate proteoglycans in the mature granulation tissue matrix (Olczyk et al., 2013) facilitate and stimulate keratinocyte proliferation and migration to restore the epidermal barrier (Diegelmann and Evans, 2004). Keratinocytes also participate in the host defense mechanisms by producing antimicrobial peptides (AMPs) such as human β-defensin and cathelicidin (hCAP18/LL-37), which also act as chemotactic agents for inflammatory cells. Along with the production of AMPs, keratinocytes express Toll-like receptors (TLRs). TLR signaling pathways mediate the recruitment of immune effector cells (Kaisho and Akira, 2006) and are essential for early skin wound healing (Chen et al., 2013). Finally, keratinocytes actively secrete several growth factors, such as keratinocyte-derived antifibrogenic factor, which initiates scar tissue remodeling (Bainbridge, 2013).

4.2.4 Remodeling

The remodeling stage lasts for several years and is characterized by transformation of the granulation tissue into scar tissue accompanied by increased mechanical strength. During this phase, some of the fibroblasts differentiate into myofibroblasts (Desmoulière et al., 2005; Farahani and Kloth, 2008; Mariappan et al., 1999; Sarrazy et al., 2011). These cells interact with the ECM via the main integrin receptors $\alpha_1\beta_1$ and $\alpha_2\beta_1$, which allows for tissue contraction (Mariappan et al., 1999; Tomasek et al., 2002).

Attenuation of inflammation and cell proliferation stop the angiogenesis process, and capillaries are disintegrated by apoptotic processes. Then a further matrix remodeling takes place, lowering the content of glycosaminoglycans and proteoglycans as well as the water content (Diegelmann and Evans, 2004; Hoffman et al., 2006), while the total collagen content increases; type I collagen becomes more abundant than collagen type III, and the number of covalent cross-links increases (Diegelmann and Evans, 2004; Gurtner et al., 2008; Reinkle and Sorg, 2012; Vedrenne et al., 2012).

4.3 Skin inflammation: care or damage

The inflammatory response plays a pivotal role in directing the outcome of the healing response and is intimately linked to the extent of scar formation. Although the mechanisms that orchestrate the differences in the scar outcome are not well understood, they possibly reflect an altered inflammatory and/or cytokine profile (Eming et al., 2007; Harty et al., 2003; Martin and Leibovich, 2005). Pathogens and/or endotoxins prolong the release of proinflammatory cytokines such as IL-1 and TNF-α, leading to an unbalanced inflammatory response (Edwards and Harding, 2004; Guo and DiPietro, 2010; Strbo et al., 2014). These processes contribute to chronicity and healing failure (Edwards and Harding, 2004; Menke et al., 2007).

Some evidence indicates that overactive neutrophil response delays wound healing and/or leads to excessive scar formation as a consequence of an exaggerated release of MMP-8 and neutrophil elastase (Canesso et al., 2014; Catalano et al., 2013; Lobmann et al., 2002; Menke et al., 2007; Yager et al., 1996) , thus generating the idea of their dual role in wound healing. In nonhealing wounds, MMPs are not adequately balanced by their endogenous tissue inhibitor of metalloproteinases (Bullen et al., 1995; Yager et al., 1996).

In pathological healing conditions, it is theorized that macrophages fail to switch to the M2 phenotype (Willenborg and Eming, 2014). In chronic venous leg ulcers the prolonged persistence of M1 macrophages may be tissue destructive (Mahdipour et al., 2011; Rodero et al., 2013; Sindrilaru et al., 2011). Thus minimizing M1 activation and promoting M2 activation in the context of chronic inflammation could represent an effective therapeutic strategy to normalize the pathways.

Excessive inflammatory conditions contribute to a deficient ECM due to the rapid degradation of collagen and other ECM components that overwhelm their synthesis by connective tissue cells. In chronic wounds, several fibroblast functions are altered. These fibroblasts show impaired proliferative and migratory capacities as well as reduced responses to growth factors (Ågren et al., 1999; Brem et al., 2007). This also results in healing failures. Furthermore, the chronic wound fibroblasts produce altered quantities of cytokines responsible for the enlarged scars (Menke et al., 2007).

LL-37 protein levels are generally low in chronic wounds, while mRNA expression levels are elevated.

4.4 Acute and chronic wounds

Wounds can be classified as acute or chronic. We still lack a univocal accepted definition of the characterizing elements (Fletcher, 2008).

Acute wounds result from surgeries or trauma such as contusions; burns (thermal, chemical, or electrical); excoriations; extractions; animal bites; or splash of objects. They normally lead to the complete healing of the injured site following the four different reparative phases in a timely and uneventful manner (Guo and Di Pietro, 2010).

When the process persists for periods longer than 4 weeks without any sign of resolution, the wound becomes chronic. It is thought that the prolonged inflammatory phase converts the acute wound into a chronic wound (Yazdanpanah et al., 2015). The cause is destruction of the ECM and essential key regulators of the healing cascade. In this context, bacterial invasion of the wound is promoted by hypoxic conditions. In addition, high levels of ROS are released, especially in situations with excessive necrotic tissues. The knowledge of the etiology and of the differences between acute and chronic wounds is fundamental for an appropriate wound management plan (Fig. 4.1).

Several other factors can contribute to the chronicity of the wound: vitamins and mineral depletion, an excess of tobacco use, and abuse of nonsteroidal antiinflammatory drugs.

Advances in the knowledge of the mechanisms occurring during epidermal homeostasis suggest that several epigenetic regulators control the genes involved in the

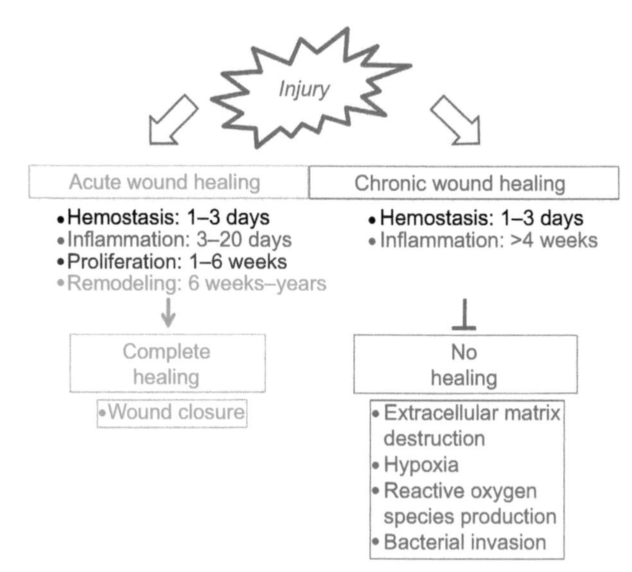

Figure 4.1 Acute and chronic wound healing. The illustration highlights the main differences between acute and chronic responses after injury.

epithelial reparative program, since they strictly control keratinocytes proliferation, differentiation and migration, and condition vessels formation during dermal renewal by orchestrating gene activation and repression (Lewis et al., 2014). The molecular basis of the epigenetic regulation mechanisms could represent a therapeutic target for the treatment of acute and chronic wounds.

4.4.1 Burns

Burns are the result of damage by fire, electric charges, chemical agents, cold liquids, or excessive sun exposure. It has been estimated that the incidence of skin burns in the United States is about 5–7 million cases/year (Hanson et al., 2010). Their treatment is crucial, since the extension and severity of the lesions in the context of the involved tissue is often abundant (Mcheik et al., 2014). To minimize the risk of systemic infection and preserve the patient's tissue, the necrotic tissue is removed and replaced by a skin autograft. For extensive burn injuries, grafting of in vitro cultured human epidermal keratinocytes in three-dimensional pluristratified layers is one option.

The long population doubling times of normal keratinocytes is a problem. To overcome this, new systems have been developed to isolate epidermal SCs (Guenou et al., 2009). These are generically classified as totipotent, pluripotent, multipotent, and adult unipotent cells on the basis of their differentiation capabilities. The first SCs are produced by the zygote during ontogenesis and are able to potentially produce an entire functional organism. The second type derives from the inner layers of the blastocyst and develops into all cells and tissue types without giving rise to an organism, while multipotent SCs derive from pluripotent SCs and differentiate into a limited range of specialized cells with specific functions. Finally the unipotent cells produce specific cell types (for example, epithelial cells only produce terminally differentiated keratinocytes); they in fact replace dead cells or cells that have lost their function in differentiated tissues.

Due to the undifferentiated state of SCs and their capacity to self-renew and differentiate into almost all tissue types, SCs are used not only in clinical practice but also in animal testing and many research fields (Koźlik and Wojcicki, 2014). They can be embryonic (ESCs), fetal (FSCs), adult (ASCs) and induced pluripotent. They are used in plastic and reconstructive surgery. Nevertheless, they are not the first choice for a reconstructive system because they are very expensive. Furthermore, the use of ESCs and FSCs is not ethically and legally approved in all countries. There is a paucity of data on their safety for human use, and the use of ESCs may be associated with an increased risk of cancer and teratoma development. ASCs are simple to isolate and due to their plasticity differentiate into a different cell type when transferred to another tissue.

SCs are also used in cellular fibrin spray formulations and are subcutaneously injected in animals trials into the stasis zone of the burned area (Koźlik and Wojcicki, 2014), promoting a decrease in cellular apoptosis and an increase of cell survival, especially if administered just after the injury (Koźlik and Wojcicki, 2014).

4.4.2 Pressure ulcers

Pressure ulcers (PUs) are mainly formed over a bony prominence that arises secondary to prolonged pressure, which prevents nutritious blood flow to the affected skin area (Black, 2015; Vélez-Díaz-Pallarés et al., 2015). Risk factors for ulcer development are immobility and emergency surgery. The Braden Scale tool is helpful in identifying patients at risk.

PUs could in part be prevented by adhering to these three principles: (1) pressure relief, (2) control of infection, and (3) adequate nutrition. Furthermore, the American College of Physicians underscores the scarcity of effective wound healing interventions. The data on the use of nonpharmacological therapy is inconclusive, especially for old people (Liu et al., 2014). Therefore, studies in older populations with PUs using sound methodology are needed.

4.4.3 Diabetic foot ulcers

It has been estimated that by 2030 there will be 360 million people with diabetes mellitus. Diabetic foot ulcers (DFUs) are one of the end-point complications of diabetes and affect about 15% of diabetic patients. Risk factors that predispose to DFUs are the male gender, a long history of diabetes, neuropathy, peripheral vascular disease, nephropathy, and rhinopathy. The diabetic foot is vulnerable to trauma due to its insensitivity. Poor control of the blood glucose level leads to a myriad of unfavorable cues in the lesion area (Gonzalez-Curiel et al., 2014; Hong et al., 2014; Hsu et al., 2014; Yazdanpanah et al., 2015).

The National Institute for Health and Clinical Excellence (a special nondepartmental health authority within the Health Development Agency for the reduction of the variation in the availability and quality of NHS treatments and care in the United Kingdom) advocates multidisciplinary approaches to counteract the sequela of adverse effects that can cause reduced quality of life, amputations, and mortality.

The measures currently used for DFUs include patient education, control of the blood glucose level, and appropriate debridement (Calderini et al., 2014). The cornerstone in wound management is pressure off-loading by the use of special insoles and boots that redistribute shearing forces. According to the literature, DFU management strategies include focused approaches starting with an appropriate and timely removal of the callus that is a predisposing factor for ulcer development (Sinwar, 2015). The infection status should be monitored regularly. The newest wound dressing systems include hydrogels, foams, and alginates loaded with antimicrobial drugs such as silver and gallium, growth factors, peptides, SCs, and/or other bioactive substances. Many of these therapies are still at an experimental stage. Moreover, the type of intervention differs on the basis of ulcer dimension, starting from simple medical management when ulcers are small or a surgical intervention for arterial reconstruction, lumbar sympathectomy, or amputation, up to an angioplasty for patients where surgery is contraindicated.

4.5 Manipulating the healing response

In burns, surgical debridement is performed to decrease the burden of necrotic tissue and bacteria. The autologous split-thickness skin grafting technique gives permanent wound closure (McHeik et al., 2013; Zilberman and Elsner, 2008).

In situations with a limited availability of donor skin, nonautologous skin grafting (allograft) is an alternative method.

Yet another alternative is the use of advanced tissue-engineered skin substitutes (Bernerd and Asselineau, 2008; Sun et al., 2006). Tissue-engineered skin is an attractive therapy due to its release of cytokines and growth factors at the wound site apart from providing protection from fluid loss and external contamination (Kroner at al., 2012). There are several commercial skin substitutes such as Celaderm, Alloderm®, Apligraf®, Orcel®, and Dermagraft®. These products may elicit host immune reactions (Ackermann et al., 2010; Gordley et al., 2009; Lev-Tov et al., 2013; Mansbridge et al., 2006; Stark et al., 2006). Other drawbacks are a short lifespan, which limits large-scale marketing (Mansbridge, 2006), and their inability to form blood vessels and stratify (Catalano et al., 2013). An interesting improvement of skin substitutes may be the integration of sweat glands or hair follicles that would offer a more natural situation. Also, ways of promoting angiogenesis would have a positive impact. The incorporation of Langerhans cells and melanocytes, in addition to keratinocytes and fibroblasts, is a desirable development. Langerhans cells monitor the skin immune reaction, and the melanocytes recreate natural pigmentation.

Therapeutic protocols have been developed using stimulation with physical agents, such as oxygen, in particular in DFU patients complicated by infection (Goldman, 2009). It has been observed that oxygen therapy can ameliorate the angiogenic response and function of immune cells (Howard et al., 2013). The molecular mechanism of action of the negative pressure wound therapy (NPWT) is not well defined. Experimental observations suggest that NPWT promotes wound healing due to a balanced modulation of antiinflammatory cytokines, mechanoreceptor and chemoreceptor-mediated cell signaling, which increase angiogenesis and the deposition of granulation tissue, and ECM remodeling (Glass et al., 2014). The application of the electric field represents the last frontier for the treatment of chronic wounds. One possible mechanism would be via the activation of epidermal SCs (Koel and Houghton, 2014).

4.6 In vitro skin test models

Skin models are valid alternative systems to in vivo animal testing of, for example, toxic or irritating effects of test materials by multiple end-point analysis, including viability (MTT-assay), membrane integrity (LDH assay), cytokine release (mainly IL-1 and IL-8), and histology (Oliver et al., 1988; Stobbe et al., 2003; Welss et al., 2004). Examples of commercial systems are the Corrositex™ system, Episkin™ (L'Orèal, Paris, France), EpiDerm (MatTek, Ashland, MA, United States), SkinEthic™ (L'Orèal), and the Epidermal Skin Test (CellSystems® Biotechnologie GmbH, Troisdorf, Germany).

4.7 Conclusions

It is clear that excessive inflammation is the leading cause for poor healing in wounds, and many modern therapies target this process. The main actors in the inflammatory process are cells and secreted cytokines, and therefore different strategies are needed to inhibit an excessive cellular response to ameliorate wound healing. The relationships among the various cell types and their effects on wound healing are still unclear. Several enigmatic events remain to be solved. As a result, several therapies to improve wound healing have failed.

A better understanding of the cellular and molecular mechanisms during the inflammatory process that control scar formation is needed. The implications would be significant for wound management in diseases where scarring is the basic pathogenic mechanism.

Another approach to the problem may consist in altering the inflammatory cascade by supplementation with exogenous substances to reduce the catabolic state. The only growth factor on the wound care market, PDGF-BB, has been proven to be effective only on a small subgroup of chronic wounds, possibly because of the corrosive proteases in many chronic wounds that destroy PDGF and other growth factors (Yager et al., 1997).

Furthermore, it has become clear that ECM components play a fundamental role in the wound healing process (Ågren and Werthén, 2007; Olczyk et al., 2014). Therefore, knowledge of the biochemical mechanisms by which ECM components modulate each stage after injury would be important for the development of new therapeutic strategies.

References

Ackermann K, Lombardi Borgia S, Korting H, Mewes K, Schäfer-Korting M. The Phenion full-thickness skin model for percutaneous absorption testing. Skin Pharmacol Physiol 2010;23:105–12.

Ågren MS, Rasmussen K, Pakkenberg B, Jørgensen B. Growth factor and proteinase profile of Vivostat® platelet-rich fibrin linked to tissue repair. Vox Sang 2014;107:37–43.

Ågren MS, Steenfos HH, Dabelsteen S, Hansen JB, Dabelsteen E. Proliferation and mitogenic response to PDGF-BB of fibroblasts isolated from venous leg ulcers is ulcer-age-dependent. J Invest Dermatol 1999;112:463–9.

Ågren MS, Werthén M. The extracellular matrix in wound healing: a closer look at therapeutics for chronic wounds. Int J Low Extrem Wounds 2007;6:82–97.

Auffray C, Fogg D, Garfa M, Elain G, Join-Lambert O, Kayal S, et al. Monitoring of blood vessels and tissues by a population of monocytes with patrolling behavior. Science 2007;317:666–70.

Auffray C, Sieweke MH, Geissmann F. Blood monocytes: development, heterogeneity, and relationship with dendritic cells. Annu Rev Immunol 2009;27:669–92.

Bainbridge P. Wound healing and the role of fibroblasts. J Wound Care 2013;22:407–12.

Bernerd F, Asselineau D. An organotypic model of skin to study photodamage and photoprotection in vitro. J Am Acad Dermatol 2008;58:S155–9.

Black J. Pressure ulcer prevention and management: a dire need for good science. Ann Intern Med 2015;162:387–8.

Blair P, Flaumenhaft R. Platelet α-granules: basic biology and clinical correlates. Blood Rev 2009;23:177–89.

Blanpain C, Fuchs E. Epidermal homeostasis: a balancing act of stem cells in the skin. Nat Rev Mol Cell Biol 2009;10:207–18.

Brancato SK, Albina JE. Wound macrophages as key regulators of repair: origin, phenotype, and function. Am J Pathol 2011;178:19–25.

Brem H, Stojadinovic O, Diegelman RF, Entero H, Lee B, Pastar I, et al. Molecular markers in patients with chronic wounds to guide surgical debridement. Mol Med 2007;13:30–9.

Bullen EC, Longaker MT, Updike DL, Benton R, Ladin D, Hou Z, et al. Tissue inhibitor of metalloproteinases-1 is decreased and activated gelatinases are increased in chronic wounds. J Invest Dermatol 1995;104:236–40.

Busti AJ, Hooper JS, Amaya CJ, Kazi S. Effects of perioperative antiinflammatory and immunomodulating therapy on surgical wound healing. Pharmacotherapy 2005;25:1566–91.

Calderini C, Cioni F, Haddoub S, Maccanelli F, Magotti MG, Tardio S. Therapeutic approach to “diabetic foot” complications. Acta Biomed 2014;85:189–204.

Canesso MC, Vieria AT, Castro TB, Schirmer BG, Cisalpino D, Martins FS, et al. Skin wound healing is accelerated and scarless in the absence of commensal microbiota. J Immunol 2014;193:5171–80.

Catalano E, Cochis A, Varoni E, Rimondini L, Azzimonti B. Tissue-engineered skin substitutes: an overview. J Artif Organs 2013;16:397–403.

Cha J, Falanga V. Stem cells in cutaneous wound healing. Clin Dermatol 2007;25:73–8.

Chen L, Guo S, Ranzer MJ, Dipietro LA. Toll-like receptor 4 has an essential role in early skin wound healing. J Invest Dermatol 2013;133:258–67.

Clemetson KJ. Platelets and primary haemostasis. Thromb Res 2012;129:220–4.

Delavary BM, van der Veer WM, van Egmond M, Niessen FB, Beelen RHJ. Macrophages in skin injury and repair. Immunobiology 2011;216:753–62.

De Rosa L, De Luca M. Cell biology: dormant and restless skin stem cells. Nature 2012;489:215–7.

Desmoulière A, Chaponnier C, Gabbiani G. Tissue repair, contraction, and the myofibroblast. Wound Repair Regen 2005;13:7–12.
Diegelmann RF, Evans MC. Wound healing: an overview of acute, fibrotic and delayed healing. Front Biosci 2004;9:283–9.
Eckes B, Nischt R, Krieg T. Cell-matrix interactions in dermal repair and scarring. Fibrogenesis Tissue Repair 2010;3:4–14.
Edwards R, Harding KG. Bacteria and wound healing. Curr Opin Infect Dis 2004;17:91–6.
Eming SA, Hammerschmidt M, Krieg T, Roers A. Interrelation of immunity and tissue repair or regeneration. Semin Cell Dev Biol 2009;20:517–27.
Eming SA, Krieg T, Davidson JM. Inflammation in wound repair: molecular and cellular mechanisms. J Invest Dermatol 2007;127:514–25.
Farahani RM, Kloth LC. The hypothesis of "biophysical matrix contraction": wound contraction revisited. Int Wound J 2008;5:477–82.
Fletcher J. Differences between acute and chronic wounds and the role of wound bed preparation. Nurs Stand 2008;22(62):64–8.
Gallo R, Kim C, Kokenyesi R, Adzick NS, Bernfield M. Syndecans-1 and -4 are induced during wound repair of neonatal but not fetal skin. J Invest Dermatol 1996;107:676–83.
Gawronska-Kozak B, Bogacki M, Rim JS, Monroe WT, Manuel JA. Scarless skin repair in immunodeficient mice. Wound Repair Regen 2006;14:265–76.
Glass GE, Murphy GF, Esmaeili A, Mai L-M, Nanchahal J. Systematic review of molecular mechanism of action of negative-pressure wound therapy. Br J Surg 2014;101:1627–36.
Goldman RJ. Hyperbaric oxygen therapy for wound healing and limb savage: a systematic review. PM R 2009;1:471–89.
Gonzalez-Curiel I, Trujillo V, Montoya-Rosales A, Rincon K, Rivas-Calderon B, deHaro-Acosta J, et al. 1,25-dihydroxyvitamin D3 induces LL-37 and HBD-2 production in keratinocytes from diabetic foot ulcers promoting wound healing: an in vitro model. PLoS One 2014;9:e111355.
Gordley K, Cole P, Hicks J, Hollier L. A comparative, long term assessment of soft tissue substitutes: AlloDerm, Enduragen, and Dermamatrix. J Plast Reconstr Aesthet Surg 2009;62:849–50.
Gordon S, Martinez FO. Alternative activation of macrophages: mechanism and functions. Immunity 2010;32:593–604.
Guenou H, Nissan X, Larcher F, Feteira J, Lemaitre G, Saidani M, et al. Human embryonic stem-cell derivatives for full reconstruction of the pluristratified epidermis: a preclinical study. Lancet 2009;374:1745–53.
Guo S, DiPietro LA. Factors affecting wound healing. J Dent Res 2010;89:219–29.
Gurtner GC, Werner S, Barrandon Y, Longaker MT. Wound repair and regeneration. Nature 2008;453:314–21.
Hanson SE, Bentz ML, Hematti P. Mesenchymal stem cell therapy for non healing cutaneous wounds. Plast Reconstr Surg 2010;125:510–6.
Harty M, Neff AW, King MW, Mescher AL. Regeneration or scarring: an immunologic perspective. Dev Dyn 2003;226:268–79.
Hinz B. Formation and function of the myofibroblast during tissue repair. J Invest Dermatol 2007;127:526–37.
Hoffman M, Harger A, Lenkowski A, Hedner U, Roberts HR, Monroe DM. Cutaneous wound healing is impaired in hemophilia B. Blood 2006;108:3053–60.
Hong S, Tian H, Lu Y, Laborde JM, Muhale FA, Wang Q, et al. Neuroprotectin/protectin D1: endogenous biosynthesis and actions on diabetic macrophages in promoting wound healing and innervation impaired by diabetes. Am J Physiol Cell Physiol 2014;307:C1058–67.

Howard M, Asmis R, Evans K, Mustoe T. Oxygen and wound care: a critical review of current therapeutic modalities and future direction. Wound Repair Regen 2013;21:503–11.

Hsu I, Parkinson LG, Shen Y, Toro A, Brown T, Zhao H, et al. Serpina3n accelerates tissue repair in a diabetic mouse model of delayed wound healing. Cell Death Dis 2014;5:e1458.

Ito M, Cotsarelis G. Is the hair follicle necessary for normal wound healing? J Invest Dermatol 2008;128:1059–61.

Ito M, Liu Y, Yang Z, Nguyen J, Liang F, Morris RJ, et al. Stem cells in the hair follicle bulge contribute to wound repair but not homeostasis of the epidermis. Nature Med 2005;11:1358–66.

Iwata Y, Yoshizaki A, Komura K, Shimizu K, Ogawa F, Hara T, et al. CD19, a response regulator of B lymphocytes, regulates wound healing through hyaluronan-induced TLR4 signaling. Am J Pathol 2009;175:649–60.

Jameson J, Cauvi G, Sharp LL, Witherdden DA, Havran WL. Gamma-delta T cell induced hyaluronan production by epithelial cells regulates inflammation. J Exp Med 2005;201:1269–79.

Jameson J, Havran WL. Skin gammadelta T-cell functions in homeostasis and wound healing. Immunol Rev 2007;215:114–22.

Kaisho T, Akira S. Toll-like receptor function and signaling. J Allergy Clin Immunol 2006;117:979–87.

Koel G, Houghton PE. Electrostimulation: current status, strength of evidence guidelines, and meta-analysis. Adv Wound Care 2014;3:118–26.

Koźlik M, Wójcicki P. The use of stem cells in plastic and reconstructive surgery. Adv Clin Exp Med 2014;23:1011–7.

Kroner E, Kaiser JS, Fischer SC, Arzt E. Bioinspired polymeric surface patterns for medical applications. J Appl Biomater Funct Mater 2012;10:287–92.

Kulidjian AA, Inman R, Issekutz TB. Rodent models of lymphocyte migration. Semin Immunol 1999;11:85–9.

Laurens N, Koolwijk P, de Maat MP. Fibrin structure and wound healing. J Thromb Haemost 2006;4:932–9.

Lev-Tov H, Li CS, Dahle S, Isseroff RR. Cellular versus acellular matrix devices in treatment of diabetic foot ulcers: study protocol for a comparative efficacy randomized controlled trial. Trials 2013;14:8.

Lewis CJ, Mardaryev AN, Sharov AA, Fessing MY, Botchkarev VA. The epigenetic regulation of wound healing. Adv Wound Care 2014;3:468–75.

Liu LQ, Moody J, Traynor M, Dyson S, Gall A. A systematic review of electrical stimulation for pressure ulcer prevention and treatment in people with spinal cord injuries. J Spinal Cord Med 2014;37:703–18.

Lobmann R, Ambrosch A, Schultz G, Waldmann K, Schiweck S, Lehnert H. Expression of matrix metalloproteinases and their inhibitors in the wounds of diabetic and non-diabetic patients. Diabetologia 2002;45:1011–6.

Lucas T, Waisman A, Ranjan R, Roes J, Krieg T, Muller W, et al. Differential roles of macrophages in diverse phases of skin repair. J Immunol 2010;184:3964–77.

Mahdipour E, Charnock JC, Mace KA. Hoxa3 promotes the differentiation of hematopoietic progenitor cells into proangiogenic Gr-1^{+}CD11b^{+} myeloid cells. Blood 2011;117:815–26.

Mansbridge J. Commercial considerations in tissue engineering. J Anat 2006;209:527–32.

Mariappan MR, Alas EA, Williams JG, Prager MD. Chitosan and chitosan sulfate have opposing effects on collagen-fibroblast interactions. Wound Repair Regen 1999;7:400–6.

Martin P, Leibovich SJ. Inflammatory cells during wound repair: the good, the bad and the ugly. Trends Cell Biol 2005;15:599–607.

McHeik JN, Barrault C, Pedretti N, Garnier J, Juchaux F, Levard G, et al. Foreskin-isolated keratinocytes provide successful extemporaneous autologous paediatric skin grafts. J Tissue Eng Regen Med 2013. http://dx.doi.org/10.1002/term.1690, in press.

Mcheik JN, Barrault C, Levard G, Morel F, Bernard FX, Lecron JC. Epidermal healing in burns: autologous keratinocyte transplantation as a standard procedure: update and perspective. Plast Reconstr Surg Glob Open 2014;2:e218.

Menke NB, Ward KR, Witten TM, Bonchev DG, Diegelmann RF. Impaired wound healing. Clin Dermatol 2007;25:19–25.

Meszaros AJ, Reichner JS, Albina JE. Macrophage phagocytosis of wound neutrophils. J Leukoc Biol 1999;65:35–42.

Midwood KS, Williams LV, Schwarzbauer JE. Tissue repair and the dynamics of the extracellular matrix. Int J Biochem Cell Biol 2004;36:1031–7.

Mills RE, Taylor KR, Podshivalova K, McKay DB, Jameson JM. Defects in skin gamma delta T cell function contribute to delayed wound repair in rapamycin-treated mice. J Immunol 2008;181:3974–83.

Mosser DM, Edwards JP. Exploring the full spectrum of macrophage activation. Nat Rev Immunol 2008;8:958–69.

Olczyk P, Komosinska-Vassev K, Winsz-Szczotka K, Stojko J, Klimek K, Kozma E. Propolis induces chondroitin/dermatan sulphate and hyaluronic acid accumulation in the skin of burned wound. Evid Based Complement Alternat Med 2013;2013:290675.

Olczyk P, Mencner Ł, Komosinska-Vassev K. The role of the extracellular matrix components in cutaneous wound healing. Biomed Res Int 2014;2014:747584.

Oliver GJ, Pemberton MA, Rhodes C. An in vitro model for identifying skin corrosive chemicals. I. Initial validation. Toxicol In Vitro 1988;2:7–17.

Park JE, Barbul A. Understanding the role of immune regulation in wound healing. Am J Surg 2004;187:11–6.

Rea S, Giles NL, Webb S, Adcroft KF, Evill LM, Strickland DH, et al. Bone marrow-derived cells in the healing burn wound—more than just inflammation. Burns 2009;35:356–64.

Reinke JM, Sorg H. Wound repair and regeneration. Eur Surg Res 2012;49:35–43.

Rodero MP, Khosrotehrani K. Skin wound healing modulation by macrophages. Int J Clin Exp Pathol 2010;3:643–53.

Rodero MP, Hodgson SS, Hollier B, Combadiere C, Khosrotehrani K. Reduced Il17a expression distinguishes a Ly6c^{lo}MHCII^{hi} macrophage population promoting wound healing. J Invest Dermatol 2013;133:783–92.

Rossi D, Zlotnik A. The biology of chemokines and their receptors. Annu Rev Immunol 2000;18:217–42.

Sarrazy V, Billet F, Micallef L, Coulomb B, Desmoulière A. Mechanisms of pathological scarring: role of myofibroblasts and current developments. Wound Repair Regen 2011;19(Suppl. 1):S10–5.

Schultz GS, Davidson JM, Kirsner RS, Bornstein P, Herman IM. Dynamic reciprocity in the wound microenvironment. Wound Repair Regen 2011;19:134–48.

Shah JM, Omar E, Pai DR, Sood S. Cellular events and biomarkers of wound healing. Indian J Plast Surg 2012;45:220–8.

Sindrilaru A, Peters T, Wieschalka S, Baican C, Baican A, Peter H, et al. An unrestrained proinflammatory M1 macrophage population induced by iron impairs wound healing in humans and mice. J Clin Invest 2011;121:985–97.

Sinno H, Prakash S. Complements and the wound healing cascade: an updated review. Plast Surg Int 2013;2013:146764.

Sinwar PD. The diabetic foot management – recent advance. Int J Surg 2015;15:27–30.

Stark HJ, Boehnke K, Mirancea N, Willhauck MJ, Pavesio A, Fusenig NE, et al. Epidermal homeostasis in long-term scaffold enforced skin equivalents. J Invest Dermatol Symp Proc 2006;11:93–105.

Stobbe JL, Drake KD, Maier KJ. Comparison of in vivo (Draize method) and in vitro (Corrositex assay) dermal corrosion values for selected industrial chemicals. Int J Toxicol 2003;22:99–107.

Strbo N, Yin N, Stojadinovic O. Innate and adaptive immune responses in wound epithelialization. Adv Wound Care 2014;3:492–501.

Sun T, Jackson S, Haycock JW, MacNeil S. Culture of skin cells in 3D rather than 2D improves their ability to survive exposure to cytotoxic agents. J Biotechnol 2006;122:372–81.

Szántó T, Joutsi-Korhonen L, Deckmyn H, Lassila R. New insights into von Willebrand disease and platelet function. Semin Thromb Hemost 2012;38:55–63.

Taylor KR, Trowbridge JM, Rudisill JA, Termeer CC, Simon JC, Gallo RL. Hyaluronan fragments stimulate endothelial recognition of injury through TLR4. J Biol Chem 2004;279:17079–84.

Tomasek JJ, Gabbiani G, Hinz B, Chaponnier C, Brown RA. Myofibroblasts and mechano-regulation of connective tissue remodelling. Nat Rev Mol Cell Biol 2002;3:349–63.

Tonnesen MG, Feng X, Clark RAF. Angiogenesis in wound healing. J Invest Dermatol Symp Proc 2000;5:40–6.

Vedrenne N, Coulomb B, Danigo A, Bonté F, Desmoulière A. The complex dialogue between (myo)fibroblasts and the extracellular matrix during skin repair processes and ageing. Pathol Biol (Paris) 2012;60:20–7.

Vélez-Díaz-Pallarés M, Lozano-Montoya I, Abraha I, Cherubini A, Soiza RL, O'Mahony D, et al. Non pharmacologic interventions to heal pressure ulcers in older patients: an overview of systematic reviews (The SENATOR-ONTOP series). J Am Med Dir Assoc 2015;16:448–69.

Welss T, Basketter DA, Schroder KR. In vitro skin irritation: facts and future. State of the art review of mechanisms and models. Toxicol In Vitro 2004;18:231–43.

Werner S, Grose R. Regulation of wound healing by growth factors and cytokines. Physiol Rev 2003;83:835–70.

Willenborg S, Eming SA. Macrophages – sensor and effectors coordinating skin damage and repair. J Dtsch Dermatol Ges 2014;12:214–21.

Willenborg S, Lucas T, van Loo G, Knipper JA, Krieg T, Haase I, et al. CCR2 recruits an inflammatory macrophage subpopulation critical for angiogenesis in tissue repair. Blood 2012;120:613–25.

Witte MB, Barbul A. Role of nitric oxide in wound repair. Am J Surg 2002;183:406–12.

Woodfin A, Voisin MB, Nourshargh S. Recent developments and complexities in neutrophil transmigration. Curr Opin Hematol 2010;17:9–17.

Wu YS, Chen SN. Apoptotic cell: linkage of inflammation and wound healing. Front Pharmacol 2014;5:1–6.

Yager DR, Zhang LY, Liang HX, Diegelmann RF, Cohen IK. Wound fluids from human pressure ulcers contain elevated matrix metalloproteinase levels and activity compared to surgical wound fluids. J Invest Dermatol 1996;107:743–8.

Yager DR, Chen SM, Ward SI, Olutoye OO, Diegelmann RF, Cohen IK. The ability of chronic wound fluids to degrade peptide growth factors is associated with increased levels of elastase activity and diminished levels of proteinase inhibitors. Wound Repair Regen 1997;5:23–32.

Yang HS, Shin J, Bhang SH, Shin JY, Park J, Im GI, et al. Enhanced skin wound healing by a sustained release of growth factors contained in platelet-rich plasma. Exp Mol Med 2011;43:622–9.

Yazdanpanah L, Nasiri M, Adarvishi S. Literature review on the management of diabetic foot ulcer. World J Diabetes 2015;6:37–53.

Yoneda A, Couchman JR. Regulation of cytoskeletal organization by syndecan transmembrane proteoglycans. Matrix Biol 2003;22:25–33.

Zilberman M, Elsner JJ. Antibiotic-eluting medical devices for various applications. J Control Release 2008;130:202–15.

Manipulating inflammation to improve healing

N. Urao, T.J. Koh
University of Illinois at Chicago, Chicago, IL, United States

5.1 Introduction

Inflammation is a critical component of the wound healing response, playing important roles in pathogen killing, angiogenesis, granulation tissue formation, collagen deposition, and wound closure [1–3]. Numerous studies have demonstrated that a dysregulated inflammatory response can lead to aberrant or failed healing and that targeting the inflammatory response is an appealing strategy for improving healing in many situations, including the closure of chronic wounds and reducing exuberant scar formation [2–5]. The purpose of this chapter is to critically review the literature on the regulation of inflammation during normal wound healing and on how dysregulation of inflammation can lead to impaired healing. We will also highlight attempts to manipulate inflammation to improve wound healing, including the development of biomaterials to manipulate inflammation.

5.2 Inflammation during wound healing

Wound healing requires the coordinated responses of many molecular pathways and a variety of cells, which regulate overlapping phases of inflammation, tissue formation, and remodeling. The inflammatory phase of wound healing begins at the moment of injury and must be properly regulated in space and time to induce an efficient healing response. The inflammatory phase is a dynamic process, involving a variety of inflammatory pathways that act as upstream signals to govern the later phases of wound healing (Fig. 5.1). Advances in leukocyte biology have led to a better understanding of inflammation during wound healing. In particular, improved knowledge of leukocyte subsets, especially monocytes/macrophages, has provided a new perspective on the pathophysiology of wound healing [6–8].

5.2.1 Inflammatory response during normal wound healing

5.2.1.1 Cells regulating inflammation

The inflammatory phase is characterized by the tissue infiltration of a series of leukocytes, including polymorphonuclear (PMN) leukocytes, monocytes/macrophages,

Wound Healing Biomaterials - Volume 1. http://dx.doi.org/10.1016/B978-1-78242-455-0.00005-7

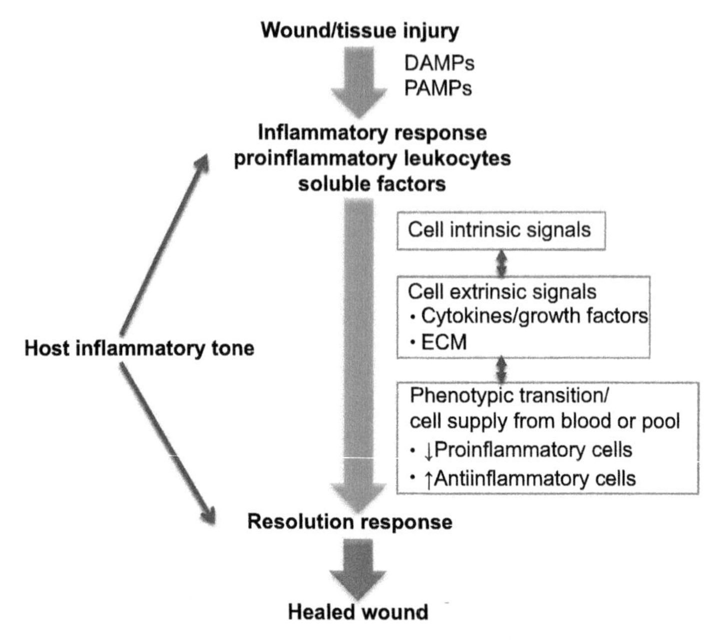

Figure 5.1 Inflammation during wound healing. Tissue injury induces an inflammatory response initiated by damage-associated molecular patterns (DAMPs) and pathogen-associated molecular patterns (PAMPs), further orchestrated by leukocytes and soluble factors and closely linked with alterations of the cell supply from blood and hematopoietic sources. The resolution of inflammation is characterized by cell-intrinsic and cell-extrinsic pathways such as cytokines/growth factors and extracellular matrix (ECM) as well as the phenotypic transition of leukocyte populations in wounds. Each of these processes is mutually interdependent, and a well-regulated inflammatory response appears to be required for the induction of the resolution response and efficient wound healing. The preexisting inflammatory tone of the host, which is often dysregulated in pathologies such as diabetes, influences both inflammatory and resolution responses, leading to poor healing.

and T cells. These leukocytes mediate essential processes for normal wound healing by fighting pathogenic organisms, removing damaged tissue and apoptotic/necrotic cells, producing growth factors, and promoting extracellular matrix (ECM) remodeling. Very early leukocyte infiltration can be mediated by tissue resident mast cells, which release vasodilative histamine and proteases as well as proinflammatory cytokines. The importance of resident mast cells in wound healing has been examined by the use of mast cell-deficient Kit mutant mice; Kit is an essential tyrosine kinase receptor driving mast cell development in mice. For example, in Kit mutant mice, neutrophil recruitment into the site of injury is reduced, supporting the idea that mast cells promote neutrophil recruitment [9]. However, Kit mutant mice exhibit phenotypes that extend beyond mast cell deficiency, and new mouse models more specifically targeting mast cells have been generated [10]. These latter mice express Cre recombinase under the control of mast cell protease genes to obtain Kit-independent mast cell-deficient mice. Using these mice, it has been demonstrated that mast cells may be dispensable for normal wound healing [11,12]. Finally, pharmacological mast cell inhibition by disodium cromoglycate reduces inflammation and scar formation in mice [13], leaving open the question of whether the alteration of specific functions of mast cells, rather than the number of mast cells, might improve wound healing.

Regardless of whether mast cells are involved, PMN leukocytes are generally considered to be the first responders following tissue damage. These cells clear debris and provide protection against infection if the body's barrier function is compromised. Neutrophils also release enzymes such as elastase and proteases as well as reactive oxygen species that can cause bystander damage to otherwise healthy tissue. In the mouse cutaneous excisional wounding model, antibody-induced neutrophil depletion accelerated wound closure in adult wild-type and diabetic mice [14]. In contrast, delayed wound closure in aged wild-type mice was further delayed by neutrophil depletion [15]. Together, these studies indicate that the influence of neutrophils on wound healing may depend on the host environment. Further investigation of diverse functions of neutrophils [16,17] and of neutrophil subsets [18] in wound healing is needed, including the complex communication between neutrophils and other immune cells that may lead to either enhanced or impaired healing.

Monocytes/macrophages follow neutrophils into the site of injury either by egressing from the blood or migrating and proliferating from their local pool. They remove damaged tissue and necrotic or apoptotic cells through phagocytosis, produce cytokines/growth factors, and present antigen to adaptive immune cells such as T cells. Macrophages play roles in both tissue damage and repair based on studies involving the selective depletion of wound macrophages in different phases of healing in different injury models [19–25]. These studies indicated that wound macrophages are involved in healing responses, including angiogenesis and collagen deposition [22,23].

Under noninflammatory conditions, peripheral tissues contain primarily tissue resident M2-like macrophages that contribute to tissue homeostasis. Upon tissue injury or infection, M1-like activation is induced by the engagement of pattern recognition receptors with damage-associated molecular patterns (DAMPs) or pathogen-associated molecular patterns (PAMPs), respectively [26]. Studies have focused on the macrophage phenotypes expressed during wound healing following tissue injury. Indeed, M1-like macrophages appear at the early stage of inflammation, and they are replaced by M2-like macrophages [6–8].

Different ontogenies of macrophages may also influence macrophage phenotypes during wound healing. In myocardial infarction, Ly6C^{hi} monocytes/M1-like macrophages and CD11c^{+} dendritic monocytes/macrophages accumulate during the initial proinflammatory phase, followed by Ly6C^{lo}/M2-like macrophages associated with antiinflammatory or healing response [27]. Ly6C^{hi} monocytes/M1-like macrophages and CD11c^{+} dendritic monocytes/macrophages were reported to be bone marrow-derived under inflammatory conditions [28]. However, it remains to be established whether Ly6C^{lo}/M2-like macrophages that accumulate in healing tissue are derived from tissue resident progenitors or from Ly6C^{hi} monocytes/M1-like macrophages during a phenotype conversion that may be regulated by the tissue environment [29]. Moreover, since Ly6C^{lo} monocytes can contribute to proinflammatory responses in some inflammatory injury models [30], the precise role(s) of monocyte subsets may be specific to each situation [29].

T lymphocytes infiltrate damaged tissue in the late inflammatory phase and remain in the tissue during the remodeling phase for weeks or longer. In classical studies, congenitally athymic nude mice that lack a normal T cell system exhibited an increased fibrotic response, suggesting that T cells may limit fibrosis [31]. A series of studies using monoclonal antibody-induced T cell depletion suggested that a subpopulation of T cells stimulates wound healing [32,33]. T cells residing in nonwounded skin of mice have been identified as $\gamma\delta$ T cells, which may help to maintain homeostasis as well as direct wound healing [34,35]. Similar skin resident T cells have been observed in wounded skin [36]. Dermal T cells may execute a variety of functions during wound healing through secreting soluble factors such as keratinocyte growth factors [34] and insulin-like growth factor (IGF) [35]. Interestingly, but not surprisingly, epidermal T cells are sensitive to metabolic conditions such as hyperglycemia [37] and obesity in mice [38] and in humans [39]. In mouse myocardial infarction, $Foxp3^+CD4^+$ regulatory T cells appear to improve healing by modulating monocyte/macrophage differentiation [40]. While the significance of T lymphocytes in human wound healing remains obscure [41], $CD4^+$ T helper 2 (T_H2) cells may promote healing through the production of the key Th2 cytokines such as IL-4, IL-5, IL-10, and IL-13 [42].

Mesenchymal stem/stromal cells (MSCs) can migrate to sites of injury from their place of residence in perivascular spaces of various tissues and/or through the circulation [43]. MSCs can be isolated from various sources such as adipose tissue, bone marrow, and peripheral blood [44–48] and expanded for therapeutic use, which has been shown to accelerate wound healing [47–50]. Moreover, endogenous MSC mobilization can be enhanced by systemic pharmacological interventions. Cytokines and growth factors such as granulocyte colony stimulating factor, granulocyte macrophage colony stimulating factor, hepatocyte growth factor, vascular endothelial growth factor (VEGF), and a combination of these have been used to promote wound healing through endogenous MSC mobilization [51], while these therapies also mobilize hematopoietic stem and progenitor cells (HSPCs) and endothelial progenitor cells (EPCs) [52]. The chemokine receptor CXCR4 is expressed on MSCs, HSPCs, and EPCs and plays important roles both in cell retention in the bone marrow (BM) and in recruitment to the wound site. The CXCR4 antagonist AMD3100, especially when combined with other mobilizing factors, has been shown to release stem and progenitor cells from their anchor bone marrow, thus driving their mobilization [52]. Combined with low-dose tacrolimus or IGF-1, AMD3100 improves skin wound healing [53] and bone fracture healing [54,55] through increased MSC and/or EPC accumulation at the wound site. MSCs may promote wound healing by transdifferentiation into multiple cell types [56]. Perhaps more significantly, MSCs have notable immunomodulatory effects on the surrounding environment following implantation and can secrete a variety of factors as well as induce factor secretion from other neighboring cells [57], especially those with antiinflammatory effects [50,58].

The supply of leukocytes from hematopoietic sources also participates in the regulation of the inflammatory response as cells infiltrating injured tissues from the blood need to be replenished. Studies on the healing of myocardial infarction as well as stroke revealed that the turnover of monocytes/macrophages in the acute inflammatory phase is rapid with an average tissue residence time of 20 h and that proinflammatory monocytes, perhaps as well as neutrophils, given their shorter lifespans, are constantly replenished by blood-borne cells at the site of damage [28]. The regulation of cell supply from the blood is dependent on multiple biological events: production and mobilization from the hematopoietic organs such as bone marrow and spleen, transendothelial migration to tissue interstitial space (cell recruitment), cell survival and proliferation, and emigration from the tissue. Regulators for these processes could occur at multiple levels from the control of gene expression to environmental cues, including adhesion molecules and receptor and cytokines/chemokines expression, as reviewed elsewhere [59]. Notably, the chemokine CCL2/MCP-1 plays essential roles in both myeloid mobilization from the bone marrow and its infiltration into the inflamed tissue [60]. At least in mice, spleen and lymphoid organs serve as monocyte reservoirs that may be deleterious for wound healing when the supply is greater than needed for efficient repair, such as in proinflammatory conditions including hypercholesterolemia and atherosclerosis [61,62]. A study demonstrated that brain injury activates hematopoietic stem cells to produce myeloid cells in mouse bone marrow [63]. Our research has shown that hindlimb ischemia injury in mice causes expansion and mobilization of hematopoietic progenitor cells in the bone marrow [64]. These studies indicate that multiple sources of inflammatory cells and hematopoietic stem and progenitor cells participate in the regulation of inflammation in peripheral tissues during wound healing, although the associated mechanisms are largely unknown.

5.2.1.2 Soluble factors regulating inflammation

Tissue damage caused by mechanical or chemical disruption, oxidative stress, or ischemia causes host cell injury or death, which results in the release of soluble factors that initiate the inflammatory cascade. Soluble factors in damaged tissue attract leukocytes that in turn produce additional soluble factors specific to each cell population. As cell phenotypes and populations shift during the healing process, the secretion or extracellular formation of soluble products shifts from proinflammatory to antiinflammatory [1,2]. In the following section, we discuss the diverse soluble factors that initiate and propagate an inflammatory response after tissue injury and summarize those factors that modulate or inhibit postinjury inflammation.

The initiation of inflammation is associated with the production of DAMPs. These molecules include cytosolic or nuclear proteins derived from necrotic cells and degraded extracellular matrices, which are normally compartmentalized and unavailable for activating inflammatory cells. Examples include uric acid [65], high mobility group box 1 protein [66], genomic [67] or mitochondrial [68] DNA, adenosine or adenosine tri-phosphate [69,70], the S100 calcium-binding protein family [71,72], hyaluronan [73], and many others [74].

Resident immune cells and/or circulating leukocytes sense DAMPs through their Toll-like receptors (TLRs) and the receptor for advanced glycosylation end product, among others [74], and propagate inflammation by producing cytokines/chemokines via signal transduction through transcription factors such as nuclear factor kappa-B, activator protein-1, and interferon regulatory factors (IRFs). DAMPs-initiated molecular pathways are extensively reviewed elsewhere [75,76]. When barriers against external microbes are disrupted, such as in skin or mucosal injury, PAMPs participate in the activation of the inflammatory signaling pathways. Peptidoglycan, lipopeptides and lipoproteins of Gram-positive bacteria, *mycoplasma* lipopeptides, and fungal zymosan are recognized by TLR2 in concert with TLR1 or TLR6 [77]. Gram-negative bacterial component lipopolysaccharide (LPS) activates TLR4 [78]. Nucleic acids from bacteria or viruses are recognized by TLR3 [79], TLR7, and TLR8 [80]. Bacterial CpG can stimulate TLR9 [81]. There is a very close relationship between the pathways for recognizing microbial products and recognizing host cell injury, and as such, DAMP receptors also detect microbial products.

Reactive oxygen species (ROS) have also been demonstrated to be a potential initiator of inflammation in addition to their well-established roles as a second messenger in cellular signaling [82]. For example, a series of experiments in zebrafish and with human neutrophils indicated that hydrogen peroxide produced in the injured site attracts neutrophils to the site through oxidation-dependent Src family kinase activation [83,84]. Moreover, studies also indicate that the responses tightly regulated by ROS may ensure an efficient wound healing response [85].

Although there is cell type- and wound environment-dependent diversity in responses to DAMPs and PAMPs, these initiators of inflammation generally result in inflammatory cytokine production from leukocytes, endothelial cells, keratinocytes, and fibroblasts. During the early stages of wound healing, monocytes and macrophages are major sources of proinflammatory cytokines such as interleukin (IL)-6, tumor necrosis factor (TNF)-α, and IL-1β [86], which are elevated as early as within 24 h following injury and are produced for several days. These cytokines play key roles in propagating the inflammatory response and when left unchecked can contribute to impaired healing [1,2].

5.2.1.3 Mechanisms of resolution of inflammation

Inflammation in damaged tissue typically resolves after several days, and immune homeostasis is gradually restored during normal healing. The major populations of leukocytes present during the later phase of inflammation are monocytes/macrophages and T cells, which produce antiinflammatory cytokines and growth factors. These include IL-10 that actively stimulates an antiinflammatory pathway and growth factors such as VEGF, fibroblast growth factor (FGF), IGF, and transforming growth factor (TGF), which are critical for vascular and tissue remodeling. Although angiogenesis is initiated by inflammatory cytokines in the early phase of inflammation, remodeling of newly formed vasculature is required to support tissue perfusion, required for normal repair and tissue regeneration. Moreover, the identification of molecules or gene products that are required for the resolution indicates that the resolution of inflammation is an active process, in which the activation of antiinflammatory signals and/or the negative regulation of inflammatory signals are necessary, rather than just a passive biological phenomenon involving the dissipation of proinflammatory mediators. Potent counterregulatory mediators termed resolvins and protectins have been reviewed [87] and have been shown to be involved in the resolution of inflammation during wound healing [88,89]. The resolution of inflammation has been a topic of intense study since many diseases such as obesity, atherosclerosis, cancer, and autoimmune diseases are associated with impairments in this process.

Cellular factors involving the resolution of inflammation include reductions in cell number via apoptosis and reverse migration and the transition from proinflammatory to antiinflammatory phenotypes typically seen in monocytes/macrophages [90–94]. Apoptosis allows for reductions in cell numbers while minimizing the production of DAMPs. Apoptotic neutrophils are ingested by macrophages, which triggers the macrophage release of antiinflammatory cytokines such as TGF-β and IL-10 [91,92]. Indeed, molecules that promote neutrophil apoptosis such as TNF-related apoptosis-inducing ligand can enhance the resolution of inflammation [95]. Besides apoptotic cell clearance by phagocytes, studies also suggest that reverse migration from the inflamed tissue contributes to a decrease in the number of inflammatory cells at local tissues [93,94,96]. Macrophages and dendritic monocytes, perhaps also neutrophils [97], can be drained from inflamed tissue to lymph nodes through C—C chemokine receptor 7 (CCR7). CCR7-mediated cell drainage to lymph nodes, where these cells potentially communicate with T lymphocytes [98], seems to promote the resolution of inflammation in atherosclerosis [99] and other inflammatory models [100].

Although alterations in the populations and functions of myeloid cells over the time course of wound healing are widely acknowledged, the regulation of these changes is still under investigation and likely occurs on multiple levels. Molecular mechanisms at the single cell level promote phenotypic changes in infiltrating neutrophils and monocytes/macrophages. In neutrophils, lipid mediators undergo class switching from proinflammatory leukotrienes and prostaglandins (PGs) to antiinflammatory lipoxins as a result of proinflammatory PGE2 stimulation [101]. Similar class switching has been observed in macrophages [102]. The intracellular biosynthesis pathway of proresolution mediators from essential fatty acids to docosahexaenoic acid by macrophages has also been described [103]. These studies indicate essential roles for lipid mediators in regulating the phenotypes of neutrophils and macrophages during the resolution of inflammation.

Monocytes/macrophages change their phenotype quickly, depending on their environment, especially when they are exposed to proinflammatory or antiinflammatory cytokines in vitro and in vivo. DAMPs and PAMPs in cooperation with Th1 cytokines such as interferon γ induce M1-type activation, whereas Th2 cytokines such as IL-4 and IL-13 induce M2-type activation. These activations are independent of the existence of T cells in vivo [26]. Moreover, M1-like and M2-like macrophages are capable of Th1 and Th2 differentiation of T cells, suggesting a major role for macrophage polarization in inflammatory responses to environmental insults [26,104]. However, IL-4 and IL-13 were undetectable in wound macrophages as well as the wound environment in mouse myocardial infarction [105] and in mouse skin wounds [106,107]. Therefore the specific factors that regulate macrophage phenotypes, especially the M2-like phenotype, in specific wound healing models remain to be elucidated.

We have demonstrated that the proinflammatory cytokine IL-1β plays an important role in regulating macrophage phenotypes during wound healing in mice and humans, such that IL-1β promotes an M1-like phenotype, and that downregulation of IL-1β production allows upregulation of peroxisome proliferator-activated receptor (PPAR) γ activity, which in turn promotes the transition to an M2-like phenotype [107,108]. In addition, an interesting in vitro study showed that IL-6, which is often considered to be a proinflammatory cytokine, enhanced M2-like macrophage polarization when acting alone, whereas costimulation with IFNγ induced an M1-like phenotype [109]. Thus downregulation of proinflammatory cytokines may play an important role in the transition from M1-like to M2-like phenotypes. Other proteins and gene products regulating the macrophage phenotypic switch in vivo include nuclear receptor subfamily 4, group a, member 1 (Nr4a1) [110] and the transcription factor interferon regulatory factor 5 (IRF5) [111].

5.2.2 Dysregulated inflammatory responses leading to impaired wound healing

As discussed, inflammation following tissue injury protects against infection and is an integral part of the healing response. Many conditions that induce impaired wound healing are associated with persistent inflammation. This may indicate that the problem of dysregulated inflammation in wound healing may not result from issues associated with the initiation of inflammation but from impairments in its resolution. In other words, deficiencies in resolution mechanisms, whether the inflammatory response is initially excessive or subnormal, may lead to nonresolving inflammation [90] as well as impaired wound healing.

The persistence of exogenous inflammatory stimuli prolongs inflammation and impairs its resolution. Excessive and/or prolonged inflammatory responses are often associated with microbe infection, which releases PAMPs and DAMPs from damaged tissue to activate pattern molecule recognition pathways such as TLR signaling. In addition, microbe contamination promotes host cell necrosis rather than apoptosis in the damaged tissue, resulting in increased proinflammatory signals and impaired apoptosis-mediated proresolution signals. Particularly in skin wounds, microbes often form biofilm that provides resistance to antibiotic and protection against host defense system. This topic has been specifically reviewed elsewhere [112–114]. In addition to microbes, the disruption of epithelium also exposes cells at the injured site to other foreign bodies, irritants, and toxic compounds, which can induce additional inflammatory responses. Once chronic inflammation is established, it can be sustained even after removing exogenous inflammatory stimuli through mechanisms mimicking autoimmune conditions or allergies [90].

In addition to sustained proinflammatory stimuli in wounds, dysregulated DAMP recognition pathways such as TLR signaling can also contribute to a persistent inflammatory response. For example, TLR expression is increased in diabetic wounds compared to nondiabetic wounds in mice [115–117] and humans [118]. Using knockout mice, TLR deficiency ameliorated the enhanced inflammatory response and improved wound healing associated in the streptozotocin-induced type 1 diabetes model [115–117]. The same group demonstrated that increased TLR expression can result from increased leukocyte infiltration in diabetic wounds and can also result from their upregulation on a per cell basis. TLR2 and TLR4 surface expression and mRNA were also increased on circulating monocytes from patients with type 1 diabetes [119] as well as type 2 diabetes [120]. However, the host status of TLR signaling might be diverse, as diabetic susceptibility to wound infection was suggested to be a result of impaired recognition of bacterial infection via the TLR pathway [121].

Chronic inflammation is characterized by a prolonged accumulation of myeloid cells such as neutrophils and macrophages, specifically those with a proinflammatory phenotype. This can be caused by impaired phenotypic transition of myeloid cells from proinflammatory to antiinflammatory/healing phenotypes [2,8,122]. A deficiency in any one of the mechanisms governing the phenotypic transition can result in sustained proinflammatory cell accumulation. For example, in a model of obesity and type 2 diabetes, increased free fatty acid levels can stimulate prostaglandin production, which enhances neutrophil survival and impairs macrophage phagocytosis [123], potentially through impairments in class switching of lipid mediators. Indeed, the biosynthetic pathway for proresolving neuroprotectin/protectin D1 is impaired in macrophages isolated from diabetic mice [124]. Under similar conditions, the dysregulation of key molecules responsible for the macrophage phenotypic switch has been demonstrated. Soluble factors present in the human and mouse diabetic wound environment are sufficient to activate the Nod-like receptor protein (NLRP)-3 inflammasome, which results in the sustained production of IL-1β by wound macrophages [125]. The sustained production of IL-1β appears to impair the upregulation of PPARγ in wound macrophages, leading to an impaired resolution of inflammation and impaired healing, suggesting that the interplay between the cytokine environment and intrinsic macrophage signaling maintains the proinflammatory state of macrophages in diabetes [107].

5.2.3 Systemic conditions that influence inflammation and wound healing

The host immune state or inflammatory tone may be an important determinant of both normal and impaired wound healing (Fig. 5.1). Low-grade, nonresolved inflammation has been identified in a wide range of chronic conditions such as obesity, diabetes, atherosclerosis, autoimmune disease, and cancer, which are thought to be associated with poor healing [126]. Moreover, a series of studies suggest that lifestyle factors such as smoking, diet, psychological stress, and sleep deprivation significantly affect immune systems and subsequently wound healing [127,128]. Finally, a subnormal immune status resulting from HIV infection or immune-suppressing drugs such as corticosteroid also likely results in impaired wound healing [126].

Proinflammatory conditions are often associated with the alteration of tissue leukocyte populations, which may depend in part on the generation of leukocytes from their progenitor cells. In obesity, for example, the infiltration of inflammatory macrophages into adipose tissue is supported by the enhanced generation of inflammatory Ly6C^{hi} monocytes in the bone marrow [129]. Similar evidence has been obtained in mouse models of diabetes [130] and hypercholesterolemia [131]. In these low-grade, chronic inflammatory conditions, homeostatic set points of the immune system may be shifted [132], which in turn may influence inflammatory responses to tissue injury during wound healing. Evidence also suggests that diet, mental stress, and sleep deprivation can potentiate activity of the hematopoietic system during inflammatory responses because the sympathetic nerve tone governs hematopoietic stem cells [128]. During normal conditions, hematopoietic stem cells sense danger signals from tissue injury [63], but aberrant activation of hematopoietic stem and progenitor cells has been reported in conditions with chronic inflammation [133–137].

A sustained supply of large numbers of proinflammatory myeloid cells from the bone marrow and the spleen may play an important role in driving the prolonged inflammatory response linked with poor wound healing associated with metabolic disease. Inflammatory monocyte mobilization from the spleen is increased in post-myocardial infarction in the mouse model of hypercholesterolemia [138]. Proinflammatory leukocytes contribute to tissue damage and proinflammatory cytokine production, which further stimulates inflammatory signaling with positive feed forward mechanisms. Indeed, the IL-1 receptor contributes to reestablishing the splenic reservoir of inflammatory monocytes in mice [28]. Since patients with chronic wounds suffer from this vicious and persistent proinflammatory cycle, novel interventions for chronic wounds could be aimed at downregulating the proinflammatory signaling network.

Systemic inflammatory conditions can influence the monocyte/macrophage system at the single cell level through epigenetic memory in the innate immune system. In vitro studies using human monocytes demonstrated that priming with the fungal cell wall component β-glucan increased proinflammatory cytokine production upon restimulation after 6 days, a phenomenon termed "trained immunity" [139,140]. Accordingly, myeloid cells recruited to wounds are intrinsically primed to be more proinflammatory in diabetic mice than those in nondiabetic mice [141]. In high-fat diet-induced obese mice, a repressive histone methylation mark, H3K27me3, is decreased at the promoter of the IL-12 gene in bone marrow progenitors, resulting in increased IL-12 production. This epigenetic signature is passed down to wound macrophages [142]. These studies indicate that systemic conditions can alter cellular phenotypes in the myeloid system, including hematopoietic progenitor cells, through epigenetic dysregulation, which in turn can lead to poor wound healing. A better understanding of hematopoietic dysfunction at the epigenetic level may provide insight into the influence of the systemic inflammatory tone on inflammatory responses during wound healing.

On the other end of the spectrum, an insufficient inflammatory response to tissue injury is associated with primary or secondary immune deficiency such as treatment with immunosuppressant drugs. Given that the initial inflammatory response governs subsequent wound healing responses, impairment of the initial inflammatory response is linked with a dysregulated inflammatory response during subsequent phases of wound healing and impaired healing [143]. In addition, it has been reported that diabetic animals exhibit an impaired early inflammatory response but persistent inflammation during skin wound healing [144]. Aging or senescence also impairs the initial inflammatory response to ischemic reperfusion of the murine heart [145]. Thus subnormal levels of the initial inflammatory response could lead to impaired healing responses.

5.3 Manipulating inflammation to improve wound healing

Dysregulated inflammatory responses are strongly associated with poor wound healing outcomes and can be caused at multiple levels. Thus therapies targeting inflammation either locally within the inflamed tissue or systemically, including sites of hematopoiesis, may lead to improved healing.

5.3.1 Targeting local inflammatory responses to improve healing

Numerous experimental studies have focused on manipulating the local biological environment to improve wound healing. This has included applying growth factors, cytokines, antibodies, and synthetic drugs targeting various components of wound healing, including the inflammatory response. In this section, we will summarize conventional clinical practice, including debridement and wound dressings, and recent experimental attempts to manipulate inflammation locally at the wound site to improve healing.

5.3.1.1 Manipulating damage-associated signals

Given that prolonged inflammatory responses result in poor wound healing outcomes, reducing damage-associated signals may lead to better outcomes. Wound debridement, typically performed surgically or chemically, is used to remove tissue debris and dead cells in wounds, which may restart acute inflammation and reduce chronic damage-associated signaling. Debridement also reduces the risk of infection because necrotic tissues can be a substrate for proliferating microbes that consume nutrients and oxygen, which are requirements for wound healing. Microbe infection itself can induce additional inflammatory signals via PAMPs, which in turn may impair inflammatory resolution and reparative processes. Preventing infection is a focus of primary care for wound patients, via debridement and adequate coverage or dressing and the use of antibiotics when required. Dressings are used not only to provide a temporary barrier and protection against exogenous stimuli but also to keep the wound moist, which may accelerate autolytic debridement, the activity of proteolytic enzymes and macrophage phagocytosis [146,147].

Biofilm formation that is resistant to basic antibiotic treatment is common in chronic wounds. Disrupting biofilm is an effective strategy to reduce microbial bioburden. Chemical compounds that have antimicrobial effects such as a silver alginate dressing [148], mechanical disruption such as surgical debridement discussed above, low-frequency ultrasonic-assisted wound debridement [149], and wireless electroceutical dressings [150] can be effective as an antibiofilm to reduce excessive inflammatory responses. It is important to note that some bacteria play an important role in the resolution of inflammation, thereby promoting healing, and thus complete sterilization of a wound may be detrimental [151].

Finally, damage-associated signals can be manipulated at the level of pattern recognition receptor (PRR) pathways such as TLR signaling. Since PRR signaling not only promotes the inflammatory response but also programs the resolution of inflammation [152], any therapy targeting PRR signaling must be fine-tuned. For example, the immunomodulatory action of hyaluronan is molecular weight-dependent [153]. Large hyaluronan polymers function as tissue integrity signals and serve to suppress the inflammatory response [154], whereas small hyaluronan polymers promote leukocyte migration in a TLR4-dependent manner [155]. Therefore the ECM can be targeted to modulate inflammatory pathways, including PRR signaling, to regulate inflammatory responses. In addition, some microorganisms actually negatively regulate these signaling PRR pathways through multiple mechanisms [156], and thus understanding how microbes manipulate PRR signaling could lead to the development of novel treatments for downregulating inflammation and improving the healing of chronic wounds.

5.3.1.2 Manipulating soluble factors and leukocyte function

Leukocyte function can be modulated by cell-extrinsic mechanisms, such as extracellular cytokines/chemokines. As the profile of cytokines/chemokines changes over the different phases of wound healing, supplementing or blocking these environmental cues at different time points can have different effects on wound healing. Proinflammatory cytokines recruit leukocytes from the blood and then activate them to propagate inflammatory responses, often through further proinflammatory cytokine production. The MCP-1/CCR2 axis regulates inflammatory monocyte recruitment to damaged tissue. In hyperinflammatory apolipoprotein E-deficient mice, monocyte-directed siRNA against CCR2 encapsulated in nanoparticles improved infarct healing [157]. Cell type and situation-specific approaches may be required because CCR2-mediated responses also regulate healing responses, such as angiogenesis, through recruiting monocytes [158] and mesenchymal progenitor cell recruitment [159]. Furthermore, $CCR2^{hi}$ monocytes were demonstrated to be a source of CX_3CR1^{hi} prohealing monocytes as the in situ transition from $CCR2^{hi}$ to CX_3CR1^{hi} monocytes in injured sites has been demonstrated in mice [160]. CX_3CR1^{hi} mononuclear phagocytes seem to be a critical component for healing in acute or chronic inflammatory conditions, as blocking these cells by systemic antibody injection or using knockout mice delayed skin wound healing [161,162]. Moreover, depletion of CX_3CR1^{hi} mononuclear phagocytes revealed that these cells integrate microbial-originated signals in chronic colitis [163]. In short, targeting of CCR2 may be effective when the host immune state is proinflammatory but not when the host immune system is well regulated. In this scenario, a systemic therapeutic approach has improved healing; the administration of anti-TNFα antibodies into the blood attenuated wound inflammation associated with reductions in monocytes in the blood and macrophages at the wound site in ob/ob mice [21].

The administration of antiinflammatory/prohealing cytokines has been used in an attempt to improve healing. The topical administration of IL-4 on experimental wounds in nondiabetic mice accelerated the rate of healing [164], although its effect on inflammation was not assessed. The systemic administration of IL-4 in the early phase (~day 5) of ligament healing in rats promoted wound healing, whereas its continuous administration (~day 11) abrogated the beneficial effect of the early phase [165]. In vitro studies showed that IL-4 downregulates a master angiogenic transcription factor, hypoxia inducible factor-1α in macrophages, and likely decreases angiogenesis in vivo [166], which has the potential to impair wound healing. Thus any wound healing treatment based on the administration of IL-4 will likely have to be healing phase-specific. With respect to another antiinflammatory cytokine, namely IL-10, the loss of IL-10 in knockout mice resulted in no significant effect in studies on healing of infarcted hearts [167] and skin wounds [168]. In contrast, treatment with exogenous human IL-10 reduced inflammation and scarring in both animal and human cutaneous wounds [169]. In contrast, IL-10 was ineffective in scar-prone dark-skinned individuals [170].

Other environmental factors that influence wound inflammation and healing include neurotransmitters, especially in diabetes where peripheral neurons are dysfunctional. Substance P is a sensory neuropeptide released from sensory fibers located in the skin. Levels of substance P are reduced in diabetic wounds associated with elevated degradation [144]. Substance P administration increased M2-type macrophages in diabetic wounds associated with improved healing [144]. In addition, a docosanoid derived from the polyunsaturated fatty acid docosahexaenoic acid (DHA) is a neuroprotective factor [171]. Interestingly, its synthesis is impaired in wounds of db/db diabetic mice, and treating wounds with neuroprotectin/protectin D1 or macrophages conditioned by this factor promoted wound healing in the diabetic mice [124]. Similarly, the topical administration of resolvin D1, which is a downstream product of DHA, accelerated wound closure and stimulated diabetic macrophage phagocytosis [89]. Thus these lipid mediators may be a target for manipulating inflammation and promoting wound healing in type 2 diabetes.

Cell-intrinsic factors regulating leukocyte activity can also be targeted. For example, local treatment with the transcription factor IRF5 via nanoparticle delivery induced macrophage transition to an antiinflammatory phenotype and improved wound healing in mice [111]. In diabetic mice, the topical pharmacological inhibition of the NLRP-3 inflammasome by glyburide also induced a switch in macrophage phenotype and downregulated the proinflammatory environment, resulting in improved healing [125]. The overexpression of Hoxa3 gene rescued diabetes-induced impairment of $CD11b^+Gr1^+$ myeloid cells in db/db mice, which are normally prohealing in wild-type mice [127]. Finally, Kruppel-like factor 4 activation by the plant-derived product Mexicanin I accelerated myeloid cell-dependent cutaneous wound healing [172].

5.3.1.3 Cell transfer or transplantation and inflammation during wound healing

The therapeutic application of various cell types, including stem and progenitor cells as well as differentiated macrophages, has been reported to improve wound healing in animal models and in some clinical trials. In both animal and human studies, these cells have been obtained from bone marrow, peripheral blood, or adipose tissue. In experimental and clinical studies, heterogeneous or more purified autologous bone marrow-derived cells, which include primarily monocytes and their progenitor cells, have shown promising but at best limited beneficial effects on healing of the ischemic heart [173], peripheral artery disease [174], and cutaneous wound healing [175,176].

Despite controversy about the fate of transferred cells, these cells likely contribute to healing through secreting growth factors [173,174].

Differentiated macrophages have been used to promote healing, but potential applications may be context-dependent. For example, exogenous M1 macrophages, but not nonactivated macrophages, reduced fibrosis and enhanced muscle fiber regeneration in murine lacerated muscles [177]. Human macrophages activated by hypoosmotic shock accelerated cardiac wound healing with increased monocyte and macrophage recruitment when they were injected into infarct myocardium of rats [178]. In addition, M2-like $CD11b^{+}Ly\text{-}6C^{lo}F4/80^{hi}$ cells isolated from a 3-day reperfusion injury improved skeletal muscle recovery after ischemia/reperfusion [179]. Furthermore, murine $Gr\text{-}1^{+}CD11b^{+}$ myeloid cells obtained from the spleen of nondiabetic mice home to the site of injury and enhance diabetic wound healing by neoangiogenesis after intravenous administration [180]. In contrast, in vitro M2 polarized macrophage administration to a cutaneous wound did not improve wound healing in both wild-type and delayed healing in diabetic db/db mice, despite those cells exhibiting an antiinflammatory phenotype [181]. These studies indicate that, consistent with the plastic nature of macrophages and with the requirement of both proinflammatory and antiinflammatory responses for proper healing, the therapeutic use of monocytes/macrophage cell therapy requires the consideration of the specific healing impairment(s) involved and the activation state of the cells prior to transfer as well as any change in phenotype that may occur in response to the evolving wound environment over the course of healing.

Adult MSCs isolated from various sites, including bone marrow, adipose tissue, and amniotic fluid, have shown potential for clinical translation for wound healing [182]. Although the mechanism through which MSCs accelerate healing is not fully understood, one of the main actions of MSCs is providing environmental cues for wound healing by secreting cytokines and growth factors [182]. For example, through the release of TNF-stimulated gene 6 protein, MSC implantation induces antiinflammatory and prohealing effects by modulating macrophage function [183], neutrophil migration [184], and dendritic cell maturation [185]. Currently, four clinical studies using MSCs for cutaneous would healing have demonstrated safety and efficacy: nonpurified autologous bone marrow cells in chronic wounds showed safety and feasibility [186]; cultured autologous MSCs from bone marrow in fibrin spray showed a strong correlation between the cell number applied and wound closure rate [187]; autologous bone marrow MSCs along with a standard wound dressing on lower extremity ulcers accelerated healing as compared with a noncell treated control group [188]; and a bone marrow MSC/collagen-based artificial dermis composite was superior to the existing skin graft in patients with intractable dermatopathies [189].

5.4 Manipulating inflammation by biomaterials

Research has focused on combining biomaterials with molecular and stem cell approaches to generate novel therapies to improve wound healing. In this section, we will briefly discuss the potential for biomaterials already in clinical therapeutic use or under development to influence inflammation during wound healing. In addition, studies indicate that inflammation can have a substantial impact on biomaterial performance and the loosely defined characteristic of "biocompatibility" [190]. Therefore we stress that a more complete understanding of inflammation caused by material–tissue interactions is warranted to better develop biomaterials to improve wound healing, particularly in different disease states that present different inflammatory environments.

5.4.1 Biomaterials applied to wound healing therapy

Biomaterials can provide structural, cellular, and biochemical support for wound healing. Some natural polymers such as polysaccharides (alginates, chitin, chitosan, heparin, chondroitin) and proteoglycans and proteins (collagen, gelatin, fibrin, keratin, silk fibroin, eggshell membrane) have been used in wound and burn management. They provide good biocompatibility, biodegradability, and similarity to macromolecules recognized by the human body [191]. Some synthetic polymers, such as biomimetic ECM micro/nanoscale fibers based on polyglycolic acid, polylactic acid, polyacrylic acid, poly-ε-caprolactone, polyvinylpyrrolidone, polyvinyl alcohol, and polyethylene glycol, exhibit in vivo and in vitro wound healing properties [191]. Synthetic polymers are appealing because it is relatively easy to modify their components to tune their mechanical properties for a wide range of tissues including cartilage, bone, vascular, nerve, and ligament [192]. Hydrogel polymers, such as polyethylene glycol diacrylate, have a high water content and can be used to cross-link natural extracellular matrices [193] while minimizing immune detection and thus foreign body reaction. Bioactive synthetic hydrogels can provide molecularly tailored biofunctions such as cell adhesion, proteolytic degradation, and growth factor-binding that may be important for improved wound healing [194]. Synthetic materials are often used to supplement natural polymers that are vulnerable to rapid degradation. Moreover, more complex scaffolds obtained from whole tissues by removing antigenic proteins and cells have been routinely used in reconstructive surgery and soft tissue replacement. These include bioprosthetic mesh, such as acellular dermal matrix derived from pig, cow, or human, which may have advantages over synthetic mesh regarding mesh infection, structural strength, and mesh explantation [195,196].

5.4.2 Biomaterials that modify inflammation to improve healing

The implantation of biomaterials can induce a host inflammatory response in response to mechanical injury as well as part of a foreign body reaction. Foreign material is recognized by myeloid cells, which attempt to clear the material from the host tissue through phagocytosis and digestion or enzymic degradation. Through this process, myeloid cells pass signals to the adaptive immune system that can contribute to rejection or chronic inflammation. An ideal biomaterial for wound healing would be one that minimizes foreign body and rejection responses but does not inhibit inflammatory responses required for the healing process. Further, a biomaterial may be designed to manipulate inflammation to maximize the healing capacity in the host (Fig. 5.2). For example, unmodified collagen scaffolds induce M2 macrophage marker expression, whereas those treated with glutaraldehyde cause increased inflammation [197]. The same group demonstrated that the scaffolds could be used to deliver cytokines

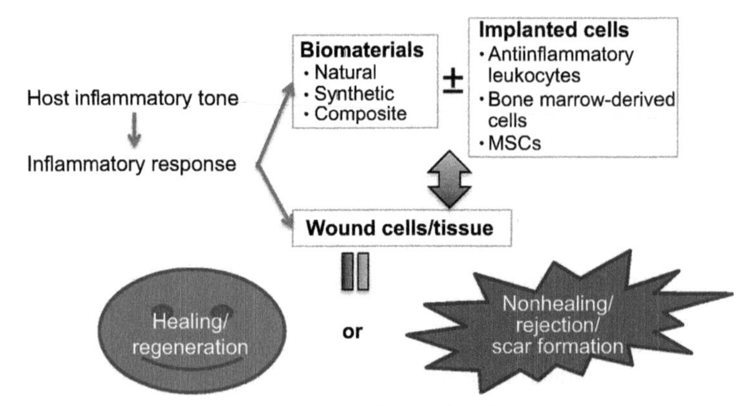

Figure 5.2 Inflammatory response in biomaterial-mediated wound care. Inflammatory responses to the therapeutic use of biomaterials and cells play important roles in wound healing outcomes such as healing or nonhealing, healing by scar or by regeneration, or rejection of materials. Natural, synthetic, or composite biomaterials are currently used in clinical practice or are being modified for better outcomes in basic and preclinical research. Biomaterials are often combined with cell therapy using antiinflammatory leukocytes, bone marrow-derived cells, and mesenchymal stem/stromal cells (MSCs). These materials inherently modify the inflammatory response in healing wounds, interacting with the host inflammatory response to regulate wound cell function and healing outcomes.

to manipulate the phenotype of wound macrophages [198]. Similarly, in a molecular size-dependent manner, hyaluronic acid-containing materials can promote phagocytosis and reduce M1 activity of macrophages [199,200] in vitro, and their topical application may enhance wound healing in vivo partly through increased M2 macrophage activity [201]. Chitosan scaffolds can regulate macrophage M1/M2 polarization, depending on different degrees of acetylation [202]. Furthermore, different biomaterials can induce proinflammatory or antiinflammatory macrophage phenotypes, depending on their composition, and thus the optimal choice of biomaterials may depend on the type of inflammatory response desired [203–206]. For example, an in situ photopolymerizable semiinterpenetrating network derived from ECM components modifies the cytokine expression profile at the wound site to favor wound healing in pigs [207]. In addition, biodegradable citric acid-derived hydrogels had an antibacterial function and promoted wound healing in rats [208]. Biomechanical properties may also play a role in modulating inflammation, as an enhanced stiffness of alginate/collagen-I scaffold can lead to increased antiinflammatory IL-10 expression of dermal fibroblasts in wound dressings [209]. In summary, these studies provide evidence that biomaterials can be designed to guide host immune responses to improve healing.

Some biomaterials have been developed to provide a controlled release of bioactive proteins at the wound site, serving both as a depot of these proteins and protecting them from the highly proteolytic environment of the wound. For example, bioscaffolds designed for the sequential release of IFNγ and IL-4 enhanced vascularization and the macrophage phenotypic switch in a murine subcutaneous implantation model [198]. Some biomaterials, such as highly porous alginate scaffolds, are designed to prevent a rapid growth factor release [210]. In addition, instead of using specific soluble factors, natural biomaterials containing a mixture of soluble factors can be used to supplement immunomodulatory and prohealing factors for healing. For example, a dehydrated human amnion/chorion composite graft released bioactive IL-4, IL-6, IL-8, and IL-10 together with growth factors such as TGFβ, bFGF, epidermal growth factor, and granulocyte-colony stimulating factor, which in turn promoted wound healing through MSC activation [211]. A bone marrow-derived extracellular matrix, which contains various growth factors including VEGF, bFGF, and platelet-derived growth factor, was obtained from a porcine sternum, and its injection with methylcellulose carrier gel improved cardiac function after rat myocardial infarction partially through reducing macrophage infiltration at a later phase [212]. Moreover, these effects were achieved at the concentration 10^3- to 10^8-fold lower than typically required to achieve a beneficial effect by a single factor [212], suggesting that natural biomaterials may be an effective source of growth factors or cytokines to improve wound healing.

Biomaterials are also being developed to improve the retention of transplanted cells for wound healing cell therapy applications [213], including MSCs, bone marrow stem cells, or monocytes/macrophages. Cell–biomaterial constructs do not guarantee enhanced functions of those transplanted cells, such as antiinflammatory effects, and introduce further complexity in terms of the action on wound healing through biomaterial's effects on cell viability, proliferation (self-renewal), and intended or spontaneous differentiation. However, some cell–biomaterial constructs have demonstrated better outcomes than cell therapy alone, which may lead to clinical translation in the near future [214,215]. It is important to note that macrophages can demonstrate unexpected phenotypes after integration with biomaterials and the host tissue environment even if they are polarized by cytokines; as such, IL-4-polarized macrophages, as compared with nonpolarized macrophages, showed increased proteolytic activity in an artificial tumor environment comprised of collagen-I gel and tumor cells [216]. Thus careful investigation will be required when extrapolating from single-cell populations to more complex systems such as cell–biomaterial composites applied to in vivo environments. These effects might be explained by modified inflammatory responses induced by cells combined with biomaterial carriers, although such speculation remains to be tested.

5.4.3 Influence of host disease status on the function of biomaterials

As discussed previously, metabolic diseases such as diabetes, obesity, hypercholesterolemia, atherosclerosis, and autoimmune diseases are linked with a low-grade chronic inflammation that likely influences the inflammatory response following tissue injury. Therefore an immune response to biomaterials might be altered in different individuals with varied preexisting immunologic tone (Fig. 5.2). For example, the interaction of materials with the host tissue interface has been studied using a prototypical adhesive material based on dendrimer/dextran in the colon [190]. The authors of this study found the existence of complex interactions that determine the overall dendrimer/dextran biomaterial compatibility, which was related to the extent and nature of immune cells in the diseased environment present before material implantation such as cancerous tissue and established colitis [190]. This study suggests the importance of assessing material–tissue interactions and their disease-driven alterations to facilitate the desired material performance in specific clinical scenarios.

5.5 Conclusions

Advances in our understanding of inflammation during wound healing offer numerous opportunities to develop novel therapeutic approaches for improving healing outcomes. Moreover, many established treatments for improving wound healing involve a favorable modification in wound inflammatory responses. Further investigation of targets in inflammatory signaling pathways is likely to provide insight into new therapeutic avenues for manipulating inflammation to improve wound healing. Biomaterials have a great potential for modifying inflammatory responses by themselves or in conjunction with the delivery of soluble factors and/or cells to provide optimal inflammatory and healing responses following tissue injury or in chronic wounds. Finally, the role of the host immune state should always be considered because it alters the local microenvironmental response in damaged tissue and can induce stem and progenitor cell dysregulation, which may further impact healing. In conclusion, we believe that continued development of biomaterials to induce the therapeutic manipulation of inflammation will lead to novel and effective approaches for improving wound healing.

5.6 Summary

- Inflammation must be regulated in time and space during wound healing to induce efficient healing of host tissues.
- Inflammation influences angiogenesis, collagen deposition, and epithelialization, all of which are required for efficient wound healing.
- The disease status of the host and associated systemic inflammation are likely to have a strong impact on the inflammatory response to additional insults, such as tissue damage.
- Manipulating inflammation is an appealing approach for novel therapies to improve wound healing.
- Research on biomaterials for wound healing should be designed to elucidate their effects on inflammation regardless of whether they are designed to modify inflammation or do so as a side effect.

References

[1] Eming SA, Krieg T, Davidson JM. Inflammation in wound repair: molecular and cellular mechanisms. J Invest Dermatol 2007;127(3):514–25. http://dx.doi.org/10.1038/sj.jid.5700701. PubMed PMID: 17299434.

[2] Koh TJ, DiPietro LA. Inflammation and wound healing: the role of the macrophage. Expert Rev Mol Med 2011;13:e23. http://dx.doi.org/10.1017/S1462399411001943. PubMed PMID: 21740602; PMCID: 3596046.

[3] Roy S, Sen CK. miRNA in wound inflammation and angiogenesis. Microcirculation 2012;19(3):224–32. http://dx.doi.org/10.1111/j.1549-8719.2011.00156.x. PubMed PMID: 22211762; PMCID: 3399420.
[4] Kempf T, Zarbock A, Vestweber D, Wollert KC. Anti-inflammatory mechanisms and therapeutic opportunities in myocardial infarct healing. J Mol Med (Berl) 2012;90(4):361–9. http://dx.doi.org/10.1007/s00109-011-0847-y. PubMed PMID: 22228177.
[5] Frangogiannis NG. Regulation of the inflammatory response in cardiac repair. Circ Res 2012;110(1):159–73. http://dx.doi.org/10.1161/CIRCRESAHA.111.243162. PubMed PMID: 22223212; PMCID: 3690135.
[6] Mahdavian Delavary B, van der Veer WM, van Egmond M, Niessen FB, Beelen RH. Macrophages in skin injury and repair. Immunobiology 2011;216(7):753–62. http://dx.doi.org/10.1016/j.imbio.2011.01.001. PubMed PMID: 21281986.
[7] Willenborg S, Eming SA. Macrophages – sensors and effectors coordinating skin damage and repair. J Dtsch Dermatol Ges 2014;12(3):214-21–3. http://dx.doi.org/10.1111/ddg.12290. PubMed PMID: 24580874.
[8] Novak ML, Koh TJ. Phenotypic transitions of macrophages orchestrate tissue repair. Am J Pathol 2013;183(5):1352–63. http://dx.doi.org/10.1016/j.ajpath.2013.06.034. PubMed PMID: 24091222; PMCID: 3969506.
[9] Egozi EI, Ferreira AM, Burns AL, Gamelli RL, Dipietro LA. Mast cells modulate the inflammatory but not the proliferative response in healing wounds. Wound Repair Regen 2003;11(1):46–54. PubMed PMID: 12581426.
[10] Rodewald HR, Feyerabend TB. Widespread immunological functions of mast cells: fact or fiction? Immunity 2012;37(1):13–24. http://dx.doi.org/10.1016/j.immuni.2012.07.007. PubMed PMID: 22840840.
[11] Antsiferova M, Martin C, Huber M, Feyerabend TB, Forster A, Hartmann K, et al. Mast cells are dispensable for normal and activin-promoted wound healing and skin carcinogenesis. J Immunol 2013;191(12):6147–55. http://dx.doi.org/10.4049/jimmunol.1301350. PubMed PMID: 24227781.
[12] Willenborg S, Eckes B, Brinckmann J, Krieg T, Waisman A, Hartmann K, et al. Genetic ablation of mast cells redefines the role of mast cells in skin wound healing and bleomycin-induced fibrosis. J Invest Dermatol 2014;134(7):2005–15. http://dx.doi.org/10.1038/jid.2014.12. PubMed PMID: 24406680.
[13] Chen L, Schrementi ME, Ranzer MJ, Wilgus TA, DiPietro LA. Blockade of mast cell activation reduces cutaneous scar formation. PLoS One 2014;9(1):e85226. http://dx.doi.org/10.1371/journal.pone.0085226. PubMed PMID: 24465509; PMCID: 3898956.
[14] Dovi JV, He LK, DiPietro LA. Accelerated wound closure in neutrophil-depleted mice. J Leukoc Biol 2003;73(4):448–55. PubMed PMID: 12660219.
[15] Nishio N, Okawa Y, Sakurai H, Isobe K. Neutrophil depletion delays wound repair in aged mice. Age (Dordr) 2008;30(1):11–9. http://dx.doi.org/10.1007/s11357-007-9043-y. PubMed PMID: 19424869; PMCID: 2276589.
[16] Mocsai A. Diverse novel functions of neutrophils in immunity, inflammation, and beyond. J Exp Med 2013;210(7):1283–99. http://dx.doi.org/10.1084/jem.20122220. PubMed PMID: 23825232; PMCID: 3698517.
[17] Nauseef WM, Borregaard N. Neutrophils at work. Nat Immunol 2014;15(7):602–11. http://dx.doi.org/10.1038/ni.2921. PubMed PMID: 24940954.
[18] Beyrau M, Bodkin JV, Nourshargh S. Neutrophil heterogeneity in health and disease: a revitalized avenue in inflammation and immunity. Open Biol 2012;2(11):120134. http://dx.doi.org/10.1098/rsob.120134. PubMed PMID: 23226600; PMCID: 3513838.

[19] Duffield JS, Forbes SJ, Constandinou CM, Clay S, Partolina M, Vuthoori S, et al. Selective depletion of macrophages reveals distinct, opposing roles during liver injury and repair. J Clin Invest 2005;115(1):56–65. http://dx.doi.org/10.1172/JCI22675. PubMed PMID: 15630444; PMCID: 539199.

[20] Goren I, Allmann N, Yogev N, Schurmann C, Linke A, Holdener M, et al. A transgenic mouse model of inducible macrophage depletion: effects of diphtheria toxin-driven lysozyme M-specific cell lineage ablation on wound inflammatory, angiogenic, and contractive processes. Am J Pathol 2009;175(1):132–47. http://dx.doi.org/10.2353/ajpath.2009.081002. PubMed PMID: 19528348; PMCID: 2708801.

[21] Goren I, Muller E, Schiefelbein D, Christen U, Pfeilschifter J, Muhl H, et al. Systemic anti-TNFα treatment restores diabetes-impaired skin repair in ob/ob mice by inactivation of macrophages. J Invest Dermatol 2007;127(9):2259–67. http://dx.doi.org/10.1038/sj.jid.5700842. PubMed PMID: 17460730.

[22] Lucas T, Waisman A, Ranjan R, Roes J, Krieg T, Muller W, et al. Differential roles of macrophages in diverse phases of skin repair. J Immunol 2010;184(7):3964–77. http://dx.doi.org/10.4049/jimmunol.0903356. PubMed PMID: 20176743.

[23] Mirza R, DiPietro LA, Koh TJ. Selective and specific macrophage ablation is detrimental to wound healing in mice. Am J Pathol 2009;175(6):2454–62. http://dx.doi.org/10.2353/ajpath.2009.090248. PubMed PMID: 19850888; PMCID: 2789630.

[24] Shechter R, London A, Varol C, Raposo C, Cusimano M, Yovel G, et al. Infiltrating blood-derived macrophages are vital cells playing an anti-inflammatory role in recovery from spinal cord injury in mice. PLoS Med 2009;6(7):e1000113. http://dx.doi.org/10.1371/journal.pmed.1000113. PubMed PMID: 19636355; PMCID: 2707628.

[25] van Amerongen MJ, Harmsen MC, van Rooijen N, Petersen AH, van Luyn MJ. Macrophage depletion impairs wound healing and increases left ventricular remodeling after myocardial injury in mice. Am J Pathol 2007;170(3):818–29. http://dx.doi.org/10.2353/ajpath.2007.060547. PubMed PMID: 17322368; PMCID: 1864893.

[26] Mills CD, Kincaid K, Alt JM, Heilman MJ, Hill AM. M-1/M-2 macrophages and the Th1/Th2 paradigm. J Immunol 2000;164(12):6166–73. PubMed PMID: 10843666.

[27] Nahrendorf M, Swirski FK, Aikawa E, Stangenberg L, Wurdinger T, Figueiredo JL, et al. The healing myocardium sequentially mobilizes two monocyte subsets with divergent and complementary functions. J Exp Med 2007;204(12):3037–47. http://dx.doi.org/10.1084/jem.20070885. PubMed PMID: 18025128; PMCID: 2118517.

[28] Leuschner F, Rauch PJ, Ueno T, Gorbatov R, Marinelli B, Lee WW, et al. Rapid monocyte kinetics in acute myocardial infarction are sustained by extramedullary monocytopoiesis. J Exp Med 2012;209(1):123–37. http://dx.doi.org/10.1084/jem.20111009. PubMed PMID: 22213805; PMCID: 3260875.

[29] Rodero MP, Licata F, Poupel L, Hamon P, Khosrotehrani K, Combadiere C, et al. In vivo imaging reveals a pioneer wave of monocyte recruitment into mouse skin wounds. PLoS One 2014;9(10):e108212. http://dx.doi.org/10.1371/journal.pone.0108212. PubMed PMID: 25272047; PMCID: 4182700.

[30] Donnelly DJ, Longbrake EE, Shawler TM, Kigerl KA, Lai W, Tovar CA, et al. Deficient CX3CR1 signaling promotes recovery after mouse spinal cord injury by limiting the recruitment and activation of Ly6C^{lo}/iNOS^{+} macrophages. J Neurosci 2011;31(27):9910–22. http://dx.doi.org/10.1523/JNEUROSCI.2114-11.2011. PubMed PMID: 21734283; PMCID: 3139517.

[31] Barbul A, Shawe T, Rotter SM, Efron JE, Wasserkrug HL, Badawy SB. Wound healing in nude mice: a study on the regulatory role of lymphocytes in fibroplasia. Surgery 1989;105(6):764–9. PubMed PMID: 2567062.

[32] Efron JE, Frankel HL, Lazarou SA, Wasserkrug HL, Barbul A. Wound healing and T-lymphocytes. J Surg Res 1990;48(5):460–3. PubMed PMID: 2352421.
[33] Barbul A, Breslin RJ, Woodyard JP, Wasserkrug HL, Efron G. The effect of in vivo T helper and T suppressor lymphocyte depletion on wound healing. Ann Surg 1989;209(4):479–83. PubMed PMID: 2522759; PMCID: 1493975.
[34] Jameson J, Ugarte K, Chen N, Yachi P, Fuchs E, Boismenu R, et al. A role for skin gammadelta T cells in wound repair. Science 2002;296(5568):747–9. http://dx.doi.org/10.1126/science.1069639. PubMed PMID: 11976459.
[35] Sharp LL, Jameson JM, Cauvi G, Havran WL. Dendritic epidermal T cells regulate skin homeostasis through local production of insulin-like growth factor 1. Nat Immunol 2005;6(1):73–9. http://dx.doi.org/10.1038/ni1152. PubMed PMID: 15592472.
[36] Toulon A, Breton L, Taylor KR, Tenenhaus M, Bhavsar D, Lanigan C, et al. A role for human skin-resident T cells in wound healing. J Exp Med 2009;206(4):743–50. http://dx.doi.org/10.1084/jem.20081787. PubMed PMID: 19307328; PMCID: 2715110.
[37] Taylor KR, Mills RE, Costanzo AE, Jameson JM. Gammadelta T cells are reduced and rendered unresponsive by hyperglycemia and chronic TNFα in mouse models of obesity and metabolic disease. PLoS One 2010;5(7):e11422. http://dx.doi.org/10.1371/journal.pone.0011422. PubMed PMID: 20625397; PMCID: 2896399.
[38] Taylor KR, Costanzo AE, Jameson JM. Dysfunctional gammadelta T cells contribute to impaired keratinocyte homeostasis in mouse models of obesity. J Invest Dermatol 2011;131(12):2409–18. http://dx.doi.org/10.1038/jid.2011.241. PubMed PMID: 21833015; PMCID: 3213272.
[39] Costanzo AE, Taylor KR, Dutt S, Han PP, Fujioka K, Jameson JM. Obesity impairs gammadelta T cell homeostasis and antiviral function in humans. PLoS One 2015;10(3):e0120918. http://dx.doi.org/10.1371/journal.pone.0120918. PubMed PMID: 25785862; PMCID: 4365046.
[40] Weirather J, Hofmann UD, Beyersdorf N, Ramos GC, Vogel B, Frey A, et al. Foxp3^{+} CD4^{+} T cells improve healing after myocardial infarction by modulating monocyte/macrophage differentiation. Circ Res 2014;115(1):55–67. http://dx.doi.org/10.1161/CIRCRESAHA.115.303895. PubMed PMID: 24786398.
[41] Hofmann U, Frantz S. Role of lymphocytes in myocardial injury, healing, and remodeling after myocardial infarction. Circ Res 2015;116(2):354–67. http://dx.doi.org/10.1161/CIRCRESAHA.116.304072. PubMed PMID: 25593279.
[42] Gause WC, Wynn TA, Allen JE. Type 2 immunity and wound healing: evolutionary refinement of adaptive immunity by helminths. Nat Rev Immunol 2013;13(8):607–14. http://dx.doi.org/10.1038/nri3476. PubMed PMID: 23827958; PMCID: 3789590.
[43] Rustad KC, Gurtner GC. Mesenchymal stem cells home to sites of injury and inflammation. Adv Wound Care (New Rochelle) 2012;1(4):147–52. http://dx.doi.org/10.1089/wound.2011.0314. PubMed PMID: 24527296; PMCID: 3623614.
[44] Hung SC, Chen NJ, Hsieh SL, Li H, Ma HL, Lo WH. Isolation and characterization of size-sieved stem cells from human bone marrow. Stem Cells 2002;20(3):249–58. http://dx.doi.org/10.1634/stemcells.20-3-249. PubMed PMID: 12004083.
[45] Jiang Y, Jahagirdar BN, Reinhardt RL, Schwartz RE, Keene CD, Ortiz-Gonzalez XR, et al. Pluripotency of mesenchymal stem cells derived from adult marrow. Nature 2002;418(6893):41–9. http://dx.doi.org/10.1038/nature00870. PubMed PMID: 12077603.
[46] Zuk PA, Zhu M, Ashjian P, De Ugarte DA, Huang JI, Mizuno H, et al. Human adipose tissue is a source of multipotent stem cells. Mol Biol Cell 2002;13(12):4279–95. http://dx.doi.org/10.1091/mbc.E02-02-0105. PubMed PMID: 12475952; PMCID: 138633.

[47] Nuschke A. Activity of mesenchymal stem cells in therapies for chronic skin wound healing. Organogenesis 2014;10(1):29–37. http://dx.doi.org/10.4161/org.27405. PubMed PMID: 24322872; PMCID: 4049892.
[48] Hanson SE. Mesenchymal stem cells: a multimodality option for wound healing. Adv Wound Care (New Rochelle) 2012;1(4):153–8. http://dx.doi.org/10.1089/wound.2011.0297. PubMed PMID: 24527297; PMCID: 3839012.
[49] Heo SC, Jeon ES, Lee IH, Kim HS, Kim MB, Kim JH. Tumor necrosis factor-α-activated human adipose tissue-derived mesenchymal stem cells accelerate cutaneous wound healing through paracrine mechanisms. J Invest Dermatol 2011;131(7):1559–67. http://dx.doi.org/10.1038/jid.2011.64. PubMed PMID: 21451545.
[50] Hocking AM. Mesenchymal stem cell therapy for cutaneous wounds. Adv Wound Care (New Rochelle) 2012;1(4):166–71. http://dx.doi.org/10.1089/wound.2011.0294. PubMed PMID: 24527299; PMCID: 3623581.
[51] Wu Y, Wang J, Scott PG, Tredget EE. Bone marrow-derived stem cells in wound healing: a review. Wound Repair Regen 2007;15(Suppl. 1):S18–26. http://dx.doi.org/10.1111/j.1524-475X.2007.00221.x. PubMed PMID: 17727462.
[52] Fruehauf S, Zeller WJ, Calandra G. Novel developments in stem cell mobilization: focus on CXCR4, vol. xiv. New York: Springer; 2012. p. 496.
[53] Lin Q, Wesson RN, Maeda H, Wang Y, Cui Z, Liu JO, et al. Pharmacological mobilization of endogenous stem cells significantly promotes skin regeneration after full-thickness excision: the synergistic activity of AMD3100 and tacrolimus. J Invest Dermatol 2014;134(9):2458–68. http://dx.doi.org/10.1038/jid.2014.162. PubMed PMID: 24682043; PMCID: 4194897.
[54] Myers TJ, Yan Y, Granero-Molto F, Weis JA, Longobardi L, Li T, et al. Systemically delivered insulin-like growth factor-I enhances mesenchymal stem cell-dependent fracture healing. Growth Factors 2012;30(4):230–41. http://dx.doi.org/10.3109/08977194.2012.683188. PubMed PMID: 22559791; PMCID: 3752908.
[55] Kumar S, Ponnazhagan S. Mobilization of bone marrow mesenchymal stem cells in vivo augments bone healing in a mouse model of segmental bone defect. Bone 2012;50(4):1012–8. http://dx.doi.org/10.1016/j.bone.2012.01.027. PubMed PMID: 22342795; PMCID: 3339043.
[56] Sasaki M, Abe R, Fujita Y, Ando S, Inokuma D, Shimizu H. Mesenchymal stem cells are recruited into wounded skin and contribute to wound repair by transdifferentiation into multiple skin cell type. J Immunol 2008;180(4):2581–7. PubMed PMID: 18250469.
[57] Gebler A, Zabel O, Seliger B. The immunomodulatory capacity of mesenchymal stem cells. Trends Mol Med 2012;18(2):128–34. http://dx.doi.org/10.1016/j.molmed.2011.10.004. PubMed PMID: 22118960.
[58] Wang Y, Chen X, Cao W, Shi Y. Plasticity of mesenchymal stem cells in immunomodulation: pathological and therapeutic implications. Nat Immunol 2014;15(11):1009–16. http://dx.doi.org/10.1038/ni.3002. PubMed PMID: 25329189.
[59] Nourshargh S, Alon R. Leukocyte migration into inflamed tissues. Immunity 2014;41(5):694–707. http://dx.doi.org/10.1016/j.immuni.2014.10.008. PubMed PMID: 25517612.
[60] Tsou CL, Peters W, Si Y, Slaymaker S, Aslanian AM, Weisberg SP, et al. Critical roles for CCR2 and MCP-3 in monocyte mobilization from bone marrow and recruitment to inflammatory sites. J Clin Invest 2007;117(4):902–9. http://dx.doi.org/10.1172/JCI29919. PubMed PMID: 17364026; PMCID: 1810572.
[61] Swirski FK, Nahrendorf M, Etzrodt M, Wildgruber M, Cortez-Retamozo V, Panizzi P, et al. Identification of splenic reservoir monocytes and their deployment to inflammatory sites. Science 2009;325(5940):612–6. http://dx.doi.org/10.1126/science.1175202. PubMed PMID: 19644120; PMCID: 2803111.

[62] Frangogiannis NG. Contribution of extramedullary organs in myocardial inflammation and remodeling: does the spleen cause cardiac melancholy? Circ Res 2014;114(2):230–2. http://dx.doi.org/10.1161/CIRCRESAHA.113.302971. PubMed PMID: 24436423.
[63] Courties G, Herisson F, Sager HB, Heidt T, Ye Y, Wei Y, et al. Ischemic stroke activates hematopoietic bone marrow stem cells. Circ Res 2015;116(3):407–17. http://dx.doi.org/10.1161/CIRCRESAHA.116.305207. PubMed PMID: 25362208.
[64] Urao N, McKinney RD, Fukai T, Ushio-Fukai M. NADPH oxidase 2 regulates bone marrow microenvironment following hindlimb ischemia: role in reparative mobilization of progenitor cells. Stem Cells 2012;30(5):923–34. http://dx.doi.org/10.1002/stem.1048. PubMed PMID: 22290850; PMCID: 3703153.
[65] Shi Y, Evans JE, Rock KL. Molecular identification of a danger signal that alerts the immune system to dying cells. Nature 2003;425(6957):516–21. http://dx.doi.org/10.1038/nature01991. PubMed PMID: 14520412.
[66] Scaffidi P, Misteli T, Bianchi ME. Release of chromatin protein HMGB1 by necrotic cells triggers inflammation. Nature 2002;418(6894):191–5. http://dx.doi.org/10.1038/nature00858. PubMed PMID: 12110890.
[67] Ishii KJ, Suzuki K, Coban C, Takeshita F, Itoh Y, Matoba H, et al. Genomic DNA released by dying cells induces the maturation of APCs. J Immunol 2001;167(5):2602–7. PubMed PMID: 11509601.
[68] Oka T, Hikoso S, Yamaguchi O, Taneike M, Takeda T, Tamai T, et al. Mitochondrial DNA that escapes from autophagy causes inflammation and heart failure. Nature 2012;485(7397):251–5. http://dx.doi.org/10.1038/nature10992. PubMed PMID: 22535248; PMCID: 3378041.
[69] Cronstein BN, Daguma L, Nichols D, Hutchison AJ, Williams M. The adenosine/neutrophil paradox resolved: human neutrophils possess both A1 and A2 receptors that promote chemotaxis and inhibit O2 generation, respectively. J Clin Invest 1990;85(4):1150–7. http://dx.doi.org/10.1172/JCI114547. PubMed PMID: 2156895; PMCID: 296546.
[70] McDonald B, Pittman K, Menezes GB, Hirota SA, Slaba I, Waterhouse CC, et al. Intravascular danger signals guide neutrophils to sites of sterile inflammation. Science 2010;330(6002):362–6. http://dx.doi.org/10.1126/science.1195491. PubMed PMID: 20947763.
[71] Ryckman C, Vandal K, Rouleau P, Talbot M, Tessier PA. Proinflammatory activities of S100: proteins S100A8, S100A9, and S100A8/A9 induce neutrophil chemotaxis and adhesion. J Immunol 2003;170(6):3233–42. PubMed PMID: 12626582.
[72] Hofmann MA, Drury S, Fu C, Qu W, Taguchi A, Lu Y, et al. RAGE mediates a novel proinflammatory axis: a central cell surface receptor for S100/calgranulin polypeptides. Cell 1999;97(7):889–901. PubMed PMID: 10399917.
[73] Jiang D, Liang J, Noble PW. Hyaluronan in tissue injury and repair. Annu Rev Cell Dev Biol 2007;23:435–61. http://dx.doi.org/10.1146/annurev.cellbio.23.090506.123337. PubMed PMID: 17506690.
[74] Kono H, Rock KL. How dying cells alert the immune system to danger. Nat Rev Immunol 2008;8(4):279–89. http://dx.doi.org/10.1038/nri2215. PubMed PMID: 18340345; PMCID: 2763408.
[75] Tang D, Kang R, Coyne CB, Zeh HJ, Lotze MT. PAMPs and DAMPs: signal 0s that spur autophagy and immunity. Immunol Rev 2012;249(1):158–75. http://dx.doi.org/10.1111/j.1600-065X.2012.01146.x. PubMed PMID: 22889221; PMCID: 3662247.
[76] Miller YI, Choi SH, Wiesner P, Fang L, Harkewicz R, Hartvigsen K, et al. Oxidation-specific epitopes are danger-associated molecular patterns recognized by pattern recognition receptors of innate immunity. Circ Res 2011;108(2):235–48. http://dx.doi.org/10.1161/CIRCRESAHA.110.223875. PubMed PMID: 21252151; PMCID: 3075542.

[77] Takeuchi O, Hoshino K, Kawai T, Sanjo H, Takada H, Ogawa T, et al. Differential roles of TLR2 and TLR4 in recognition of gram-negative and gram-positive bacterial cell wall components. Immunity 1999;11(4):443–51. PubMed PMID: 10549626.
[78] Chow JC, Young DW, Golenbock DT, Christ WJ, Gusovsky F. Toll-like receptor-4 mediates lipopolysaccharide-induced signal transduction. J Biol Chem 1999;274(16):10689–92. PubMed PMID: 10196138.
[79] Alexopoulou L, Holt AC, Medzhitov R, Flavell RA. Recognition of double-stranded RNA and activation of NF-κB by toll-like receptor 3. Nature 2001;413(6857):732–8. http://dx.doi.org/10.1038/35099560. PubMed PMID: 11607032.
[80] Heil F, Hemmi H, Hochrein H, Ampenberger F, Kirschning C, Akira S, et al. Species-specific recognition of single-stranded RNA via toll-like receptor 7 and 8. Science 2004;303(5663):1526–9. http://dx.doi.org/10.1126/science.1093620. PubMed PMID: 14976262.
[81] Hemmi H, Takeuchi O, Kawai T, Kaisho T, Sato S, Sanjo H, et al. Toll-like receptor recognizes bacterial DNA. Nature 2000;408(6813):740–5. http://dx.doi.org/10.1038/35047123. PubMed PMID: 11130078.
[82] Urao N, Ushio-Fukai M. Redox regulation of stem/progenitor cells and bone marrow niche. Free Radic Biol Med 2013;54:26–39. http://dx.doi.org/10.1016/j.freeradbiomed.2012.10.532. PubMed PMID: 23085514; PMCID: 3637653.
[83] Niethammer P, Grabher C, Look AT, Mitchison TJ. A tissue-scale gradient of hydrogen peroxide mediates rapid wound detection in zebrafish. Nature 2009;459(7249):996–9. http://dx.doi.org/10.1038/nature08119. PubMed PMID: 19494811; PMCID: 2803098.
[84] Yoo SK, Starnes TW, Deng Q, Huttenlocher A. Lyn is a redox sensor that mediates leukocyte wound attraction in vivo. Nature 2011;480(7375):109–12. http://dx.doi.org/10.1038/nature10632. PubMed PMID: 22101434; PMCID: 3228893.
[85] van der Vliet A, Janssen-Heininger YM. Hydrogen peroxide as a damage signal in tissue injury and inflammation: murderer, mediator, or messenger? J Cell Biochem 2014;115(3):427–35. http://dx.doi.org/10.1002/jcb.24683. PubMed PMID: 24122865; PMCID: 4363740.
[86] Mirza RE, Koh TJ. Contributions of cell subsets to cytokine production during normal and impaired wound healing. Cytokine 2015;71(2):409–12. http://dx.doi.org/10.1016/j.cyto.2014.09.005. PubMed PMID: 25281359; PMCID: 4297569.
[87] Serhan CN, Chiang N, Dalli J. The resolution code of acute inflammation: novel pro-resolving lipid mediators in resolution. Semin Immunol 2015;27(3):200–15. http://dx.doi.org/10.1016/j.smim.2015.03.004. PubMed PMID: 25857211.
[88] Hellmann J, Tang Y, Spite M. Proresolving lipid mediators and diabetic wound healing. Curr Opin Endocrinol Diabetes Obes 2012;19(2):104–8. http://dx.doi.org/10.1097/MED.0b013e3283514e00. PubMed PMID: 22374140; PMCID: 4038027.
[89] Tang Y, Zhang MJ, Hellmann J, Kosuri M, Bhatnagar A, Spite M. Proresolution therapy for the treatment of delayed healing of diabetic wounds. Diabetes 2013;62(2):618–27. http://dx.doi.org/10.2337/db12-0684. PubMed PMID: 23043160; PMCID: 3554373.
[90] Nathan C, Ding A. Nonresolving inflammation. Cell 2010;140(6):871–82. http://dx.doi.org/10.1016/j.cell.2010.02.029. PubMed PMID: 20303877.
[91] Bratton DL, Henson PM. Neutrophil clearance: when the party is over, clean-up begins. Trends Immunol 2011;32(8):350–7. http://dx.doi.org/10.1016/j.it.2011.04.009. PubMed PMID: 21782511; PMCID: 3151332.
[92] Kennedy AD, DeLeo FR. Neutrophil apoptosis and the resolution of infection. Immunol Res 2009;43(1–3):25–61. http://dx.doi.org/10.1007/s12026-008-8049-6. PubMed PMID: 19066741.

[93] Mathias JR, Perrin BJ, Liu TX, Kanki J, Look AT, Huttenlocher A. Resolution of inflammation by retrograde chemotaxis of neutrophils in transgenic zebrafish. J Leukoc Biol 2006;80(6):1281–8. http://dx.doi.org/10.1189/jlb.0506346. PubMed PMID: 16963624.
[94] Robertson AL, Holmes GR, Bojarczuk AN, Burgon J, Loynes CA, Chimen M, et al. A zebrafish compound screen reveals modulation of neutrophil reverse migration as an anti-inflammatory mechanism. Sci Transl Med 2014;6(225):225ra29. http://dx.doi.org/10.1126/scitranslmed.3007672. PubMed PMID: 24574340; PMCID: 4247228.
[95] McGrath EE, Marriott HM, Lawrie A, Francis SE, Sabroe I, Renshaw SA, et al. TNF-related apoptosis-inducing ligand (TRAIL) regulates inflammatory neutrophil apoptosis and enhances resolution of inflammation. J Leukoc Biol 2011;90(5):855–65. http://dx.doi.org/10.1189/jlb.0211062. PubMed PMID: 21562052; PMCID: 3644175.
[96] Woodfin A, Voisin MB, Beyrau M, Colom B, Caille D, Diapouli FM, et al. The junctional adhesion molecule JAM-C regulates polarized transendothelial migration of neutrophils in vivo. Nat Immunol 2011;12(8):761–9. http://dx.doi.org/10.1038/ni.2062. PubMed PMID: 21706006; PMCID: 3145149.
[97] Beauvillain C, Cunin P, Doni A, Scotet M, Jaillon S, Loiry ML, et al. CCR7 is involved in the migration of neutrophils to lymph nodes. Blood 2011;117(4):1196–204. http://dx.doi.org/10.1182/blood-2009-11-254490. PubMed PMID: 21051556.
[98] Leiriao P, del Fresno C, Ardavin C. Monocytes as effector cells: activated Ly-6C^{high} mouse monocytes migrate to the lymph nodes through the lymph and cross-present antigens to CD8$^+$T cells. Eur J Immunol 2012;42(8):2042–51. http://dx.doi.org/10.1002/eji.201142166. PubMed PMID: 22585535.
[99] Trogan E, Feig JE, Dogan S, Rothblat GH, Angeli V, Tacke F, et al. Gene expression changes in foam cells and the role of chemokine receptor CCR7 during atherosclerosis regression in ApoE-deficient mice. Proc Natl Acad Sci USA 2006;103(10):3781–6. http://dx.doi.org/10.1073/pnas.0511043103. PubMed PMID: 16537455; PMCID: 1450154.
[100] Winter S, Rehm A, Wichner K, Scheel T, Batra A, Siegmund B, et al. Manifestation of spontaneous and early autoimmune gastritis in CCR7-deficient mice. Am J Pathol 2011;179(2):754–65. http://dx.doi.org/10.1016/j.ajpath.2011.04.012. PubMed PMID: 21801869; PMCID: 3157176.
[101] Levy BD, Clish CB, Schmidt B, Gronert K, Serhan CN. Lipid mediator class switching during acute inflammation: signals in resolution. Nat Immunol 2001;2(7):612–9. http://dx.doi.org/10.1038/89759. PubMed PMID: 11429545.
[102] Norris PC, Gosselin D, Reichart D, Glass CK, Dennis EA. Phospholipase A2 regulates eicosanoid class switching during inflammasome activation. Proc Natl Acad Sci USA 2014;111(35):12746–51. http://dx.doi.org/10.1073/pnas.1404372111. PubMed PMID: 25139986; PMCID: 4156727.
[103] Serhan CN, Yang R, Martinod K, Kasuga K, Pillai PS, Porter TF, et al. Maresins: novel macrophage mediators with potent antiinflammatory and proresolving actions. J Exp Med 2009;206(1):15–23. http://dx.doi.org/10.1084/jem.20081880. PubMed PMID: 19103881; PMCID: 2626672.
[104] Mosser DM, Edwards JP. Exploring the full spectrum of macrophage activation. Nat Rev Immunol 2008;8(12):958–69. http://dx.doi.org/10.1038/nri2448. PubMed PMID: 19029990; PMCID: 2724991.
[105] Frangogiannis NG, Mendoza LH, Lindsey ML, Ballantyne CM, Michael LH, Smith CW, et al. IL-10 is induced in the reperfused myocardium and may modulate the reaction to injury. J Immunol 2000;165(5):2798–808. PubMed PMID: 10946312.
[106] Daley JM, Brancato SK, Thomay AA, Reichner JS, Albina JE. The phenotype of murine wound macrophages. J Leukoc Biol 2010;87(1):59–67. PubMed PMID: 20052800; PMCID: 2801619.

[107] Mirza RE, Fang MM, Novak ML, Urao N, Sui A, Ennis WJ, et al. Macrophage PPARγ and impaired wound healing in type 2 diabetes. J Pathol 2015;236(4):433–44. http://dx.doi.org/10.1002/path.4548. PubMed PMID: 25875529.
[108] Mirza RE, Fang MM, Ennis WJ, Koh TJ. Blocking interleukin-1β induces a healing-associated wound macrophage phenotype and improves healing in type 2 diabetes. Diabetes 2013;62(7):2579–87. http://dx.doi.org/10.2337/db12-1450. PubMed PMID: 23493576; PMCID: 3712034.
[109] Fernando MR, Reyes JL, Iannuzzi J, Leung G, McKay DM. The pro-inflammatory cytokine, interleukin-6, enhances the polarization of alternatively activated macrophages. PLoS One 2014;9(4):e94188. http://dx.doi.org/10.1371/journal.pone.0094188. PubMed PMID: 24736635; PMCID: 3988054.
[110] Hilgendorf I, Gerhardt LM, Tan TC, Winter C, Holderried TA, Chousterman BG, et al. Ly-6C^{high} monocytes depend on Nr4a1 to balance both inflammatory and reparative phases in the infarcted myocardium. Circ Res 2014;114(10):1611–22. http://dx.doi.org/10.1161/CIRCRESAHA.114.303204. PubMed PMID: 24625784; PMCID: 4017349.
[111] Courties G, Heidt T, Sebas M, Iwamoto Y, Jeon D, Truelove J, et al. In vivo silencing of the transcription factor IRF5 reprograms the macrophage phenotype and improves infarct healing. J Am Coll Cardiol 2013;63(15):1556–66. http://dx.doi.org/10.1016/j.jacc.2013.11.023. PubMed PMID: 24361318.
[112] Martin JM, Zenilman JM, Lazarus GS. Molecular microbiology: new dimensions for cutaneous biology and wound healing. J Invest Dermatol 2010;130(1):38–48. http://dx.doi.org/10.1038/jid.2009.221. PubMed PMID: 19626034.
[113] Percival SL, Hill KE, Williams DW, Hooper SJ, Thomas DW, Costerton JW. A review of the scientific evidence for biofilms in wounds. Wound Repair Regen 2012;20(5):647–57. http://dx.doi.org/10.1111/j.1524-475X.2012.00836.x. PubMed PMID: 22985037.
[114] Zhao G, Usui ML, Lippman SI, James GA, Stewart PS, Fleckman P, et al. Biofilms and inflammation in chronic wounds. Adv Wound Care (New Rochelle) 2013;2(7):389–99. http://dx.doi.org/10.1089/wound.2012.0381. PubMed PMID: 24527355; PMCID: 3763221.
[115] Devaraj S, Tobias P, Jialal I. Knockout of toll-like receptor-4 attenuates the pro-inflammatory state of diabetes. Cytokine 2011;55(3):441–5. http://dx.doi.org/10.1016/j.cyto.2011.03.023. PubMed PMID: 21498084.
[116] Dasu MR, Thangappan RK, Bourgette A, DiPietro LA, Isseroff R, Jialal I. TLR2 expression and signaling-dependent inflammation impair wound healing in diabetic mice. Lab Invest 2010;90(11):1628–36. http://dx.doi.org/10.1038/labinvest.2010.158. PubMed PMID: 20733560.
[117] Dasu MR, Jialal I. Amelioration in wound healing in diabetic toll-like receptor-4 knockout mice. J Diabetes Complications 2013;27(5):417–21. http://dx.doi.org/10.1016/j.jdiacomp.2013.05.002. PubMed PMID: 23773694; PMCID: 3770740.
[118] Dasu MR, Martin SJ. Toll-like receptor expression and signaling in human diabetic wounds. World J Diabetes 2014;5(2):219–23. http://dx.doi.org/10.4239/wjd.v5.i2.219. PubMed PMID: 24748934; PMCID: 3990321.
[119] Devaraj S, Dasu MR, Rockwood J, Winter W, Griffen SC, Jialal I. Increased toll-like receptor (TLR) 2 and TLR4 expression in monocytes from patients with type 1 diabetes: further evidence of a proinflammatory state. J Clin Endocrinol Metab 2008;93(2):578–83. http://dx.doi.org/10.1210/jc.2007-2185. PubMed PMID: 18029454; PMCID: 2243229.
[120] Dasu MR, Devaraj S, Park S, Jialal I. Increased toll-like receptor (TLR) activation and TLR ligands in recently diagnosed type 2 diabetic subjects. Diabetes Care 2010;33(4):861–8. http://dx.doi.org/10.2337/dc09-1799. PubMed PMID: 20067962; PMCID: 2845042.

[121] Nguyen KT, Seth AK, Hong SJ, Geringer MR, Xie P, Leung KP, et al. Deficient cytokine expression and neutrophil oxidative burst contribute to impaired cutaneous wound healing in diabetic, biofilm-containing chronic wounds. Wound Repair Regen 2013;21(6):833–41. http://dx.doi.org/10.1111/wrr.12109. PubMed PMID: 24118295.

[122] Dey A, Allen J, Hankey-Giblin PA. Ontogeny and polarization of macrophages in inflammation: blood monocytes versus tissue macrophages. Front Immunol 2014;5:683. http://dx.doi.org/10.3389/fimmu.2014.00683. PubMed PMID: 25657646; PMCID: 4303141.

[123] Hellmann J, Zhang MJ, Tang Y, Rane M, Bhatnagar A, Spite M. Increased saturated fatty acids in obesity alter resolution of inflammation in part by stimulating prostaglandin production. J Immunol 2013;191(3):1383–92. http://dx.doi.org/10.4049/jimmunol.1203369. PubMed PMID: 23785121; PMCID: 3720716.

[124] Hong S, Tian H, Lu Y, Laborde JM, Muhale FA, Wang Q, et al. Neuroprotectin/protectin D1: endogenous biosynthesis and actions on diabetic macrophages in promoting wound healing and innervation impaired by diabetes. Am J Physiol Cell Physiol 2014;307(11):C1058–67. http://dx.doi.org/10.1152/ajpcell.00270.2014. PubMed PMID: 25273880; PMCID: 4254953.

[125] Mirza RE, Fang MM, Weinheimer-Haus EM, Ennis WJ, Koh TJ. Sustained inflammasome activity in macrophages impairs wound healing in type 2 diabetic humans and mice. Diabetes 2014;63(3):1103–14. http://dx.doi.org/10.2337/db13-0927. PubMed PMID: 24194505; PMCID: 3931398.

[126] Nahrendorf M, Pittet MJ, Swirski FK. Monocytes: protagonists of infarct inflammation and repair after myocardial infarction. Circulation 2010;121(22):2437–45. http://dx.doi.org/10.1161/CIRCULATIONAHA.109.916346. PubMed PMID: 20530020; PMCID: 2892474.

[127] Mahdipour E, Charnock JC, Mace KA. Hoxa3 promotes the differentiation of hematopoietic progenitor cells into proangiogenic Gr-1^{+}CD11b^{+} myeloid cells. Blood 2011;117(3):815–26. http://dx.doi.org/10.1182/blood-2009-12-259549. PubMed PMID: 20974673.

[128] Nahrendorf M, Swirski FK. Lifestyle effects on hematopoiesis and atherosclerosis. Circ Res 2015;116(5):884–94. http://dx.doi.org/10.1161/CIRCRESAHA.116.303550. PubMed PMID: 25722442; PMCID: 4347940.

[129] Nagareddy PR, Kraakman M, Masters SL, Stirzaker RA, Gorman DJ, Grant RW, et al. Adipose tissue macrophages promote myelopoiesis and monocytosis in obesity. Cell Metab 2014;19(5):821–35. http://dx.doi.org/10.1016/j.cmet.2014.03.029. PubMed PMID: 24807222; PMCID: 4048939.

[130] Nagareddy PR, Murphy AJ, Stirzaker RA, Hu Y, Yu S, Miller RG, et al. Hyperglycemia promotes myelopoiesis and impairs the resolution of atherosclerosis. Cell Metab 2013;17(5):695–708. http://dx.doi.org/10.1016/j.cmet.2013.04.001. PubMed PMID: 23663738; PMCID: 3992275.

[131] Swirski FK, Libby P, Aikawa E, Alcaide P, Luscinskas FW, Weissleder R, et al. Ly-6C^{hi} monocytes dominate hypercholesterolemia-associated monocytosis and give rise to macrophages in atheromata. J Clin Invest 2007;117(1):195–205. http://dx.doi.org/10.1172/JCI29950. PubMed PMID: 17200719; PMCID: 1716211.

[132] Medzhitov R. Origin and physiological roles of inflammation. Nature 2008;454(7203):428–35. http://dx.doi.org/10.1038/nature07201. PubMed PMID: 18650913.

[133] Baldridge MT, King KY, Boles NC, Weksberg DC, Goodell MA. Quiescent haematopoietic stem cells are activated by IFN-γ in response to chronic infection. Nature 2010;465(7299):793–7. http://dx.doi.org/10.1038/nature09135. PubMed PMID: 20535209; PMCID: 2935898.

[134] Seijkens T, Hoeksema MA, Beckers L, Smeets E, Meiler S, Levels J, et al. Hypercholesterolemia-induced priming of hematopoietic stem and progenitor cells aggravates atherosclerosis. FASEB J 2014;28(5):2202–13. http://dx.doi.org/10.1096/fj.13-243105. PubMed PMID: 24481967.

[135] Oduro Jr KA, Liu F, Tan Q, Kim CK, Lubman O, Fremont D, et al. Myeloid skewing in murine autoimmune arthritis occurs in hematopoietic stem and primitive progenitor cells. Blood 2012;120(11):2203–13. http://dx.doi.org/10.1182/blood-2011-11-391342. PubMed PMID: 22855602; PMCID: 3447779.

[136] Wu WC, Sun HW, Chen HT, Liang J, Yu XJ, Wu C, et al. Circulating hematopoietic stem and progenitor cells are myeloid-biased in cancer patients. Proc Natl Acad Sci USA 2014;111(11):4221–6. http://dx.doi.org/10.1073/pnas.1320753111. PubMed PMID: 24591638; PMCID: 3964061.

[137] Griseri T, McKenzie BS, Schiering C, Powrie F. Dysregulated hematopoietic stem and progenitor cell activity promotes interleukin-23-driven chronic intestinal inflammation. Immunity 2012;37(6):1116–29. http://dx.doi.org/10.1016/j.immuni.2012.08.025. PubMed PMID: 23200826; PMCID: 3664922.

[138] Panizzi P, Swirski FK, Figueiredo JL, Waterman P, Sosnovik DE, Aikawa E, et al. Impaired infarct healing in atherosclerotic mice with Ly-6C^{hi} monocytosis. J Am Coll Cardiol 2010;55(15):1629–38. http://dx.doi.org/10.1016/j.jacc.2009.08.089. PubMed PMID: 20378083; PMCID: 2852892.

[139] Saeed S, Quintin J, Kerstens HH, Rao NA, Aghajanirefah A, Matarese F, et al. Epigenetic programming of monocyte-to-macrophage differentiation and trained innate immunity. Science 2014;345(6204):1251086. http://dx.doi.org/10.1126/science.1251086. PubMed PMID: 25258085.

[140] Cheng SC, Quintin J, Cramer RA, Shepardson KM, Saeed S, Kumar V, et al. mTOR- and HIF-1α-mediated aerobic glycolysis as metabolic basis for trained immunity. Science 2014;345(6204):1250684. http://dx.doi.org/10.1126/science.1250684. PubMed PMID: 25258083; PMCID: 4226238.

[141] Bannon P, Wood S, Restivo T, Campbell L, Hardman MJ, Mace KA. Diabetes induces stable intrinsic changes to myeloid cells that contribute to chronic inflammation during wound healing in mice. Dis Model Mech 2013;6(6):1434–47. http://dx.doi.org/10.1242/dmm.012237. PubMed PMID: 24057002; PMCID: 3820266.

[142] Gallagher KA, Joshi A, Carson WF, Schaller M, Allen R, Mukerjee S, et al. Epigenetic changes in bone marrow progenitor cells influence the inflammatory phenotype and alter wound healing in type 2 diabetes. Diabetes 2015;64(4):1420–30. http://dx.doi.org/10.2337/db14-0872. PubMed PMID: 25368099; PMCID: 4375075.

[143] Weinheimer-Haus EM, Mirza RE, Koh TJ. Nod-like receptor protein-3 inflammasome plays an important role during early stages of wound healing. PLoS One 2015;10(3):e0119106. http://dx.doi.org/10.1371/journal.pone.0119106. PubMed PMID: 25793779; PMCID: 4368510.

[144] Leal EC, Carvalho E, Tellechea A, Kafanas A, Tecilazich F, Kearney C, et al. Substance P promotes wound healing in diabetes by modulating inflammation and macrophage phenotype. Am J Pathol 2015;185(6):1638–48. http://dx.doi.org/10.1016/j.ajpath.2015.02.011. PubMed PMID: 25871534.

[145] Bujak M, Kweon HJ, Chatila K, Li N, Taffet G, Frangogiannis NG. Aging-related defects are associated with adverse cardiac remodeling in a mouse model of reperfused myocardial infarction. J Am Coll Cardiol 2008;51(14):1384–92. http://dx.doi.org/10.1016/j.jacc.2008.01.011. PubMed PMID: 18387441; PMCID: 3348616.

[146] Madhok BM, Vowden K, Vowden P. New techniques for wound debridement. Int Wound J 2013;10(3):247–51. http://dx.doi.org/10.1111/iwj.12045. PubMed PMID: 23418808.

[147] Powers JG, Morton LM, Phillips TJ. Dressings for chronic wounds. Dermatol Ther 2013;26(3):197–206. http://dx.doi.org/10.1111/dth.12055. PubMed PMID: 23742280.

[148] Percival SL, Slone W, Linton S, Okel T, Corum L, Thomas JG. The antimicrobial efficacy of a silver alginate dressing against a broad spectrum of clinically relevant wound isolates. Int Wound J 2011;8(3):237–43. http://dx.doi.org/10.1111/j.1742-481X.2011.00774.x. PubMed PMID: 21470369.

[149] Crone S, Garde C, Bjarnsholt T, Alhede M. A novel in vitro wound biofilm model used to evaluate low-frequency ultrasonic-assisted wound debridement. J Wound Care 2015. http://dx.doi.org/10.12968/jowc.2015.24.2.64. 24(2):64, 6–9, 72, PubMed PMID: 25647434.

[150] Banerjee J, Ghatak PD, Roy S, Khanna S, Hemann C, Deng B, et al. Silver-zinc redox-coupled electroceutical wound dressing disrupts bacterial biofilm. PLoS One 2015;10(3):e0119531. http://dx.doi.org/10.1371/journal.pone.0119531. PubMed PMID: 25803639; PMCID: 4372374.

[151] Scales BS, Huffnagle GB. The microbiome in wound repair and tissue fibrosis. J Pathol 2013;229(2):323–31. http://dx.doi.org/10.1002/path.4118. PubMed PMID: 23042513; PMCID: 3631561.

[152] Murray PJ, Smale ST. Restraint of inflammatory signaling by interdependent strata of negative regulatory pathways. Nat Immunol 2012;13(10):916–24. http://dx.doi.org/10.1038/ni.2391. PubMed PMID: 22990889.

[153] Ruppert SM, Hawn TR, Arrigoni A, Wight TN, Bollyky PL. Tissue integrity signals communicated by high-molecular weight hyaluronan and the resolution of inflammation. Immunol Res 2014;58(2–3):186–92. http://dx.doi.org/10.1007/s12026-014-8495-2. PubMed PMID: 24614953; PMCID: 4106675.

[154] Petrey AC, de la Motte CA. Hyaluronan, a crucial regulator of inflammation. Front Immunol 2014;5:101. http://dx.doi.org/10.3389/fimmu.2014.00101. PubMed PMID: 24653726; PMCID: 3949149.

[155] Muto J, Morioka Y, Yamasaki K, Kim M, Garcia A, Carlin AF, et al. Hyaluronan digestion controls DC migration from the skin. J Clin Invest 2014;124(3):1309–19. http://dx.doi.org/10.1172/JCI67947. PubMed PMID: 24487587; PMCID: 3934161.

[156] Kondo T, Kawai T, Akira S. Dissecting negative regulation of toll-like receptor signaling. Trends Immunol 2012;33(9):449–58. http://dx.doi.org/10.1016/j.it.2012.05.002. PubMed PMID: 22721918.

[157] Majmudar MD, Keliher EJ, Heidt T, Leuschner F, Truelove J, Sena BF, et al. Monocyte-directed RNAi targeting CCR2 improves infarct healing in atherosclerosis-prone mice. Circulation 2013;127(20):2038–46. http://dx.doi.org/10.1161/CIRCULATIONAHA.112.000116. PubMed PMID: 23616627; PMCID: 3661714.

[158] Willenborg S, Lucas T, van Loo G, Knipper JA, Krieg T, Haase I, et al. CCR2 recruits an inflammatory macrophage subpopulation critical for angiogenesis in tissue repair. Blood 2012;120(3):613–25. http://dx.doi.org/10.1182/blood-2012-01-403386. PubMed PMID: 22577176.

[159] Ishikawa M, Ito H, Kitaori T, Murata K, Shibuya H, Furu M, et al. MCP/CCR2 signaling is essential for recruitment of mesenchymal progenitor cells during the early phase of fracture healing. PLoS One 2014;9(8):e104954. http://dx.doi.org/10.1371/journal.pone.0104954. PubMed PMID: 25133509; PMCID: 4136826.

[160] Dal-Secco D, Wang J, Zeng Z, Kolaczkowska E, Wong CH, Petri B, et al. A dynamic spectrum of monocytes arising from the in situ reprogramming of $CCR2^+$ monocytes at a site of sterile injury. J Exp Med 2015;212(4):447–56. http://dx.doi.org/10.1084/jem.20141539. PubMed PMID: 25800956; PMCID: 4387291.

[161] Ishida Y, Gao JL, Murphy PM. Chemokine receptor CX3CR1 mediates skin wound healing by promoting macrophage and fibroblast accumulation and function. J Immunol 2008;180(1):569–79. PubMed PMID: 18097059.
[162] Clover AJ, Kumar AH, Caplice NM. Deficiency of CX3CR1 delays burn wound healing and is associated with reduced myeloid cell recruitment and decreased sub-dermal angiogenesis. Burns 2011;37(8):1386–93. http://dx.doi.org/10.1016/j.burns.2011.08.001. PubMed PMID: 21924836.
[163] Longman RS, Diehl GE, Victorio DA, Huh JR, Galan C, Miraldi ER, et al. CX_3CR1^+ mononuclear phagocytes support colitis-associated innate lymphoid cell production of IL-22. J Exp Med 2014;211(8):1571–83. http://dx.doi.org/10.1084/jem.20140678. PubMed PMID: 25024136; PMCID: 4113938.
[164] Salmon-Ehr V, Ramont L, Godeau G, Birembaut P, Guenounou M, Bernard P, et al. Implication of interleukin-4 in wound healing. Lab Invest 2000;80(8):1337–43. PubMed PMID: 10950124.
[165] Chamberlain CS, Leiferman EM, Frisch KE, Wang S, Yang X, Brickson SL, et al. The influence of interleukin-4 on ligament healing. Wound Repair Regen 2011;19(3):426–35. http://dx.doi.org/10.1111/j.1524-475X.2011.00682.x. PubMed PMID: 21518087; PMCID: 3147289.
[166] Dehne N, Tausendschon M, Essler S, Geis T, Schmid T, Brune B. IL-4 reduces the proangiogenic capacity of macrophages by down-regulating HIF-1α translation. J Leukoc Biol 2014;95(1):129–37. http://dx.doi.org/10.1189/jlb.0113045. PubMed PMID: 24006507.
[167] Zymek P, Nah DY, Bujak M, Ren G, Koerting A, Leucker T, et al. Interleukin-10 is not a critical regulator of infarct healing and left ventricular remodeling. Cardiovasc Res 2007;74(2):313–22. http://dx.doi.org/10.1016/j.cardiores.2006.11.028. PubMed PMID: 17188669; PMCID: 1924681.
[168] Eming SA, Werner S, Bugnon P, Wickenhauser C, Siewe L, Utermohlen O, et al. Accelerated wound closure in mice deficient for interleukin-10. Am J Pathol 2007;170(1): 188–202. http://dx.doi.org/10.2353/ajpath.2007.060370. PubMed PMID: 17200193; PMCID: 1762712.
[169] Kieran I, Knock A, Bush J, So K, Metcalfe A, Hobson R, et al. Interleukin-10 reduces scar formation in both animal and human cutaneous wounds: results of two preclinical and phase II randomized control studies. Wound Repair Regen 2013;21(3):428–36. http://dx.doi.org/10.1111/wrr.12043. PubMed PMID: 23627460.
[170] Kieran I, Taylor C, Bush J, Rance M, So K, Boanas A, et al. Effects of interleukin-10 on cutaneous wounds and scars in humans of African continental ancestral origin. Wound Repair Regen 2014;22(3):326–33. http://dx.doi.org/10.1111/wrr.12178. PubMed PMID: 24844332.
[171] Hong S, Gronert K, Devchand PR, Moussignac RL, Serhan CN. Novel docosatrienes and 17S-resolvins generated from docosahexaenoic acid in murine brain, human blood, and glial cells. Autacoids in anti-inflammation. J Biol Chem 2003;278(17):14677–87. http://dx.doi.org/10.1074/jbc.M300218200. PubMed PMID: 12590139.
[172] Ou L, Shi Y, Dong W, Liu C, Schmidt TJ, Nagarkatti P, et al. Kruppel-like factor KLF4 facilitates cutaneous wound healing by promoting fibrocyte generation from myeloid-derived suppressor cells. J Invest Dermatol 2015;135(5):1425–34. http://dx.doi.org/10.1038/jid.2015.3. PubMed PMID: 25581502; PMCID: 4402119.
[173] Tongers J, Losordo DW, Landmesser U. Stem and progenitor cell-based therapy in ischaemic heart disease: promise, uncertainties, and challenges. Eur Heart J 2011;32(10):1197–206. http://dx.doi.org/10.1093/eurheartj/ehr018. PubMed PMID: 21362705; PMCID: 3094549.

[174] Raval Z, Losordo DW. Cell therapy of peripheral arterial disease: from experimental findings to clinical trials. Circ Res 2013;112(9):1288–302. http://dx.doi.org/10.1161/CIRCRESAHA.113.300565. PubMed PMID: 23620237; PMCID: 3838995.

[175] Cha J, Falanga V. Stem cells in cutaneous wound healing. Clin Dermatol 2007;25(1):73–8. http://dx.doi.org/10.1016/j.clindermatol.2006.10.002. PubMed PMID: 17276204.

[176] Wu Y, Zhao RC, Tredget EE. Concise review: bone marrow-derived stem/progenitor cells in cutaneous repair and regeneration. Stem Cells 2010;28(5):905–15. http://dx.doi.org/10.1002/stem.420. PubMed PMID: 20474078; PMCID: 2964514.

[177] Novak ML, Weinheimer-Haus EM, Koh TJ. Macrophage activation and skeletal muscle healing following traumatic injury. J Pathol 2014;232(3):344–55. http://dx.doi.org/10.1002/path.4301. PubMed PMID: 24255005; PMCID: 4019602.

[178] Leor J, Rozen L, Zuloff-Shani A, Feinberg MS, Amsalem Y, Barbash IM, et al. Ex vivo activated human macrophages improve healing, remodeling, and function of the infarcted heart. Circulation 2006;114(Suppl. 1):I94–100. http://dx.doi.org/10.1161/CIRCULATIONAHA.105.000331. PubMed PMID: 16820652.

[179] Niedermayer W, Schaefer J, Held K, Schwarzkopf HJ, Ulber F, Wustenfeld E. Changes of renal perfusion and renal vascular resistance in diastolic arterial counterpulsation. Z Kreislaufforsch 1968;57(5):466–74. PubMed PMID: 5678696.

[180] Tong X, Lv G, Huang J, Min Y, Yang L, Lin PC. Gr-1^{+}CD11b^{+} myeloid cells efficiently home to site of injury after intravenous administration and enhance diabetic wound healing by neoangiogenesis. J Cell Mol Med 2014;18(6):1194–202. http://dx.doi.org/10.1111/jcmm.12265. PubMed PMID: 24645717; PMCID: 4112018.

[181] Jetten N, Roumans N, Gijbels MJ, Romano A, Post MJ, de Winther MP, et al. Wound administration of M2-polarized macrophages does not improve murine cutaneous healing responses. PLoS One 2014;9(7):e102994. http://dx.doi.org/10.1371/journal.pone.0102994. PubMed PMID: 25068282; PMCID: 4113363.

[182] Isakson M, de Blacam C, Whelan D, McArdle A, Clover AJ. Mesenchymal stem cells and cutaneous wound healing: current evidence and future potential. Stem Cells Int 2015;2015:831095. http://dx.doi.org/10.1155/2015/831095. PubMed PMID: 26106431.

[183] Choi H, Lee RH, Bazhanov N, Oh JY, Prockop DJ. Anti-inflammatory protein TSG-6 secreted by activated MSCs attenuates zymosan-induced mouse peritonitis by decreasing TLR2/NF-κB signaling in resident macrophages. Blood 2011;118(2):330–8. http://dx.doi.org/10.1182/blood-2010-12-327353. PubMed PMID: 21551236; PMCID: 3138686.

[184] Dyer DP, Thomson JM, Hermant A, Jowitt TA, Handel TM, Proudfoot AE, et al. TSG-6 inhibits neutrophil migration via direct interaction with the chemokine CXCL8. J Immunol 2014;192(5):2177–85. http://dx.doi.org/10.4049/jimmunol.1300194. PubMed PMID: 24501198; PMCID: 3988464.

[185] Liu Y, Yin Z, Zhang R, Yan K, Chen L, Chen F, et al. MSCs inhibit bone marrow-derived DC maturation and function through the release of TSG-6. Biochem Biophys Res Commun 2014;450(4):1409–15. http://dx.doi.org/10.1016/j.bbrc.2014.07.001. PubMed PMID: 25014173.

[186] Badiavas EV, Falanga V. Treatment of chronic wounds with bone marrow-derived cells. Arch Dermatol 2003;139(4):510–6. http://dx.doi.org/10.1001/archderm.139.4.510. PubMed PMID: 12707099.

[187] Falanga V, Iwamoto S, Chartier M, Yufit T, Butmarc J, Kouttab N, et al. Autologous bone marrow-derived cultured mesenchymal stem cells delivered in a fibrin spray accelerate healing in murine and human cutaneous wounds. Tissue Eng 2007;13(6):1299–312. http://dx.doi.org/10.1089/ten.2006.0278. PubMed PMID: 17518741.

[188] Dash NR, Dash SN, Routray P, Mohapatra S, Mohapatra PC. Targeting nonhealing ulcers of lower extremity in human through autologous bone marrow-derived mesenchymal stem cells. Rejuvenation Res 2009;12(5):359–66. http://dx.doi.org/10.1089/rej.2009.0872. PubMed PMID: 19929258.

[189] Yoshikawa T, Mitsuno H, Nonaka I, Sen Y, Kawanishi K, Inada Y, et al. Wound therapy by marrow mesenchymal cell transplantation. Plast Reconstr Surg 2008;121(3):860–77. http://dx.doi.org/10.1097/01.prs.0000299922.96006.24. PubMed PMID: 18317135.

[190] Oliva N, Carcole M, Beckerman M, Seliktar S, Hayward A, Stanley J, et al. Regulation of dendrimer/dextran material performance by altered tissue microenvironment in inflammation and neoplasia. Sci Transl Med 2015;7(272):272ra11. http://dx.doi.org/10.1126/scitranslmed.aaa1616. PubMed PMID: 25632035.

[191] Mogosanu GD, Grumezescu AM. Natural and synthetic polymers for wounds and burns dressing. Int J Pharm 2014;463(2):127–36. http://dx.doi.org/10.1016/j.ijpharm.2013.12.015. PubMed PMID: 24368109.

[192] Hanson S, D'Souza RN, Hematti P. Biomaterial-mesenchymal stem cell constructs for immunomodulation in composite tissue engineering. Tissue Eng Part A 2014;20(15–16):2162–8. http://dx.doi.org/10.1089/ten.tea.2013.0359. PubMed PMID: 25140989.

[193] Zhu J. Bioactive modification of poly(ethylene glycol) hydrogels for tissue engineering. Biomaterials 2010;31(17):4639–56. http://dx.doi.org/10.1016/j.biomaterials.2010.02.044. PubMed PMID: 20303169; PMCID: 2907908.

[194] Zhu J, Marchant RE. Design properties of hydrogel tissue-engineering scaffolds. Expert Rev Med Devices 2011;8(5):607–26. http://dx.doi.org/10.1586/erd.11.27. PubMed PMID: 22026626; PMCID: 3206299.

[195] Clemens MW, Selber JC, Liu J, Adelman DM, Baumann DP, Garvey PB, et al. Bovine versus porcine acellular dermal matrix for complex abdominal wall reconstruction. Plast Reconstr Surg 2013;131(1):71–9. http://dx.doi.org/10.1097/PRS.0b013e3182729e58. PubMed PMID: 22965235.

[196] Baumann DP, Butler CE. Bioprosthetic mesh in abdominal wall reconstruction. Semin Plast Surg 2012;26(1):18–24. http://dx.doi.org/10.1055/s-0032-1302461. PubMed PMID: 23372454; PMCID: 3348742.

[197] Spiller KL, Anfang RR, Spiller KJ, Ng J, Nakazawa KR, Daulton JW, et al. The role of macrophage phenotype in vascularization of tissue engineering scaffolds. Biomaterials 2014;35(15):4477–88. http://dx.doi.org/10.1016/j.biomaterials.2014.02.012. PubMed PMID: 24589361; PMCID: 4000280.

[198] Spiller KL, Nassiri S, Witherel CE, Anfang RR, Ng J, Nakazawa KR, et al. Sequential delivery of immunomodulatory cytokines to facilitate the M1-to-M2 transition of macrophages and enhance vascularization of bone scaffolds. Biomaterials 2015;37:194–207. http://dx.doi.org/10.1016/j.biomaterials.2014.10.017. PubMed PMID: 25453950; PMCID: 4312192.

[199] He H, Zhang S, Tighe S, Son J, Tseng SC. Immobilized heavy chain-hyaluronic acid polarizes lipopolysaccharide-activated macrophages toward M2 phenotype. J Biol Chem 2013;288(36):25792–803. http://dx.doi.org/10.1074/jbc.M113.479584. PubMed PMID: 23878196; PMCID: 3764786.

[200] Franz S, Allenstein F, Kajahn J, Forstreuter I, Hintze V, Moller S, et al. Artificial extracellular matrices composed of collagen I and high-sulfated hyaluronan promote phenotypic and functional modulation of human pro-inflammatory M1 macrophages. Acta Biomater 2013;9(3):5621–9. http://dx.doi.org/10.1016/j.actbio.2012.11.016. PubMed PMID: 23168224.

[201] Tolg C, Telmer P, Turley E. Specific sizes of hyaluronan oligosaccharides stimulate fibroblast migration and excisional wound repair. PloS One 2014;9(2):e88479. http://dx.doi.org/10.1371/journal.pone.0088479. PubMed PMID: 24551108; PMCID: 3923781.
[202] Vasconcelos DP, Fonseca AC, Costa M, Amaral IF, Barbosa MA, Aguas AP, et al. Macrophage polarization following chitosan implantation. Biomaterials 2013;34(38):9952–9. http://dx.doi.org/10.1016/j.biomaterials.2013.09.012. PubMed PMID: 24074837.
[203] Badylak SF, Valentin JE, Ravindra AK, McCabe GP, Stewart-Akers AM. Macrophage phenotype as a determinant of biologic scaffold remodeling. Tissue Eng Part A 2008;14(11):1835–42. http://dx.doi.org/10.1089/ten.tea.2007.0264. PubMed PMID: 18950271.
[204] Brown BN, Valentin JE, Stewart-Akers AM, McCabe GP, Badylak SF. Macrophage phenotype and remodeling outcomes in response to biologic scaffolds with and without a cellular component. Biomaterials 2009;30(8):1482–91. http://dx.doi.org/10.1016/j.biomaterials.2008.11.040. PubMed PMID: 19121538; PMCID: 2805023.
[205] McGarvey JR, Pettaway S, Shuman JA, Novack CP, Zellars KN, Freels PD, et al. Targeted injection of a biocomposite material alters macrophage and fibroblast phenotype and function following myocardial infarction: relation to left ventricular remodeling. J Pharmacol Exp Ther 2014;350(3):701–9. http://dx.doi.org/10.1124/jpet.114.215798. PubMed PMID: 25022514; PMCID: 4152878.
[206] Wolf MT, Dearth CL, Ranallo CA, LoPresti ST, Carey LE, Daly KA, et al. Macrophage polarization in response to ECM coated polypropylene mesh. Biomaterials 2014;35(25):6838–49. http://dx.doi.org/10.1016/j.biomaterials.2014.04.115. PubMed PMID: 24856104; PMCID: 4347831.
[207] Kleinbeck KR, Faucher LD, Kao WJ. Biomaterials modulate interleukin-8 and other inflammatory proteins during reepithelialization in cutaneous partial-thickness wounds in pigs. Wound Repair Regen 2010;18(5):486–98. http://dx.doi.org/10.1111/j.1524-475X.2010.00609.x. PubMed PMID: 20731797; PMCID: 3279954.
[208] Xie Z, Aphale NV, Kadapure TD, Wadajkar AS, Orr S, Gyawali D, et al. Design of antimicrobial peptides conjugated biodegradable citric acid derived hydrogels for wound healing. J Biomed Mater Res A 2015;103(12):3907–18. http://dx.doi.org/10.1002/jbm.a.35512. PubMed PMID: 26014899.
[209] Branco da Cunha C, Klumpers DD, Li WA, Koshy ST, Weaver JC, Chaudhuri O, et al. Influence of the stiffness of three-dimensional alginate/collagen-I interpenetrating networks on fibroblast biology. Biomaterials 2014;35(32):8927–36. http://dx.doi.org/10.1016/j.biomaterials.2014.06.047. PubMed PMID: 25047628.
[210] Wang L, Shansky J, Borselli C, Mooney D, Vandenburgh H. Design and fabrication of a biodegradable, covalently crosslinked shape-memory alginate scaffold for cell and growth factor delivery. Tissue Eng Part A 2012;18(19–20):2000–7. http://dx.doi.org/10.1089/ten.TEA.2011.0663. PubMed PMID: 22646518; PMCID: 3463276.
[211] Koob TJ, Rennert R, Zabek N, Massee M, Lim JJ, Temenoff JS, et al. Biological properties of dehydrated human amnion/chorion composite graft: implications for chronic wound healing. Int Wound J 2013;10(5):493–500. http://dx.doi.org/10.1111/iwj.12140. PubMed PMID: 23902526; PMCID: 4228928.
[212] Ravi S, Caves JM, Martinez AW, Xiao J, Wen J, Haller CA, et al. Effect of bone marrow-derived extracellular matrix on cardiac function after ischemic injury. Biomaterials 2012;33(31):7736–45. http://dx.doi.org/10.1016/j.biomaterials.2012.07.010. PubMed PMID: 22819498; PMCID: 3849033.

[213] Roche ET, Hastings CL, Lewin SA, Shvartsman DE, Brudno Y, Vasilyev NV, et al. Comparison of biomaterial delivery vehicles for improving acute retention of stem cells in the infarcted heart. Biomaterials 2014;35(25):6850–8. http://dx.doi.org/10.1016/j.biomaterials.2014.04.114. PubMed PMID: 24862441; PMCID: 4051834.
[214] Atala A, Kasper FK, Mikos AG. Engineering complex tissues. Sci Transl Med 2012;4(160):160rv12. http://dx.doi.org/10.1126/scitranslmed.3004890. PubMed PMID: 23152327.
[215] Tibbitt MW, Anseth KS. Dynamic microenvironments: the fourth dimension. Sci Transl Med 2012;4(160):160ps24. http://dx.doi.org/10.1126/scitranslmed.3004804. PubMed PMID: 23152326.
[216] Linde N, Gutschalk CM, Hoffmann C, Yilmaz D, Mueller MM. Integrating macrophages into organotypic co-cultures: a 3D in vitro model to study tumor-associated macrophages. PLoS One 2012;7(7):e40058. http://dx.doi.org/10.1371/journal.pone.0040058. PubMed PMID: 22792213; PMCID: 3391227.

Modelling wound healing

Y.H. Martin, F.V. Lali, A.D. Metcalfe
Blond McIndoe Research Foundation, Queen Victoria Hospital, East Grinstead, United Kingdom; The University of Brighton, Brighton, United Kingdom

6.1 Introduction

Wound healing in humans has a number of unique processes that are linked to parameters including age, sex, physiology and the microenvironment of the wound. The ability to perform controlled, clinical experimentation on the mechanism of action and subsequent therapeutic intervention of wound repair is limited. This has led investigators to look for novel approaches to begin to unravel the mechanisms involved in human wound healing. These approaches range from in vitro characterisation to in vivo assessments in other organisms and to consented patient volunteer and clinical trials. It is challenging to design experiments that capture the complex interactions that are orchestrated during wound healing. A number of in vitro assays and surgical approaches that can lead to important insights into the mechanism of wound repair have been developed. In this chapter, we will consider a number of different model systems that have been developed for characterising wound healing, some of which have led to clinical therapeutic studies.

6.2 In vitro models of wound healing

In vitro models to study wound healing range from simple monolayer scratch assays to increasingly complex tissue engineered skin with stratified epithelia to full-thickness ex vivo human skin models.

Scratch models are widely employed to study keratinocyte and fibroblast migration. Confluent monolayers are scratched using pipette tips or other blunt objects (Fig. 6.1). The wound closes as cells migrate into the scratched area, which is measured over time using image analysis software (Schreier et al., 1993). To differentiate between the effect of proliferation and migration, cells can be mitotically inactivated using reagents such as mitomycin C, which will result in the scratch being repopulated solely by migrating cells. Typically, these assays are carried out on tissue-culture grade polystyrene surfaces. These surfaces can be coated by extracellular matrix (ECM) molecules to mimic physiological substrates (Li et al., 2004a,b).

Scratch assays are the simplest form of in vitro assays employed to study wound healing. They are particularly suitable to study mechanisms that regulate cell behaviour, such as signalling pathways and posttranscriptional regulation, as well as

Wound Healing Biomaterials - Volume 1. http://dx.doi.org/10.1016/B978-1-78242-455-0.00006-9

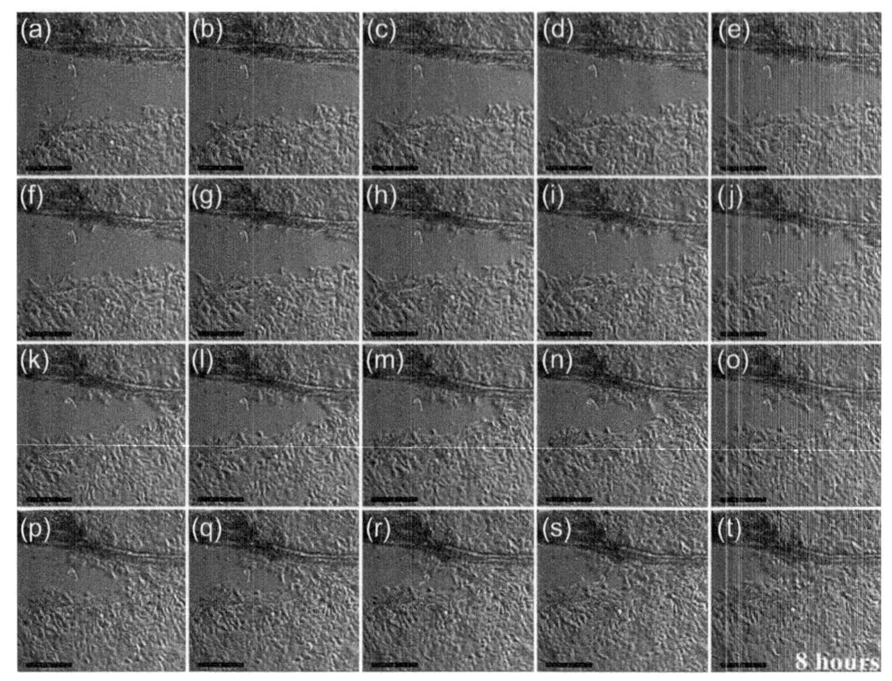

Figure 6.1 A series of images (a–t) showing a scratch in a layer of human keratinocytes grown on plastic closing over an 8-h period (interval length between images was 25 min).

potentially novel therapeutic agents, which can modulate this behaviour (Pullar et al., 2003; Yang et al., 2011). Modified versions of the scratch wound model exist. For example, several scratches can be combined in a single assay, which may more closely mimic in vivo cell signalling (Turchi et al., 2002; Bertero et al., 2011). In addition, some researchers do not create a scratch but rather grow cells to confluency around a plug, which can then be removed to assess migration into the void space. Importantly, damage-associated molecular patterns are not released as is the case when cell monolayers are scratched. Such assays can be further modified to assess paracrine signalling between keratinocytes and dermal fibroblasts by growing keratinocytes in tissue culture inserts suspended over fibroblasts grown in the well below (Goulet et al., 1996). Differences in keratinocyte cell behaviour in the presence and absence of fibroblasts can then be assessed in detail (Loo and Halliwell, 2012; Loo et al., 2011).

Although two-dimensional monolayer experiments are ideal for analysing individual cellular functions such as migration, mechanistically at the single cell level, wound healing is much more complex than that. This has led to the development of more sophisticated approaches to dissecting more than the migratory processes involved in wound repair. Three-dimensional skin models, which allow epithelial cells to stratify, are a more sophisticated approach to studying the behaviour of keratinocytes. To achieve this, keratinocytes are expanded to confluency on porous filters prior to being raised to the air–liquid interface, which subsequently facilitates stratification (Prunieras et al., 1983; Rosdy and Clauss, 1990; Poumay and Coquette, 2007;

Frankart et al., 2012). Such assays are particularly useful to study the effects of chemicals on skin irritation or sensitisation (Coquette et al., 2003).

Decellularised dermis can be used as a more physiologically relevant matrix to assess keratinocyte cell behaviour than porous filters. For this purpose, fibroblasts are removed with detergents, freezing or lethal irradiation, leaving behind intact acellular ECM. Keratinocytes seeded on top of the dermal template can then be scratched or wounded using punch biopsies, and the effect of different dermal substitutes or other treatments such as growth factors can be assessed (Xie et al., 2010; van Kilsdonk et al., 2013). Epithelial outgrowth can also be studied from a skin punch biopsy explanted onto the dermal substrate (Forsberg et al., 2006).

More complex tissue engineered three-dimensional skin models can be fabricated to contain several cutaneous cell types or cutaneous microenvironments to model wound healing. The dermal component in such models may take the form of a collagen gel, seeded with keratinocytes, which can be used to specifically monitor contraction during wound healing while also studying epithelialisation (Bell et al., 1983; Amadeu et al., 2003; Schneider et al., 2008). To model angiogenesis in wound healing, three-dimensional microenvironmental niches can be created to assess the angiogenic response of endothelial or other cell types to such environmental stimuli. These can consist of individual ECM components, such as fibrin or collagen, complex ECM formulations, such as matrigel, or combinations thereof. Such assays can then be used to study the effect of the niche environment, added growth factors or peptides in combination with other cutaneous cells (Potter et al., 2006; Demidova-Rice et al., 2011; Feng et al., 2013). Commercially available skin substitutes have also been used to model cell behaviour to predict in vivo outcomes (Falanga et al., 2002).

To fully recapitulate the effect of intact cutaneous structures and physiologically relevant ratios of different cells, ex vivo models of human skin can be employed. Simple models exist where whole skin punch biopsies are seeded onto tissue culture plastic, and cell outgrowth is measured. This model has been used to assess the effects of topically applied pharmaceutical compounds as well as different pH microenvironments on cell outgrowth (Harris et al., 2009; Sharpe et al., 2009). In a more sophisticated version of this model, full-thickness and partial-thickness intact skin is maintained in culture for several weeks. Such skin can then be wounded using punch biopsies or burned prior to assessing wound healing, which has been reported to recapitulate cell behaviour in line with clinical observation (Kratz et al., 1994; Emanuelsson and Kratz, 1997; Kratz, 1998; Rizzo et al., 2012; Xu et al., 2012; Balaji et al., 2014). This type of model is frequently used in conjunction with other in vitro models described to corroborate cell behaviour studies in monolayer culture with data gained using intact skin. Such studies have been used to compare the effect of signalling pathways in normal and aberrant wound healing (Stojadinovic et al., 2005) as well as assessing the effect of different compounds on wound healing (Pullar et al., 2006; Tomic-Canic et al., 2007; Pastar et al., 2012).

6.3 In vivo models of wound healing

In vitro and ex vivo models of wound healing have been useful in dissecting the molecular pathways utilised in specific skin cells or limited cocultures. These in vitro models are, however, often too simplified to recapitulate the complex interplay between the resident and recruited cell types and the stem cell niches, which impact the wound healing process. In vivo models are therefore indispensable in studying the mechanisms of wound repair.

6.4 The rodent model

Small rodents, including mice and rats, have been used widely in wound healing studies because of a number of advantages compared to the porcine model (Morris et al., 1997; Mogford and Mustoe, 2001; Dorsett-Martin, 2004; Galiano et al., 2004). These are mainly due to their low cost of husbandry and the opportunity of using several animals in studies, increasing statistical power. Additionally, the comprehensive knowledge of the mouse genome and availability of transgenic models provide for detailed molecular studies into the mechanism of skin pathologies and healing phenotypes.

The skin of the mouse differs from human skin in a number of respects. It is more hairy and in most cases needs shaving before wounds are made. There are hairless mouse strains that are immunocompetent and suitable for wound healing experiments (Danielsen et al., 2016). The mouse skin is loose over the underlying muscle, and the contractile force of the panniculus carnosus muscle in the subcutaneous tissue promotes wound healing by contraction. This level of contraction is not characteristic of the human skin. Therefore to model the normal epithelialisation seen in human skin, mouse skin wounds are often fixed with a splint or suturing to underlying tissue to reduce the contraction and permit wound closure by epithelialisation (Wang et al., 2013). It is also possible to create a wound where mechanical contraction potentially provided by the underlying muscle is reduced by creating a full-thickness wound by extending beyond the panniculus carnosus muscle (Eming et al., 2007).

6.5 The pig model

The pig model of skin wound healing is thought to most closely approximate that of human skin wound healing, with the caveat that it perhaps does not have precisely the same immune response as the human. Structurally, the thickness of the epidermis and dermis and their relative ratios are close to that in humans (Sullivan et al., 2001). Furthermore, the vascular network and appendages such as hair follicles closely resemble those found in the human skin, and a further level of similarity is borne in the content of elastin and collagen (Sullivan et al., 2001; Swindle et al., 2012). Humans and pigs also respond similarly to growth factors (Swindle et al., 2012). According to Debeer et al. (2013), some of the main histological differences between porcine and human skin are that the pig has a reduced granular cell layer, reduced amounts of melanin and melanocytes, a wider distribution of apocrine sweat glands and more sparse and smaller sebaceous glands. Pig skin lacks eccrine sweat glands. In humans eccrine sweat glands are important for the formation of new epidermis during wound healing (Rittié et al., 2013). Furthermore, Bode et al. (2010) discuss the immunological responses of the pig, which are thought to show some differences when compared to the human response. The authors consider these to have a minor impact (with the possible exception of gamma delta T cells (γδT) and a greater role of the innate immune system).

Typically, the White Yorkshire pig has been used in many wound healing studies, and the Göttingen mini pig is also increasingly being used in preclinical safety evaluations (Bode et al., 2010). The main drawbacks of the pig model are the prohibitive costs of husbandry and the fact that it does not completely replicate the scarring mechanisms observed in humans. The Red Duroc pig has been used as a better model of scarring, but even this does not fully replicate the hypertrophic scarring observed in humans (Zhu et al., 2003, 2007).

6.6 Types of wounds

6.6.1 Tape stripping

This is a method of detaching the top layers of the stratified epithelium using adhesive tape. It easily removes the cornified layer of dead cells and the successive deeper layer of the epithelium without breaching the basement membrane. As a result, no vascular damage or blood loss results, but the water barrier function of skin is lost. This provides a good model for the loss of barrier function, and this is measured by transepidermal water loss (TEWL) (Gao et al., 2013). Alongside imaging of skin topography, TEWL can be useful in the comparison of skin from different anatomical sites (Kleesz et al., 2012). Tape-stripped skin repairs easily by the process of epidermal proliferation and turnover whereby the underlying propagating cells initially proliferate and then differentiate to replace the top layers.

6.6.2 Suction blister wound model

The suction blister model permits the investigation of the restoration of the epidermal barrier function and blood flow, thought to be closely associated with the level of inflammation in the wound (Koivukangas and Oikarinen, 2003). The suction blister model has had extensive use for understanding the basic biology of epidermal wound healing and the healing of burn injuries. The process involves a prolonged vacuum being used on the skin to disrupt the dermo–epidermal junction, separating the dermis from the epidermis but leaving the basal lamina undisturbed (Koskela et al., 2009). The method is sensitive and is able to detect changes in collagen synthesis due to various diseases, topical or systemic therapies, the effects of lifestyle such as smoking or matrix metalloproteinase (MMP) inhibitors on epidermal healing (Ågren et al., 2001; Koivukangas and Oikarinen, 2003; Sørensen et al., 2009). Other applications of the suction blister technique include measurements of pharmacological agents or their derivatives from interstitial fluid and assays of various enzymes and cytokines (Koivukangas and Oikarinen, 2003). The company Electronic Diversities (Flinksburg, MD, USA) provides complete instrumentation that produces standardised blisters (Alexis et al., 1999).

6.6.3 Laser

Erbium:Yttrium-Aluminium-Garnet laser produces more uniform epidermal wounds than the suction blister method (Ferraq et al., 2012). Laser wounds are made in less than 1 min while suction blisters take 1–2 h to raise. The healing kinetics was similar for the laser- and suction blister-induced epidermal wounds when compared with optical coherence tomography and histology (Ferraq et al., 2012).

6.6.4 Minimal puncture wounds

Standardised puncture wounds can easily be inflicted using spring-loaded lancet devices (Olerud et al., 1995; Varol and Anderson, 2010). As opposed to the three aforementioned epidermal wound models, the puncture wounds extend into the dermis and bleed. Wound healing can be evaluated macroscopically with magnification (×10), laser Doppler perfusion imaging and microscopically (Varol and Anderson, 2010).

6.6.5 Split-thickness and full-thickness wounds and grafts

Split-thickness wounds are those which cause loss of the epidermis and the top layer of the dermis. Usually, this involves the papillary dermis and the top of the rete ridges, but the bases of the skin appendages remain intact. As a result, the split-thickness wound retains most of the follicular and other stem cell niches, potentially allowing it to repair without scarring. A split-thickness wound is the wound type created at the donor site of autologous skin grafts. Full-thickness wounds, on the other hand, cause injury to the entire depth of both the epidermis and dermis.

6.6.6 Incisional and excisional wounds

Full-thickness wounds are caused by trauma, which arises from physical damage to the skin. They are categorised as incisional or excisional. Incisional wounds such as those made by surgical blades create damage to the tissue structure, parting the tissue along the line of the incision but without the loss of tissue volume. The mechanics of fibroblast contraction can be investigated by this methodology.

Excisional wounds, on the other hand, result in the loss of tissue volume, an open wound bed and a clear outer boundary. While both incisional and excisional wounds undergo the same phases of wound healing, excisional wounds undergo an extensive proliferative phase when a granulation tissue forms as it is vascularised and remodelled. The full-thickness wound finally closes by epithelialisation over the granulation tissue. Myofibroblast-driven contraction also plays a role in excisional wound healing and is proportional to wound size.

Incisional and excisional wounds are supported to heal in two distinct ways. Primary intention is used to bring the edges of incisional wounds together by suturing or similar intervention. This allows the wound to heal quickly by minimising the gap between the edges of incisional wounds. On the other hand, excisional wounds are supported by standard dressing and left to heal spontaneously by the process of granulation and epithelialisation, so-called secondary intention. Healing by secondary intention in full-thickness wounds often results in scarring, which may need surgical intervention in some patients.

6.6.7 Human skin graft and skin substitute models

Animal models have been essential for the development of novel therapies for wound healing and in burns. The use of sprayed autologous or allogeneic keratinocytes to augment healing of severe burns is now common practise in many clinical units (Wood et al., 2006; Atiyeh and Costagliola, 2007; James et al., 2010; De Corte et al., 2012; Auxenfans et al., 2013). Further refinements and improvements to this technique are demonstrating efficacy in small and large animal models (Gustafson et al., 2007; Ananta et al., 2011; Eldardiri et al., 2012; Pontiggia et al., 2013), information which is essential for development towards clinical application.

A number of research groups have developed novel in vivo wound healing models in a genetically modified human context (Scheid et al., 2000; Escamez et al., 2004; Carretero et al., 2006). The model developed by Escámez et al. (2004) is based on the regeneration of human skin on the back of immunocompromised athymic nude mice by transplantation of a bioengineered skin equivalent seeded with human cells. This model system provides a suitable in vivo tool to test gene transfer strategies for human skin repair (Escamez et al., 2004). This study and others also serve as a useful platform for studies in genetically modified mice, allowing the evaluation of therapeutic approaches for impaired wound healing.

Preclinical testing of human therapeutics such as monoclonal antibodies has been limited in murine models due to species differences in pharmacokinetics and biologic responses (Waldron-Lynch et al., 2012). To overcome these constraints these authors developed a murine skin transplant model in humanised mice and used it to test human monoclonal antibody therapies (Waldron-Lynch et al., 2012).

These humanised-animal model approaches will likely become more frequently used as the technology develops.

6.6.8 In vivo models of delayed wound healing

Certain wounds fail to heal, typically by the epithelial layer never closing. The open wound bed remains exposed and infection is usually associated. Some nonhealing wounds may have a primary infectious aetiology in which case treatment of the infection is an essential part of the management (Doig et al., 2012). Several local and systemic factors contribute to healing time. Infection, a major local factor, prolongs inflammation and tissue damage. The increased activity of MMPs often leads to the degradation of growth factors and the provisional matrix. The delay in healing may also be due to recurrent trauma to the wound, probably due to the anatomical location, such as a joint (Wray, 1983). The type of wound may also predispose it to a delay in healing. Following burns, the loss of critical stem cell niches and basal layer cells limits the epithelialisation potential, exposing the wound to infection and inflammatory complications. Full-thickness burns are especially difficult to heal for this reason.

6.6.8.1 Ischaemic wounds

Long-term, nonhealing wounds are characteristically ischaemic, and surgical debridement promotes healing by removing the dead skin and exposing a vascularised wound bed that permits graft take. Although tissue ischaemia can be simulated in vitro in a hypoxic chamber, wound ischaemia is among the complex phenomena that require in vivo modelling. The best characterised in vivo models for the ischaemic wound are the skin flap, the ischaemic hindlimb and the rabbit ear models.

The McFarlane flap is among the earliest and widely used of these models in which a rectangular skin flap is made on the back side of a small rodent (McFarlane et al., 1965). This flap of skin is continuous with the rest of the body at only one short edge, and a region of necrotic ischaemia develops distal to this edge. Wounds made in this ischaemic zone provide a good model for the healing of ischaemic wounds. To stop neovascularisation from the wound bed, a silicone splint is sutured over the wound bed, separating the flap and depriving it of vascular supply.

The McFarlane flap has been extensively adapted. One of these adaptations was the bipedicle flap of Schwarz et al. (1995) in which both short edges remain attached to the back of the animal, and the long edges are sutured together on the underside, making a closed tube. It has less necrosis than the original McFarlane flap but develops a gradient of ischaemia from the pedicles to the centre, allowing the study of healing characteristics of wounds made in the ischaemic zone.

Besides the flap model, ischaemic wound areas are generated by ligation of the vasculature, which has been described for the hindlimb and the rabbit ear. The Ischaemic Hind Limb model is induced by ligation of the iliac and femoral arteries of the animal (usually rat), such as described by Paek et al. (2002). The contralateral limb acts as a control, and wounds made in congruent locations can be studied for the effect of ischaemia. The ischaemic rabbit ear model is also well established and has some noteworthy benefits in modelling ischaemic human wound healing. The first of these is the practical consideration that backlighting allows for noninvasive examination of the ear vasculature. Second, the cartilaginous rabbit ear does not heal by contraction, like the typical rodent loose skin. The model is therefore closer to human skin. Before wounding, ischaemic insult is established by ligation of arteries supplying the ear (Ahn and Mustoe, 1990).

6.6.8.2 Diabetic wound models

Diabetes is one of the major diseases associated with poor wound healing. There are a number of rodent models for diabetic wounds. The benefit of the rodent model is that disease can be either induced in the wild-type animal or spontaneous disease allowed to develop in animals with predisposing genetic backgrounds. Diabetes can be induced in the wild-type strain using streptozotocin, which is toxic to the pancreatic islet cells. A second diabetogenic, alloxan, is also used (Tengrup et al., 1988), but because of the differences in the mechanism of action between streptozotocin and alloxan (Gai et al., 2004) there are species variations in toxicity that should be considered in their choice. The C57/BL6 mouse has two mutant strains with deregulated leptin signalling, which become obese within weeks of birth and spontaneously develop diabetes and the characteristic delay in wound healing. The two genetic strains used are the ob/ob mice, which are deficient in leptin (Galiano et al., 2004), and the db/db, which have a nonfunctional mutant leptin receptor in the hypothalamus (Osborn et al., 2010).

The specific role of local hyperglycaemia in the diabetic wound has been contested, but a clear systemic change in diabetes is associated with poor healing (Velander et al., 2008). The debate as to the utility of these models and how they work is still ongoing. For instance, the impaired healing observed may just correlate with leptin deficiency than hyperglycaemia, or it may simply reflect the obesity of the animals, and perhaps as a consequence of this increased growth also affects the mechanobiology around the healing wound in these models. In a similar vein, burn wounds often present with similar metabolic deregulation associated with hyperglycaemia, insulin resistance and elevated inflammatory markers in blood (Gauglitz et al., 2009).

6.6.8.3 Burn wound model

The severity of thermal burns to the skin depends on the depth of injury and the area of coverage. Burn wounds can be caused by a scald or contact with a hot conductive material. These thermal burns are reproduced in animal models by exposure of a brass plate to hot water to generate partial-thickness burns or hot oil to produce full-thickness burns (Davis et al., 1990; Middelkoop et al., 2004; Branski et al., 2008). A range of similar conductive materials and temperatures are used in various laboratories to produce the burn wound.

6.7 Assessment of healing outcomes

6.7.1 Scarring

Epidermal wounds induced by tape stripping, suction or laser and split-thickness wounds regenerate/repair without complications. However, when a full-thickness wound heals in the mammalian adult, it leaves behind a scar tissue, which does not entirely restore the structure and function of the unwounded tissue (Metcalfe and Ferguson, 2007). These wounds heal quickly at the expense of regeneration to minimise the complications of infection in the exposed wound (Metcalfe and Ferguson, 2007).

It has been shown that in both incisional and excisional wounds, scarring depends on the depth of the wound with full-thickness wounds predisposing to prominent scars (Dunkin et al., 2007). During the remodelling phase, myofibroblasts synthesise type I collagen, replacing type III collagen, which is expressed in the early phases of wound healing (Clore et al., 1979). The excessive deposition of poorly organised collagen promotes scarring. Both incisional and excisional wounds often heal with a scar lacking in appendages such as hair follicles and sweat glands. A number of problem scar types that have been characterised include contractures at joints, hypertrophic scars, keloids and altered pigmentation at the scar site. Each of these scar types can have medical, lifestyle or aesthetic implications for the patient, and they are indicators of poor outcome. These outcomes cannot be attributed to a single aetiology or wound type. The need for research to understand their aetiology and how to improve outcomes remains an important theme in wound biology. Many researchers attempt to reproduce these scar types in animal models for investigating interventions.

6.7.2 Contractures

These are scars that limit the free movement of joints and muscle, compromising locomotion and physical dexterity. Several functionally limiting events can occur after a deep burn, and even with good initial surgical management, contractures are common (Kamolz et al., 2009; Fufa et al., 2014). As these authors point out, acute burn care is often managed by multidisciplinary, specialised burn units, but postburn contractures may be referred to hand or limb surgeons, who should be familiar with nonsurgical and operative approaches to improve function and the expected outcomes of such contractures (Fufa et al., 2014). The most common and functionally limiting are contractures of the hand and digits as well as of the skin of the upper arm and shoulder. Skin contractures are also often accompanied by debilitating chronic pain, requiring a pain medication that can in extreme cases lead to other undesirable side effects and unwanted dependency on the drugs administered to control the pain.

The management of contractures often includes complete surgical excision of scar tissue and soft tissue defect resurfacing; this is most commonly achieved with full-thickness skin grafts (Kamolz et al., 2009; Fufa et al., 2014). The surgical release of contracture can result in exposing tendons or joints, at which point tissue transfer may have to be considered (Fufa et al., 2014). It is important that rehabilitation of the joints starts early postoperatively to prevent any recontracture after surgery (Fufa et al., 2014).

6.7.3 Hypertrophic scars

These are scars that form within the boundary of the wound but extend above the level of the surrounding unwounded epidermis.

6.7.4 Keloid scars

Keloid scarring is a form of aberrant wound healing that results in the growth and expansion of an exuberant scar beyond the confines of the original skin injury. Keloid scars are characterised by excessive ECM disposition, prolonged fibroblast proliferation, increases in inflammatory cell infiltration and angiogenesis (Varmeh et al., 2011). At present, there is no satisfactory treatment for keloid, and progressive disease can lead to severe disfigurements and bodily dysfunction.

The mechanisms involved in keloid formation are poorly understood but are believed to be influenced by genetic predisposition, disrupted growth factor signalling and immunological dysfunction, all resulting in overproduction and aberrant collagen deposition (Shih and Bayat, 2010).

Keloid has also been described as a benign dermal tumour that develops from an aberrant wound healing process. The idea that the processes of wound healing and tumourigenesis share many similar mechanisms is an old one first proposed by Virchow in the 1860s, and cancer has been described as an "overhealing wound" (Schafer

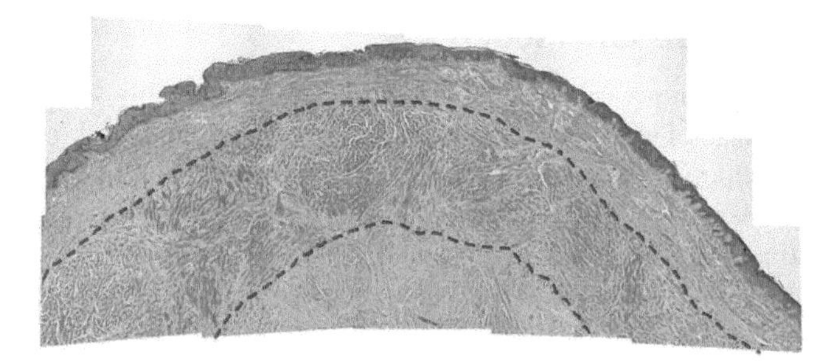

Figure 6.2 Composite set of images of a proportion of a histological section through an ear keloid. Note the central area of condensed collagen (within the dashed lines). Masson's trichrome stain.

and Werner, 2008). As Varmeh et al. (2011) point out, keloid fibroblasts at some point stop proliferating, and thus the scar remains benign. Characterising this control of proliferation in the keloid, however, could be important for therapeutic intervention (Fig. 6.2).

6.7.5 Animal models for hypertrophic and keloid scars

The pig model of healing is thought by many to most closely resemble human skin (Sullivan et al., 2001), but it is limited in its ability to reproduce hypertrophic scarring. The only strain that has been characterised for hypertrophic scarring is the Red Duroc pig, which was shown to present a scar of this type lasting up to 5 months (Zhu et al., 2003). The limitations of using the pig (among them the cost of husbandry) have encouraged the development and use of other models of hypertrophic scarring. These are mainly models using the more representative Tight Skin mouse (Tsk) or by xenotransplantation into immunocompromised athymic rodents.

Full-thickness excision wounds in the Tsk mouse can transiently reproduce human hypertrophic scarring (Ehrlich and Needle, 1983). They have limited use, however, because the scars are not long lasting. Surgical scars in the rabbit ear remain elevated for months (Morris et al., 1997), providing a useful model for hypertrophic scarring. Besides these options, hypertrophic and keloid scarring can be modelled by xenotransplantation of established human scars. To avoid rejection, recipient animals have to be immunosuppressed, immunocompromised (Shetlar et al., 1985) or transplantation is targeted into immunologically privileged sites (Hochman et al., 2005). The existence of inbred strains of athymic nude mice obviated the need to use immunosuppressive comedication. Shetlar et al. (1985) first used the nude mouse to study a human keloid scar implanted subcutaneously. Typically, these keloid implants were deepithelialised and diced before implantation. The limitations of the mouse model are in the small animal size and therefore the implant size and surgical manipulation for studies. A strain of athymic rats has been bred and characterised for use as a larger animal model to overcome the size limitations of the mouse (Festing et al., 1978). These rats have been successfully used to implant keloid scars sandwiched into a flap supplied by a single pedicle. This closely resembles the keloid scar found on human skin.

6.7.6 Skin fibrosis and models

Skin fibrosis occurs in a variety of human diseases, most notably systemic sclerosis (SSc). The end stage of scleroderma, like dermal scarring of human skin, is characterised by the deposition of excess collagen in the dermis with a loss of the differentiated structures, such as hair follicles, sweat glands etc. and adipose tissue.

The initiating factors for this process and the early stages are believed to occur through vascular injury and immune dysfunction with a dysregulated inflammatory response. However, because of the insidious onset of the disease, this stage is rarely observed in humans and remains poorly understood. Animal models have provided a means to examine these early stages and to isolate and understand the effect of perturbations in signalling pathways, chemokines and cytokines.

Long et al. (2014) describe the Tight Skin 2 (Tsk2) mouse model of SSc. The Tsk2 mutation is distinct from the original Tsk mutation, which is an autosomal dominant mutation on mouse chromosome 2; Tsk2 maps to mouse chromosome 1. This mouse strain has many features of SSc, including tight skin, excessive collagen deposition, alterations in the ECM, increased elastic fibres and the presence of antinuclear antibodies with age (Long et al., 2014). These authors compared how fibrosis affected wound healing in Tsk2/+ mice compared with their wild-type littermates. The potential sources of this response were investigated by wounding Tsk2/+ mice that were genetically deficient either for a known fibrosis mediator (the NLRP3 inflammasome) or for elastic fibres in the skin, using a fibulin-5 knock-out. Long et al. (2014) found that the loss of elastic fibres restored normal wound healing in the Tsk2/+ mouse and that the loss of the NLRP3 inflammasome had no effect. This led these workers to postulate that elastic fibre alteration was the main reason for delayed wound healing in the Tsk2/+ mouse (Long et al., 2014). They also speculated that therapies promoting collagen deposition in the tissue matrix without elastin deposition might promote wound healing in SSc and other fibrotic diseases (Long et al., 2014).

6.7.7 The effects of mechanics on healing

The mechanical environment of a wound site affects the rate and quality of wound healing. Mechanical stress influences wound healing by affecting the behaviour of cells within the dermis, but it remains unclear how these mechanical forces affect the process of epithelialisation (Evans et al., 2013). Tensile forces are known to affect the behaviour of cells within epithelia, however, and the material properties of extracellular matrices, such as substrate stiffness, have been shown to affect the morphology, proliferation, differentiation and migration of many different cell types (Engler et al., 2004; Evans et al., 2013).

Several studies have established that skin wounds, which are under mechanical tension, are more prone to heal with the formation of a scar. Surgeons therefore aim to reduce tension at incision sites postsurgery. Simple methods, such as adhesive tape, have proved to be successful in improving wound healing; novel devices are under preclinical development, and some are in phase I clinical trials (Atkinson et al., 2005; Gurtner et al., 2011; Wong et al., 2013).

6.7.8 *Mammalian models of regeneration*

Mammals are generally considered to be poor regenerators, but there are a few mammalian models that display a robust ability to regenerate (Muneoka et al., 2008; Wu et al., 2013). Two such model systems are the regenerating ear tips of the MRL/MpJ mouse and digit regeneration seen in both humans and mice.

A number of studies have shown that mammalian foetal skin regenerates perfectly, but adult skin repairs by the formation of scar tissue (Ferguson and O'Kane, 2004). Wound healing and tissue regeneration are likely to be processes that begin along the same pathway but because of subtly different microenvironmental cues can diverge to result in either scar formation or scar-free regeneration. This idea evolved from studies showing that within the same mammalian model, both repair and regeneration can occur (Beare et al., 2006; Metcalfe and Ferguson, 2007; Wu et al., 2013; Simkin et al., 2013). The influencing factors for this are likely to be numerous and possibly include inflammatory cell components, nerve ingrowth, cytokine signalling, thrombin activation, complement factors, the presence of progenitor and dedifferentiated cell populations and other parameters still to be discovered (Clark et al., 1998; Metcalfe and Ferguson, 2007; Kumar and Brockes, 2012; Buckley et al., 2012; Seifert and Maden, 2014; Gawronska-Kozak et al., 2014; Gourevitch et al., 2014).

Utilising these model systems of mammalian regeneration and comparing them with equivalent established model systems in lower animals may hold the key to beginning to understand scar-free healing. These model systems may also inform the development of more robust preclinical approaches for wound healing and for assisting in the translation of therapeutics, which modify the healing process clinically.

6.8 Translational medicine

6.8.1 *Incisional acute wounding models and scar revision*

Cutaneous scarring affects up to 100 million people per annum (Kieran et al., 2013). There is no effective scar reducing/preventing therapeutic developed to date. Skin scarring is associated with psychosocial distress and has a negative effect on quality of life (So et al., 2011). Scar revision does not erase a scar but helps to make it less noticeable and more acceptable to the patient.

Shah et al. (1992, 1994) demonstrated that a neutralizing antibody to transforming growth factor (TGF)-β1 reduced skin scarring in a rat incisional wound healing model. The TGF-β family of cytokines plays a key role in scarring (Ferguson and O'Kane, 2004; Metcalfe and Ferguson, 2007), and TGF-β3 is known to improve scar appearance across a range of mammalian species. Bush et al. (2010) describe a comprehensive set of preclinical and clinical trials using intradermal avotermin (TGF-β3). As part of this, So et al. (2011) performed a phase II study to assess the efficacy of intradermal avotermin for the improvement of scar appearance following scar-revision surgery. Primary endpoint data from the combined surgical groups showed that avotermin significantly improved scar appearance as compared with placebo (So et al., 2011). Avotermin advanced to phase III trials but unfortunately failed to meet the trial endpoints.

Additional approaches to scar reduction in the skin have included using other growth factors, such as basic fibroblast growth factor, which is thought to reduce scarring and promote wound healing by inhibiting the TGF-β1/sons of mothers against decapentaplegic homologue protein-dependent pathway (Shi et al., 2013).

Interleukin (IL)-10 is an antiinflammatory and antifibrotic cytokine. In the embryo it is important for scarless wound repair. Preclinical studies in rat incisional wounds revealed that IL-10 resulted in healing with decreased inflammation, better scar histology and better macroscopic scar appearance (Kieran et al., 2013). These authors describe phase II randomised controlled clinical trials where four incisions were made on each arm of 175 healthy volunteers. These four incisions received four different concentrations of recombinant human IL-10 (rhIL-10), which were matched with four control incisions that received either standard care or placebo. IL-10-treated human incisions at low concentrations healed with better macroscopic scar appearance and less red scars. These studies were carried out in predominantly Caucasian volunteers (Kieran et al., 2013).

Scars in humans of African continental ancestry heal with an exaggerated inflammatory response and a generally wider scar (Kieran et al., 2014). These authors investigated the effects of IL-10 on cutaneous scarring in volunteers of African ancestral origin in an exploratory, single-centre, within-subject, double-blind randomised controlled trial. A range of doses of IL-10 were administered. Subjects showed a trend toward favouring treatment with 5ng/100μL/linear cm IL-10 between 5 and 9 months postexcision (Kieran et al., 2014). No concentration of IL-10 produced a statistically significant improvement in scarring as compared with placebo. As Ågren and Danielsen (2014) observed, the discrepancy in response to IL-10 in relation to skin type is striking and indicates some fundamental biological differences in wound healing mechanisms that are currently unknown. The outcomes of scar-reducing therapies have yet to prove their translation, but the methods and approaches used to try to unravel the wound healing cascades should be recognised (Ågren and Danielsen, 2014). As Eskes et al. (2012) quite rightly point out, randomised clinical trials should ensure adequate allocation concealment and blinding of outcome assessors, alongside an application of intention-to-treat analysis and patient-oriented outcome measures.

6.8.2 Connexin 43 treatment of chronic wounds

Other wound healing approaches on chronic wounds or excisional wound models for scar reduction approaches that have been in various phase I and phase II clinical trials have included Connexin 43 (Cx 43) targeting. Preclinical and clinical studies demonstrated that ACT1, a peptide mimetic of the carboxyl-terminus of Cx 43, formulated as Granexin™ gel (FirstString Research, Mount Pleasant, SC, USA), has shown to be safe and effective for reducing scar formation and promoting faster healing of incisional and excisional wounds with markedly less wound gaping (Ghatnekar et al., 2009; Ongstad et al., 2013). In a prospective, multicentre clinical trial, adults with chronic venous leg ulcers were randomised to treatment with an ACT1 gel formulation plus conventional standard of care (Ghatnekar et al., 2014). A significantly greater reduction in the mean percent ulcer area from baseline to 12 weeks was associated with the incorporation of ACT1 therapy when compared to compression bandage therapy alone (Ghatnekar et al., 2014). Likewise, ACT1 in hydroxyethyl cellulose gel reduced the ulcer area as compared to standard of care alone after 12 weeks of treatment of diabetic foot ulcers in a randomised controlled trial (Grek et al., 2015). No placebo gel was used.

Alternative approaches targeting Cx 43 include Nexagon® (CoDa Therapeutics, San Diego, CA, USA), which is a natural, unmodified oligonucleotide (30-mer) that downregulates Cx 43. By modulating early biological responses in tissue repair, Nexagon® is believed to modulate the inflammatory response and decreases wound spread to accelerate healing, leading to reduced inflammation, swelling and scarring.

6.8.3 Likely future trends for modelling wound healing

Though there has been extensive experimental work performed on different species in an attempt to recapitulate human wound healing, these models are still not completely characterised. This is because of the complex interactions of a number of factors between the species under investigation that are used to try and predict translational mechanisms that will influence the clinical outcome. These factors include overlapping events in the wound healing cascade, cells from multiple sources, newly deposited and maturing ECM microenvironments, chemicals, enzymes and growth factors, to name just a few of the key players involved. Therapeutic interventions include the manipulation of one or more of these phases or factors to try and positively influence the healing outcome. For therapeutic interventions, stratification of the patient population will become an increasingly important part of the advance beyond phase I and II clinical trials.

Additionally, new approaches, including multidimensional mathematical modelling, will become an increasingly important tool for quantitatively defining, then testing, the nature of the often complex interactions that occur during the wound healing process (Cumming et al., 2010; Friedman and Xue, 2011; Menon et al., 2012; Murphy et al., 2012).

A greater understanding of the mechanisms of scar reduction therapies and scarless repair mechanisms will likely lead to the development of improved animal models with a greater ability to define targets for scar prevention. Key to this will be understanding similarities and differences in the wound repair process across models and species. In some situations within the same mammalian model both repair and regeneration occur; whether this is influenced by inflammatory cell components or other parameters still has to be ascertained. It is likely that these subtle differences in mechanism, inflammatory cell influence as well as emerging model systems may hold the key to reducing scarring and lead to the development of better models of wound healing.

References

Ågren, M.S., Danielsen, P.L., 2014. Antiscarring pharmaceuticals: lost in translation? Wound Repair Regen. 22, 293–294.

Ågren, M.S., Mirastschijski, U., Karlsmark, T., Saarialho-Kere, U.K., 2001. Topical synthetic inhibitor of matrix metalloproteinases delays epidermal regeneration of human wounds. Exp. Dermatol. 10, 337–348.

Ahn, S.T., Mustoe, T.A., 1990. Effects of ischemia on ulcer wound healing: a new model in the rabbit ear. Ann. Plast. Surg. 24, 17–23.

Alexis, A.F., Wilson, D.C., Todhunter, J.A., Stiller, M.J., 1999. Reassessment of the suction blister model of wound healing: introduction of a new higher pressure device. Int. J. Dermatol. 38, 613–617.

Amadeu, T.P., Coulomb, B., Desmouliere, A., Costa, A.M., 2003. Cutaneous wound healing: myofibroblastic differentiation and in vitro models. Int. J. Low Extrem. Wounds 2, 60–68.

Ananta, M., Mudera, V., Brown, R.A., 2011. A rapid fabricated living dermal equivalent for skin tissue engineering: an in-vivo evaluation in an acute wound model. Tissue Eng. Part A 18, 353–361.

Atiyeh, B.S., Costagliola, M., 2007. Cultured epithelial autograft (CEA) in burn treatment: three decades later. Burns 33, 405–413.

Atkinson, J.A., McKenna, K.T., Barnett, A.G., McGrath, D.J., Rudd, M., 2005. A randomized, controlled trial to determine the efficacy of paper tape in preventing hypertrophic scar formation in surgical incisions that traverse Langer's skin tension lines. Plast. Reconstr. Surg. 116, 1648–1656.

Auxenfans, C., Shipkov, H., Bach, C., Catherine, Z., Lacroix, P., Bertin-Maghit, M., Damour, O., Braye, F., 2013. Cultured allogenic keratinocytes for extensive burns: a retrospective study over 15 years. Burns 40 (1), 82–88.

Balaji, S., Moles, C.M., Bhattacharya, S.S., LeSaint, M., Dhamija, Y., Le, L.D., King, A., Kidd, M., Bouso, M.F., Shaaban, A., Crombleholme, T.M., Bollyky, P., Keswani, S.G., 2014. Comparison of interleukin 10 homologs on dermal wound healing using a novel human skin ex vivo organ culture model. J. Surg. Res. 190, 358–366.

Beare, A.H., Metcalfe, A.D., Ferguson, M.W., 2006. Location of injury influences the mechanisms of both regeneration and repair within the MRL/MpJ mouse. J. Anat. 209, 547–559.

Bell, E., Sher, S., Hull, B., Merrill, C., Rosen, S., Chamson, A., Asselineau, D., Dubertret, L., Coulomb, B., Lapiere, C., Nusgens, B., Neveux, Y., 1983. The reconstitution of living skin. J. Invest. Dermatol. 81, 2s–10s.

Bertero, T., Gastaldi, C., Bourget-Ponzio, I., Imbert, V., Loubat, A., Selva, E., Busca, R., Mari, B., Hofman, P., Barbry, P., Meneguzzi, G., Ponzio, G., Rezzonico, R., 2011. miR-483-3p controls proliferation in wounded epithelial cells. FASEB J. 25, 3092–3105.

Bode, G., Clausing, P., Gervais, F., Loegsted, J., Luft, J., Nogues, V., Sims, J., 2010. The utility of the minipig as an animal model in regulatory toxicology. J. Pharmacol. Toxicol. Methods 62, 196–220.

Branski, L.K., Mittermayr, R., Herndon, D.N., Norbury, W.B., Masters, O.E., Hofmann, M., Traber, D.L., Redl, H., Jeschke, M.G., 2008. A porcine model of full-thickness burn, excision and skin autografting. Burns 34, 1119–1127.

Buckley, G., Wong, J., Metcalfe, A.D., Ferguson, M.W., 2012. Denervation affects regenerative responses in MRL/MpJ and repair in C57BL/6 ear wounds. J. Anat. 220, 3–12.

Bush, J., So, K., Mason, T., Occleston, N.L., O'Kane, S., Ferguson, M.W., 2010. Therapies with emerging evidence of efficacy: avotermin for the improvement of scarring. Dermatol. Res. Pract. 2010.

Carretero, M., Escamez, M.J., Prada, F., Mirones, I., Garcia, M., Holguin, A., Duarte, B., Podhajcer, O., Jorcano, J.L., Larcher, F., Del Rio, M., 2006. Skin gene therapy for acquired and inherited disorders. Histol. Histopathol. 21, 1233–1247.

Clark, L.D., Clark, R.K., Heber-Katz, E., 1998. A new murine model for mammalian wound repair and regeneration. Clin. Immunol. Immunopathol. 88, 35–45.

Clore, J.N., Cohen, I.K., Diegelmann, R.F., 1979. Quantitation of collagen types I and III during wound healing in rat skin. Proc. Soc. Exp. Biol. Med. 161, 337–340.

Coquette, A., Berna, N., Vandenbosch, A., Rosdy, M., De Wever, B., Poumay, Y., 2003. Analysis of interleukin-1alpha (IL-1alpha) and interleukin-8 (IL-8) expression and release in in vitro reconstructed human epidermis for the prediction of in vivo skin irritation and/or sensitization. Toxicol. In Vitro 17, 311–321.

Cumming, B.D., McElwain, D.L., Upton, Z., 2010. A mathematical model of wound healing and subsequent scarring. J. R. Soc. Interface 7, 19–34.

Danielsen, P.L., Lerche, C.M., Wulf, H.C., Jorgensen, L.M., Liedberg, A.-S., Hansson, C., Ågren, M.S., 2016. Acute ultraviolet radiation perturbs epithelialization but not biomechanical strength of full-thickness cutaneous wounds. Photochem. Photobiol. 92, 187–192.

Davis, S.C., Mertz, P.M., Eaglstein, W.H., 1990. Second-degree burn healing: the effect of occlusive dressings and a cream. J. Surg. Res. 48, 245–248.

De Corte, P., Verween, G., Verbeken, G., Rose, T., Jennes, S., De Coninck, A., Roseeuw, D., Vanderkelen, A., Kets, E., Haddow, D., Pirnay, J.P., 2012. Feeder layer- and animal product-free culture of neonatal foreskin keratinocytes: improved performance, usability, quality and safety. Cell Tissue Bank. 13, 175–189.

Debeer, S., Le Luduec, J.B., Kaiserlian, D., Laurent, P., Nicolas, J.F., Dubois, B., Kanitakis, J., 2013. Comparative histology and immunohistochemistry of porcine versus human skin. Eur. J. Dermatol. 23, 456–466.

Demidova-Rice, T.N., Geevarghese, A., Herman, I.M., 2011. Bioactive peptides derived from vascular endothelial cell extracellular matrices promote microvascular morphogenesis and wound healing in vitro. Wound Repair Regen. 19, 59–70.

Doig, K.D., Holt, K.E., Fyfe, J.A., Lavender, C.J., Eddyani, M., Portaels, F., Yeboah-Manu, D., Pluschke, G., Seemann, T., Stinear, T.P., 2012. On the origin of *Mycobacterium ulcerans*, the causative agent of Buruli ulcer. BMC Genomics 13, 258.

Dorsett-Martin, W.A., 2004. Rat models of skin wound healing: a review. Wound Repair Regen. 12, 591–599.

Dunkin, C.S., Pleat, J.M., Gillespie, P.H., Tyler, M.P., Roberts, A.H., McGrouther, D.A., 2007. Scarring occurs at a critical depth of skin injury: precise measurement in a graduated dermal scratch in human volunteers. Plast. Reconstr. Surg. 119, 1722–1732.

Ehrlich, H.P., Needle, A.L., 1983. Wound healing in tight-skin mice: delayed closure of excised wounds. Plast. Reconstr. Surg. 72, 190–198.

Eldardiri, M., Martin, Y., Roxburgh, J., Lawrence-Watt, D.J., Sharpe, J.R., 2012. Wound contraction is significantly reduced by the use of microcarriers to deliver keratinocytes and fibroblasts in an in vivo pig model of wound repair and regeneration. Tissue Eng. Part A 18, 587–597.

Emanuelsson, P., Kratz, G., 1997. Characterization of a new in vitro burn wound model. Burns 23, 32–36.

Eming, S.A., Werner, S., Bugnon, P., Wickenhauser, C., Siewe, L., Utermohlen, O., Davidson, J.M., Krieg, T., Roers, A., 2007. Accelerated wound closure in mice deficient for interleukin-10. Am. J. Pathol. 170, 188–202.

Engler, A., Bacakova, L., Newman, C., Hategan, A., Griffin, M., Discher, D., 2004. Substrate compliance versus ligand density in cell on gel responses. Biophys. J. 86, 617–628.

Escamez, M.J., Garcia, M., Larcher, F., Meana, A., Munoz, E., Jorcano, J.L., Del Rio, M., 2004. An in vivo model of wound healing in genetically modified skin-humanized mice. J. Invest. Dermatol. 123, 1182–1191.

Eskes, A.M., Brölmann, F.E., Sumpio, B.E., Mayer, D., Moore, Z., Ågren, M.S., Hermans, M., Cutting, K., Legemate, D.A., Ubbink, D.T., Vermeulen, H., 2012. Fundamentals of randomized clinical trials in wound care: design and conduct. Wound Repair Regen. 20, 449–455.

Evans, N.D., Oreffo, R.O., Healy, E., Thurner, P.J., Man, Y.H., 2013. Epithelial mechanobiology, skin wound healing, and the stem cell niche. J. Mech. Behav. Biomed. Mater. 28, 397–409.

Falanga, V., Isaacs, C., Paquette, D., Downing, G., Kouttab, N., Butmarc, J., Badiavas, E., Hardin-Young, J., 2002. Wounding of bioengineered skin: cellular and molecular aspects after injury. J. Invest. Dermatol. 119, 653–660.

Feng, X., Tonnesen, M.G., Mousa, S.A., Clark, R.A., 2013. Fibrin and collagen differentially but synergistically regulate sprout angiogenesis of human dermal microvascular endothelial cells in 3-dimensional matrix. Int. J. Cell Biol. 2013, 231279.

Ferguson, M.W., O'Kane, S., 2004. Scar-free healing: from embryonic mechanisms to adult therapeutic intervention. Philos. Trans. R. Soc. Lond. B Biol. Sci. 359, 839–850.

Ferraq, Y., Black, D.R., Theunis, J., Mordon, S., 2012. Superficial wounding model for epidermal barrier repair studies: comparison of Erbium:YAG laser and the suction blister method. Lasers Surg. Med. 44, 525–532.

Festing, M.F., May, D., Connors, T.A., Lovell, D., Sparrow, S., 1978. An athymic nude mutation in the rat. Nature 274, 365–366.

Forsberg, S., Saarialho-Kere, U., Rollman, O., 2006. Comparison of growth-inhibitory agents by fluorescence imaging of human skin re-epithelialization in vitro. Acta Derm. Venereol. 86, 292–299.

Frankart, A., Malaisse, J., De Vuyst, E., Minner, F., de Rouvroit, C.L., Poumay, Y., 2012. Epidermal morphogenesis during progressive in vitro 3D reconstruction at the air-liquid interface. Exp. Dermatol. 21, 871–875.

Friedman, A., Xue, C., 2011. A mathematical model for chronic wounds. Math. Biosci. Eng. 8, 253–261.

Fufa, D.T., Chuang, S.S., Yang, J.Y., 2014. Postburn contractures of the hand. J. Hand Surg. Am. 39, 1869–1876.

Gai, W., Schott-Ohly, P., Schulte im, W.S., Gleichmann, H., 2004. Differential target molecules for toxicity induced by streptozotocin and alloxan in pancreatic islets of mice in vitro. Exp. Clin. Endocrinol. Diabetes 112, 29–37.

Galiano, R.D., Michaels, J., Dobryansky, M., Levine, J.P., Gurtner, G.C., 2004. Quantitative and reproducible murine model of excisional wound healing. Wound Repair Regen. 12, 485–492.

Gao, Y., Wang, X., Chen, S., Li, S., Liu, X., 2013. Acute skin barrier disruption with repeated tape stripping: an in vivo model for damage skin barrier. Skin Res. Technol. 19, 162–168.

Gauglitz, G.G., Herndon, D.N., Kulp, G.A., Meyer III, W.J., Jeschke, M.G., 2009. Abnormal insulin sensitivity persists up to three years in pediatric patients post-burn. J. Clin. Endocrinol. Metab. 94, 1656–1664.

Gawronska-Kozak, B., Grabowska, A., Kopcewicz, M., Kur, A., 2014. Animal models of skin regeneration. Reprod. Biol. 14, 61–67.

Ghatnekar, G.S., Grek, C.L., Armstrong, D.G., Desai, S.C., Gourdie, R.G., 2014. The effect of a connexin43-based peptide on the healing of chronic venous leg ulcers: a multicenter, randomized trial. J. Invest. Dermatol. 135, 289–298.

Ghatnekar, G.S., O'Quinn, M.P., Jourdan, L.J., Gurjarpadhye, A.A., Draughn, R.L., Gourdie, R.G., 2009. Connexin43 carboxyl-terminal peptides reduce scar progenitor and promote regenerative healing following skin wounding. Regen. Med. 4, 205–223.

Goulet, F., Poitras, A., Rouabhia, M., Cusson, D., Germain, L., Auger, F.A., 1996. Stimulation of human keratinocyte proliferation through growth factor exchanges with dermal fibroblasts in vitro. Burns 22, 107–112.

Gourevitch, D., Kossenkov, A.V., Zhang, Y., Clark, L., Chang, C., Showe, L.C., Heber-Katz, E., 2014. Inflammation and its correlates in regenerative wound healing: an alternate perspective. Adv. Wound Care (New Rochelle) 3, 592–603.

Grek, C.L., Prasad, G.M., Viswanathan, V., Armstrong, D.G., Gourdie, R.G., Ghatnekar, G.S., 2015. Topical administration of a connexin43-based peptide augments healing of chronic neuropathic diabetic foot ulcers: a multicenter, randomized trial. Wound Repair Regen. 23, 203–212.

Gurtner, G.C., Dauskardt, R.H., Wong, V.W., Bhatt, K.A., Wu, K., Vial, I.N., Padois, K., Korman, J.M., Longaker, M.T., 2011. Improving cutaneous scar formation by controlling the mechanical environment: large animal and phase I studies. Ann. Surg. 254, 217–225.

Gustafson, C.J., Birgisson, A., Junker, J., Huss, F., Salemark, L., Johnson, H., Kratz, G., 2007. Employing human keratinocytes cultured on macroporous gelatin spheres to treat full thickness-wounds: an in vivo study on athymic rats. Burns 33, 726–735.

Harris, K.L., Bainbridge, N.J., Jordan, N.R., Sharpe, J.R., 2009. The effect of topical analgesics on ex vivo skin growth and human keratinocyte and fibroblast behavior. Wound Repair Regen. 17, 340–346.

Hochman, B., Vilas Boas, F.C., Mariano, M., Ferreiras, L.M., 2005. Keloid heterograft in the hamster (*Mesocricetus auratus*) cheek pouch, Brazil. Acta Cir. Bras. 20, 200–212.

James, S.E., Booth, S., Dheansa, B., Mann, D.J., Reid, M.J., Shevchenko, R.V., Gilbert, P.M., 2010. Sprayed cultured autologous keratinocytes used alone or in combination with meshed autografts to accelerate wound closure in difficult-to-heal burns patients. Burns 36, e10–e20.

Kamolz, L.P., Kitzinger, H.B., Karle, B., Frey, M., 2009. The treatment of hand burns. Burns 35, 327–337.

Kieran, I., Knock, A., Bush, J., So, K., Metcalfe, A., Hobson, R., Mason, T., O'Kane, S., Ferguson, M., 2013. Interleukin-10 reduces scar formation in both animal and human cutaneous wounds: results of two preclinical and phase II randomized control studies. Wound Repair Regen. 21, 428–436.

Kieran, I., Taylor, C., Bush, J., Rance, M., So, K., Boanas, A., Metcalfe, A., Hobson, R., Goldspink, N., Hutchison, J., Ferguson, M., 2014. Effects of interleukin-10 on cutaneous wounds and scars in humans of African continental ancestral origin. Wound Repair Regen. 22, 326–333.

Kleesz, P., Darlenski, R., Fluhr, J.W., 2012. Full-body skin mapping for six biophysical parameters: baseline values at 16 anatomical sites in 125 human subjects. Skin Pharmacol. Physiol. 25, 25–33.

Koivukangas, V., Oikarinen, A., 2003. Suction blister model of wound healing. Methods Mol. Med. 78, 255–261.

Koskela, M., Gaddnas, F., Ala-Kokko, T.I., Laurila, J.J., Saarnio, J., Oikarinen, A., Koivukangas, V., 2009. Epidermal wound healing in severe sepsis and septic shock in humans. Crit. Care 13, R100.

Kratz, G., 1998. Modeling of wound healing processes in human skin using tissue culture. Microsc. Res. Tech. 42, 345–350.

Kratz, G., Lake, M., Gidlund, M., 1994. Insulin like growth factor-1 and -2 and their role in the re-epithelialisation of wounds; interactions with insulin like growth factor binding protein type 1. Scand. J. Plast. Reconstr. Surg. Hand Surg. 28, 107–112.

Kumar, A., Brockes, J.P., 2012. Nerve dependence in tissue, organ, and appendage regeneration. Trends Neurosci. 35, 691–699.

Li, W., Fan, J., Chen, M., Guan, S., Sawcer, D., Bokoch, G.M., Woodley, D.T., 2004a. Mechanism of human dermal fibroblast migration driven by type I collagen and platelet-derived growth factor-BB. Mol. Biol. Cell 15, 294–309.

Li, W., Henry, G., Fan, J., Bandyopadhyay, B., Pang, K., Garner, W., Chen, M., Woodley, D.T., 2004b. Signals that initiate, augment, and provide directionality for human keratinocyte motility. J. Invest. Dermatol. 123, 622–633.

Long, K.B., Burgwin, C.M., Huneke, R., Artlett, C.M., Blankenhorn, E.P., 2014. Tight skin 2 mice exhibit delayed wound healing caused by increased elastic fibers in fibrotic skin. Adv. Wound Care (New Rochelle) 3, 573–581.

Loo, A.E., Halliwell, B., 2012. Effects of hydrogen peroxide in a keratinocyte-fibroblast co-culture model of wound healing. Biochem. Biophys. Res. Commun. 423, 253–258.

Loo, A.E., Ho, R., Halliwell, B., 2011. Mechanism of hydrogen peroxide-induced keratinocyte migration in a scratch-wound model. Free Radic. Biol. Med. 51, 884–892.

McFarlane, R.M., Deyoung, G., Henry, R.A., 1965. The design of a pedicle flap in the rat to study necrosis and its prevention. Plast. Reconstr. Surg. 35, 177–182.

Menon, S.N., Flegg, J.A., McCue, S.W., Schugart, R.C., Dawson, R.A., McElwain, D.L., 2012. Modelling the interaction of keratinocytes and fibroblasts during normal and abnormal wound healing processes. Proc. Biol. Sci. 279, 3329–3338.

Metcalfe, A.D., Ferguson, M.W., 2007. Bioengineering skin using mechanisms of regeneration and repair. Biomaterials 28, 5100–5113.

Middelkoop, E., van den Bogaerdt, A.J., Lamme, E.N., Hoekstra, M.J., Brandsma, K., Ulrich, M.M., 2004. Porcine wound models for skin substitution and burn treatment. Biomaterials 25, 1559–1567.

Mogford, J., Mustoe, T.A., 2001. Experimental models of wound healing. In: Falanga, V. (Ed.), Cutaneous Wound Healing. Martin Dunitz Ltd, London, pp. 109–122.

Morris, D.E., Wu, L., Zhao, L.L., Bolton, L., Roth, S.I., Ladin, D.A., Mustoe, T.A., 1997. Acute and chronic animal models for excessive dermal scarring: quantitative studies. Plast. Reconstr. Surg. 100, 674–681.

Muneoka, K., Allan, C.H., Yang, X., Lee, J., Han, M., 2008. Mammalian regeneration and regenerative medicine. Birth Defects Res. C, Embryo Today 84, 265–280.

Murphy, K.E., McCue, S.W., McElwain, D.L., 2012. Clinical strategies for the alleviation of contractures from a predictive mathematical model of dermal repair. Wound Repair Regen. 20, 194–202.

Olerud, J.E., Odland, G.F., Burgess, E.M., Wyss, C.R., Fisher, L.D., Matsen III, F.A., 1995. A model for the study of wounds in normal elderly adults and patients with peripheral vascular disease or diabetes mellitus. J. Surg. Res. 59, 349–360.

Ongstad, E.L., O'Quinn, M.P., Ghatnekar, G.S., Yost, M.J., Gourdie, R.G., 2013. A connexin43 mimetic peptide promotes regenerative healing and improves mechanical properties in skin and heart. Adv. Wound Care (New Rochelle) 2, 55–62.

Osborn, O., Sanchez-Alavez, M., Brownell, S.E., Ross, B., Klaus, J., Dubins, J., Beutler, B., Conti, B., Bartfai, T., 2010. Metabolic characterization of a mouse deficient in all known leptin receptor isoforms. Cell Mol. Neurobiol. 30, 23–33.

Paek, R., Chang, D.S., Brevetti, L.S., Rollins, M.D., Brady, S., Ursell, P.C., Hunt, T.K., Sarkar, R., Messina, L.M., 2002. Correlation of a simple direct measurement of muscle pO(2) to a clinical ischemia index and histology in a rat model of chronic severe hindlimb ischemia. J. Vasc. Surg. 36, 172–179.

Pastar, I., Khan, A.A., Stojadinovic, O., Lebrun, E.A., Medina, M.C., Brem, H., Kirsner, R.S., Jimenez, J.J., Leslie, C., Tomic-Canic, M., 2012. Induction of specific microRNAs inhibits cutaneous wound healing. J. Biol. Chem. 287, 29324–29335.

Pontiggia, L., Klar, A., Bottcher-Haberzeth, S., Biedermann, T., Meuli, M., Reichmann, E., 2013. Optimizing in vitro culture conditions leads to a significantly shorter production time of human dermo-epidermal skin substitutes. Pediatr. Surg. Int. 29, 249–256.

Potter, M.J., Linge, C., Cussons, P., Dye, J.F., Sanders, R., 2006. An investigation to optimize angiogenesis within potential dermal replacements. Plast. Reconstr. Surg. 117, 1876–1885.

Poumay, Y., Coquette, A., 2007. Modelling the human epidermis in vitro: tools for basic and applied research. Arch. Dermatol. Res. 298, 361–369.

Prunieras, M., Regnier, M., Woodley, D., 1983. Methods for cultivation of keratinocytes with an air-liquid interface. J. Invest. Dermatol. 81, 28s–33s.

Pullar, C.E., Chen, J., Isseroff, R.R., 2003. PP2A activation by beta2-adrenergic receptor agonists: novel regulatory mechanism of keratinocyte migration. J. Biol. Chem. 278, 22555–22562.

Pullar, C.E., Grahn, J.C., Liu, W., Isseroff, R.R., 2006. Beta2-adrenergic receptor activation delays wound healing. FASEB J. 20, 76–86.

Rittié, L., Sachs, D.L., Orringer, J.S., Voorhees, J.J., Fisher, G.J., 2013. Eccrine sweat glands are major contributors to reepithelialization of human wounds. Am. J. Pathol. 182, 163–171.

Rizzo, A.E., Beckett, L.A., Baier, B.S., Isseroff, R.R., 2012. The linear excisional wound: an improved model for human ex vivo wound epithelialization studies. Skin Res. Technol. 18, 125–132.

Rosdy, M., Clauss, L.C., 1990. Terminal epidermal differentiation of human keratinocytes grown in chemically defined medium on inert filter substrates at the air-liquid interface. J. Invest. Dermatol. 95, 409–414.

Schäfer, M., Werner, S., 2008. Cancer as an overhealing wound: an old hypothesis revisited. Nat. Rev. Mol. Cell Biol. 9, 628–638.

Scheid, A., Meuli, M., Gassmann, M., Wenger, R.H., 2000. Genetically modified mouse models in studies on cutaneous wound healing. Exp. Physiol. 85, 687–704.

Schneider, A., Garlick, J.A., Egles, C., 2008. Self-assembling peptide nanofiber scaffolds accelerate wound healing. PLoS One 3, e1410.

Schreier, T., Degen, E., Baschong, W., 1993. Fibroblast migration and proliferation during in vitro wound healing. A quantitative comparison between various growth factors and a low molecular weight blood dialysate used in the clinic to normalize impaired wound healing. Res. Exp. Med. (Berl) 193, 195–205.

Schwarz, D.A., Lindblad, W.J., Rees, R.R., 1995. Altered collagen metabolism and delayed healing in a novel model of ischemic wounds. Wound Repair Regen. 3, 204–212.

Seifert, A.W., Maden, M., 2014. New insights into vertebrate skin regeneration. Int. Rev. Cell Mol. Biol. 310, 129–169.

Shah, M., Foreman, D.M., Ferguson, M.W., 1992. Control of scarring in adult wounds by neutralising antibody to transforming growth factor beta. Lancet 339, 213–214.

Shah, M., Foreman, D.M., Ferguson, M.W., 1994. Neutralising antibody to TGF-beta 1,2 reduces cutaneous scarring in adult rodents. J. Cell Sci. 107 (Pt 5), 1137–1157.

Sharpe, J.R., Harris, K.L., Jubin, K., Bainbridge, N.J., Jordan, N.R., 2009. The effect of pH in modulating skin cell behaviour. Br. J. Dermatol. 161, 671–673.

Shetlar, M.R., Shetlar, C.L., Hendricks, L., Kischer, C.W., 1985. The use of athymic nude mice for the study of human keloids. Proc. Soc. Exp. Biol. Med. 179, 549–552.

Shi, H.X., Lin, C., Lin, B.B., Wang, Z.G., Zhang, H.Y., Wu, F.Z., Cheng, Y., Xiang, L.J., Guo, D.J., Luo, X., Zhang, G.Y., Fu, X.B., Bellusci, S., Li, X.K., Xiao, J., 2013. The anti-scar effects of basic fibroblast growth factor on the wound repair in vitro and in vivo. PLoS One 8, e59966.

Shih, B., Bayat, A., 2010. Genetics of keloid scarring. Arch. Dermatol. Res. 302, 319–339.

Simkin, J., Han, M., Yu, L., Yan, M., Muneoka, K., 2013. The mouse digit tip: from wound healing to regeneration. Methods Mol. Biol. 1037, 419–435.

So, K., McGrouther, D.A., Bush, J.A., Durani, P., Taylor, L., Skotny, G., Mason, T., Metcalfe, A., O'Kane, S., Ferguson, M.W., 2011. Avotermin for scar improvement following scar revision surgery: a randomized, double-blind, within-patient, placebo-controlled, phase II clinical trial. Plast. Reconstr. Surg. 128, 163–172.

Sørensen, L.T., Zillmer, R., Ågren, M., Ladelund, S., Karlsmark, T., Gottrup, F., 2009. Effect of smoking, abstention, and nicotine patch on epidermal healing and collagenase in skin transudate. Wound Repair Regen. 17, 347–353.

Stojadinovic, O., Brem, H., Vouthounis, C., Lee, B., Fallon, J., Stallcup, M., Merchant, A., Galiano, R.D., Tomic-Canic, M., 2005. Molecular pathogenesis of chronic wounds: the role of beta-catenin and c-myc in the inhibition of epithelialization and wound healing. Am. J. Pathol. 167, 59–69.

Sullivan, T.P., Eaglstein, W.H., Davis, S.C., Mertz, P., 2001. The pig as a model for human wound healing. Wound Repair Regen. 9, 66–76.

Swindle, M.M., Makin, A., Herron, A.J., Clubb Jr., F.J., Frazier, K.S., 2012. Swine as models in biomedical research and toxicology testing. Vet. Pathol. 49, 344–356.

Tengrup, I., Hallmans, G., Ågren, M.S., 1988. Granulation tissue formation and metabolism of zinc and copper in alloxan-diabetic rats. Scand. J. Plast. Reconstr. Surg. Hand Surg. 22, 41–45.

Tomic-Canic, M., Mamber, S.W., Stojadinovic, O., Lee, B., Radoja, N., McMichael, J., 2007. Streptolysin O enhances keratinocyte migration and proliferation and promotes skin organ culture wound healing in vitro. Wound Repair Regen. 15, 71–79.

Turchi, L., Chassot, A.A., Rezzonico, R., Yeow, K., Loubat, A., Ferrua, B., Lenegrate, G., Ortonne, J.P., Ponzio, G., 2002. Dynamic characterization of the molecular events during in vitro epidermal wound healing. J. Invest. Dermatol. 119, 56–63.

van Kilsdonk, J.W., van den Bogaard, E.H., Jansen, P.A., Bos, C., Bergers, M., Schalkwijk, J., 2013. An in vitro wound healing model for evaluation of dermal substitutes. Wound Repair Regen. 21, 890–896.

Varmeh, S., Egia, A., McGrouther, D., Tahan, S.R., Bayat, A., Pandolfi, P.P., 2011. Cellular senescence as a possible mechanism for halting progression of keloid lesions. Genes Cancer 2, 1061–1066.

Varol, A.L., Anderson, C.D., 2010. A minimally invasive human in vivo cutaneous wound model for the evaluation of innate skin reactivity and healing status. Arch. Dermatol. Res. 302, 383–393.

Velander, P., Theopold, C., Hirsch, T., Bleiziffer, O., Zuhaili, B., Fossum, M., Hoeller, D., Gheerardyn, R., Chen, M., Visovatti, S., Svensson, H., Yao, F., Eriksson, E., 2008. Impaired wound healing in an acute diabetic pig model and the effects of local hyperglycemia. Wound Repair Regen. 16, 288–293.

Waldron-Lynch, F., Deng, S., Preston-Hurlburt, P., Henegariu, O., Herold, K.C., 2012. Analysis of human biologics with a mouse skin transplant model in humanized mice. Am. J. Transplant. 12, 2652–2662.

Wang, X., Ge, J., Tredget, E.E., Wu, Y., 2013. The mouse excisional wound splinting model, including applications for stem cell transplantation. Nat. Protoc. 8, 302–309.

Wong, V.W., Beasley, B., Zepeda, J., Dauskardt, R.H., Yock, P.G., Longaker, M.T., Gurtner, G.C., 2013. A mechanomodulatory device to minimize incisional scar formation. Adv. Wound Care (New Rochelle) 2, 185–194.

Wood, F.M., Kolybaba, M.L., Allen, P., 2006. The use of cultured epithelial autograft in the treatment of major burn wounds: eleven years of clinical experience. Burns 32, 538–544.

Wray, R.C., 1983. Force required for wound closure and scar appearance. Plast. Reconstr. Surg. 72, 380–382.

Wu, Y., Wang, K., Karapetyan, A., Fernando, W.A., Simkin, J., Han, M., Rugg, E.L., Muneoka, K., 2013. Connective tissue fibroblast properties are position-dependent during mouse digit tip regeneration. PLoS One 8, e54764.

Xie, Y., Rizzi, S.C., Dawson, R., Lynam, E., Richards, S., Leavesley, D.I., Upton, Z., 2010. Development of a three-dimensional human skin equivalent wound model for investigating novel wound healing therapies. Tissue Eng. Part C, Methods 16, 1111–1123.

Xu, W., Jong, H.S., Jia, S., Zhao, Y., Galiano, R.D., Mustoe, T.A., 2012. Application of a partial-thickness human ex vivo skin culture model in cutaneous wound healing study. Lab. Invest. 92, 584–599.

Yang, X., Wang, J., Guo, S.L., Fan, K.J., Li, J., Wang, Y.L., Teng, Y., Yang, X., 2011. miR-21 promotes keratinocyte migration and re-epithelialization during wound healing. Int. J. Biol. Sci. 7, 685–690.

Zhu, K.Q., Carrougher, G.J., Gibran, N.S., Isik, F.F., Engrav, L.H., 2007. Review of the female Duroc/Yorkshire pig model of human fibroproliferative scarring. Wound Repair Regen. 15 (Suppl. 1), S32–S39.

Zhu, K.Q., Engrav, L.H., Gibran, N.S., Cole, J.K., Matsumura, H., Piepkorn, M., Isik, F.F., Carrougher, G.J., Muangman, P.M., Yunusov, M.Y., Yang, T.M., 2003. The female, red Duroc pig as an animal model of hypertrophic scarring and the potential role of the cones of skin. Burns 29, 649–664.

Part Two

Therapeutics and tissue regeneration for wound healing

Stem cell therapies for wounds

A. Abdullahi, S. Amini-Nik, M.G. Jeschke
University of Toronto, Toronto, ON, Canada

7.1 Introduction

A number of disease conditions like burns and diabetes are associated with impaired wound healing. It is estimated that 5 million people out of the 21 million people suffering from diabetes will develop complications like chronic wounds that will fail to heal (Association, 2014). With the rise in obesity and diabetes, these numbers are slated to double by 2023, putting huge strains on healthcare systems. Because of ischemia, infection, neuropathy, and impaired wound repair the incidence of amputations are also slated to increase further, reducing the quality of life. Wound therapy has moved from the aim of managing the symptoms of chronic wounds to realizing that a more proactive therapy is essential for an optimal healing process. Quite a few discoveries and breakthroughs have been made in tissue engineering (Jeschke et al., 2004) and wound care (Tsourdi et al., 2013). Essential events in the area of stem cell research have significantly influenced our view of wound repair and skin regeneration. Stem cell therapy has revolutionized our thinking in that it is now possible to envision the restoration of damaged or nonfunctional tissue back to its functional state. The idea behind stem cell therapy is that by implanting stem cells in the damaged tissue, they can provide necessary cues for regeneration and healing (Cha and Falanga, 2007). This chapter will help answer many of the questions about stem cells and their therapeutic role in injury. What source of tissue should be used? What adhesion and scaffold materials are required for stem cell transplantation? And finally, how the stem cells are delivered to injured tissues will be addressed, among other questions.

7.2 Wound healing

Following injury significant skin tissue damage occurs that results in the formation of a wound characterized by the disruption of both the original skin tissue architecture and homeostasis (Bielefeld et al., 2013; Gurtner et al., 2008; Stocum and Cameron, 2011). Upon formation of the wound, a number of extracellular and intracellular events are activated to restore tissue integrity and homeostasis. However, the extent to which these events are initiated varies and is dependent on a number of factors such as severity, nature of the injury, and size of the wound (Gurtner et al., 2008). Despite these variations due to the nature of the wound, the overall process and sequence of events involved in wound healing and repair are conserved.

Wound Healing Biomaterials - Volume 1. http://dx.doi.org/10.1016/B978-1-78242-455-0.00007-0

A highly complex and orchestrated sequence of events, wound healing is comprised of three overlapping key biochemical and cellular phases (see Fig. 7.1): (1) inflammation, (2) proliferation, and (3) remodeling (Bielefeld et al., 2013). First, platelets contribute to hemostasis and bolus of growth factors such as transforming growth factor (TGF)-β and platelet-derived growth factors. The inflammatory stage is characterized by infiltration of neutrophils, macrophages, and lymphocytes and the release of cytokines into the wound bed (Amini-Nik et al., 2014; Bielefeld et al., 2013; Koh and Dipietro, 2011; Singer and Clark, 1999). This phase of the wound healing cascade begins within minutes and lasts for days. The recruitment of these immune cells is pivotal for adequate wound closure as they phagocytoze dead cells as well as pathogens (Koh and Dipietro, 2011; Singer and Clark, 1999). In addition to clearing out pathogens, these cells also secrete important cytokines and growth factors such as tumor necrosis factor (TNF)-α, interleukin (IL)-1, and fibroblast growth factor (FGF) that promote the chemotactic migration and proliferation of fibroblasts, keratinocytes, and stem cells to the wound (Gurtner et al., 2008).The secretion of these chemical mediators and arrival of these cells initiate the proliferation phase of the wound healing cascade and provide a niche that orchestrates the outcome of healing.

Hallmark events of the proliferation phase, which starts about 2 days following injury, include angiogenesis, formation of new granulation tissue, and epithelialization (Singer and Clark, 1999). Upon the onset of the proliferation phase, fibroblasts begin to migrate into the wound bed and proliferate. Fibroblasts subsequently initiate new fibronectin and collagen deposition, forming a temporary matrix for epithelial cells to migrate on (Bielefeld et al., 2013; Gurtner et al., 2008). A number of elements are also released during this phase, such as the matrix metalloproteinases (MMPs), which help to degrade the basement membrane to facilitate and clear a path for keratinocyte migration on the wound bed (Martin, 1997). These key cells assist in reestablishing a new functional epidermal layer. A number of growth factors are also released during this phase such as keratinocyte growth factor, TGF-α, TGF-β, epidermal growth factor (EGF), and heparin-binding EGF, which regulate the proliferation and motility of keratinocytes (Bielefeld et al., 2013; Gurtner et al., 2008; Werner and Grose, 2003). The release of these chemical mediators along with FGF-2 and vascular endothelial growth factor also help to set the stage for angiogenesis, which is the formation of new capillaries and small blood vessels (Werner and Grose, 2003), possibly initiated by Tie2 lineage cells (Sarkar et al., 2012). This signals the end of the proliferative stage, and the healing process is transitioned into the final phase of the wound healing cascade.

The remodeling phase lasts from weeks to years and is characterized by the reorganization of collagen deposited earlier. Although initially dominated by type III collagen, as the wound bed matures it is replaced with the more organized type I collagen in cutaneous wounds (Gurtner et al., 2008). This replacement of collagen types is mediated primarily via the proteolytic activity of the MMPs (Armstrong and Jude, 2002). This reorganization of collagen confers greater tensile strength in the new scar tissue that is formed; however, the original tensile strength of the uninjured skin is never fully regained (Singer and Clark, 1999). Moreover, the involvement of skeletal muscle stem cells (Pax7 + cells) in the healing of rodents' skin and their fate change to myofibroblastic-like

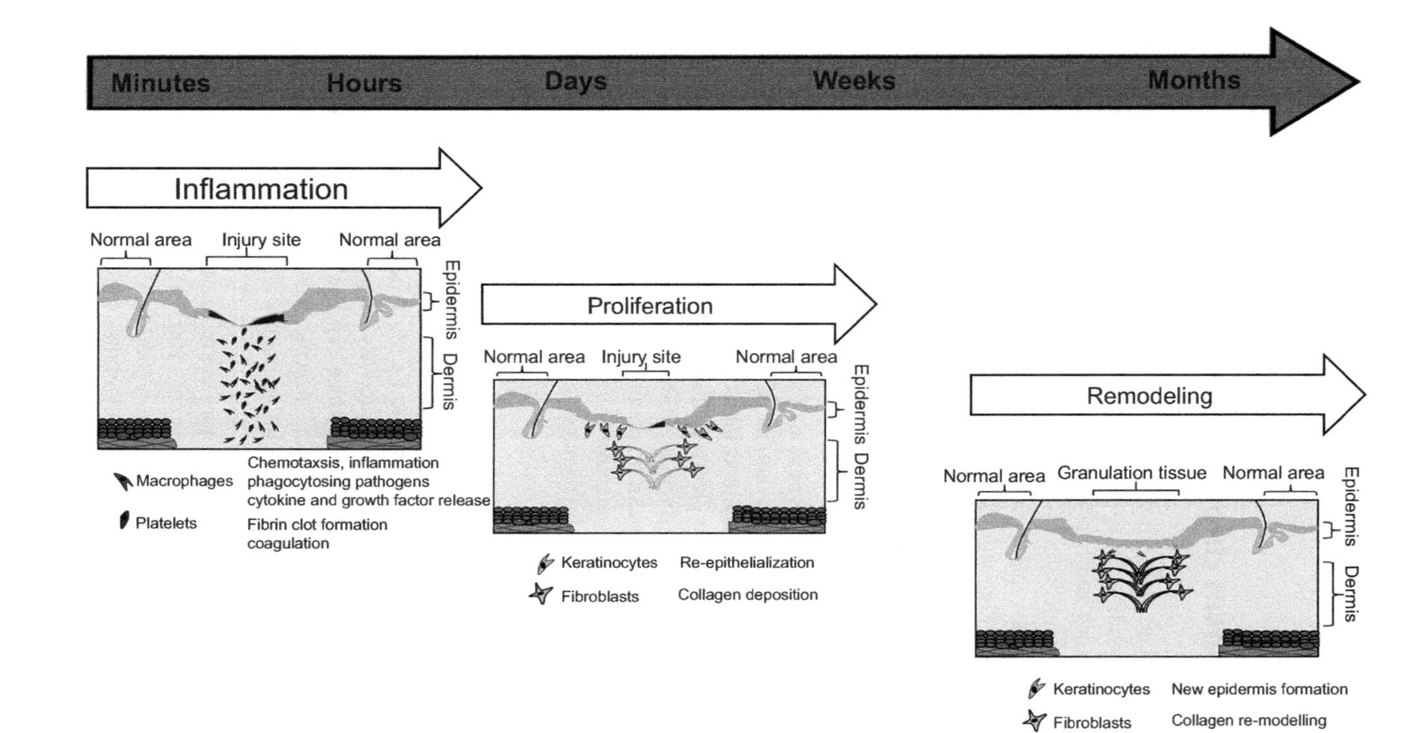

Figure 7.1 Schematic representation of the three phases of wound healing postinjury. The figure illustrates an excisional full-thickness biopsy wound in rodent skin and highlights the three distinct yet overlapping stages of skin healing.

cells may provide another source of strength (Amini-Nik et al., 2011). Thus if the healing cascade follows the aforementioned phases and occurs at the precise allocated times the wound heals without difficulty; however, if this fails, the healing process becomes compromised, shifting an acute wound into a chronic wound that fails to resolve.

7.3 Acute versus chronic wounds

In healthy individuals with no underlying conditions, an acute wound should heal within weeks, depending on age. However, if the wound does not follow the normal healing trajectory and becomes fixed in one of the wound healing phases, the wound becomes chronic. Chronic wounds are thus defined as wounds that have failed to proceed through the repair process in an orderly and timely process to reestablish a sustained anatomic and functional barrier (Lazarus et al., 1994). In particular, chronic wounds are thought to regress in the inflammatory phase, unable to progress through the wound healing process (Hunt et al., 2000; Lazarus et al., 1994). Therefore an understanding of the causative factors that facilitate chronic wounds is imperative to help guide caregivers in selecting the appropriate care for those suffering from this condition.

The pathophysiological transformation of an acute wound to a chronic wound has not been fully uncovered. However, a number of local and systemic factors like increased free radical production, ischemia, and infections have all been implicated in altering or disrupting one or more of the phases of the wound healing process and in the development of chronic wounds (Ågren et al, 2000; Mast and Schultz, 1996). For instance, defective cytokine and growth factor production, impaired cellular infiltration, enhanced expression of proteinases, and defective extracellular matrix (ECM) components have all been found in chronic wounds (Ågren et al., 2000; Mast and Schultz, 1996).

7.3.1 Defective cytokine/growth factor production

Macrophages and fibroblasts are recruited to the wound site during healing for cytokine production and the formation of the ECM. When it comes to differentiating between acute and chronic wounds, the activity as well as the level of production of cytokines/growth factors that they produce can help to identify if the wound has become a chronic wound. For instance, in acute wound healing there is a precise balance between the production of proinflammatory and antiinflammatory cytokines in the wound bed (Mast and Schultz, 1996). However, in chronic wounds there is an exaggerated production of proinflammatory cytokines, which have devastating effects on the wound bed environment (Lerman et al., 2003; Mast and Schultz, 1996; Wetzler et al., 2000). In fact, it has been show that wound fluid derived from chronic wounds shows an enrichment of proinflammatory cytokines such as TNF-α and IL-1β (Zykova et al., 2000). The elevated levels of these proinflammatory cytokines are detrimental for wound closure as they inhibit growth and alter the morphology of fibroblasts (Werner and Grose, 2003). In contrast, other important inflammatory cytokines like

IL-6 that help to recruit fibroblasts and keratinocytes to the wound bed are significantly impaired (Werner and Grose, 2003).

7.3.2 Abnormal cell activity and impaired tissue remodeling

In conjunction with impaired production of necessary chemokines and proteins, mainly collagen, which are vital for optimal wound closure, macrophages and fibroblasts also show a reduced capacity to migrate to the wound bed in chronic wounds (Wetzler et al., 2000; Lerman et al., 2003). Impaired fibroblast proliferation in diabetic patients has been shown to result in reduced levels of collagen deposition, thereby affecting tissue remodeling and ECM integrity (Boulton, 2004; Lerman et al., 2003; Loots et al., 1998; Seibold et al., 1985). To further complicate matters, chronic wounds show elevated levels of MMPs, which degrade the already reduced collagen depot in the wound bed (Boulton, 2004; Seibold et al., 1985). Furthermore, the environment of chronic wounds is rich with reactive oxygen species (ROS) that is released by infiltrating neutrophils and macrophages (Moseley et al., 2004; Wlaschek and Scharffetter-Kochanek, 2005). These free radicals have devastating effects as they can damage cell membranes, blood vessels supplying the wound bed, and ECM proteins (Moseley et al., 2004; Wenk et al., 1999). Thus the combination of these factors leads to a decrease in proliferation, inadequate vascularization, and the accumulation of necrotic tissue due to ischemia.

Unlike deficient healing, in conditions like excessive scars, aberrations of physiologic wound healing lead to the formation of keloids or hypertrophic scars. As explained, the transformation of a wound clot into granulation tissue requires a delicate balance between ECM deposition and degradation. In the case of excessive healing, this balance favors increased fibroblast activity with greater and more sustained ECM deposition. Current evidences suggest that a more prolonged inflammatory period, with excessive infiltration of immune cells, may contribute to excessive healing (Amini-Nik et al., 2014). In fact, it has been shown that in hypertrophic scars the number of inflammatory cells, such as macrophages, is correlated with cellularity of the tissue and with the levels of the protein beta-catenin. Beta-catenin is a protein that is involved in the proliferation phase of wound healing, and excessive beta-catenin expression leads to excessive healing, eg, hypertrophic scars (Cheon et al., 2002). As such, it is suggested that the association between the number of inflammatory cells and beta-catenin levels contributes to the development of hypertrophic scars.

7.4 Burns

Burn injury represents one of the most debilitating forms of trauma and constitutes a major global public health problem. Burns rank among the fourth most common types of trauma worldwide, following traffic accidents, falls, and interpersonal violence (Peck, 2012). Difficulty in creating clinically effective treatment options stems partly from the fact that burn injury is multifaceted. Besides the acute effects on skin, the hypermetabolic response postthermal injury can result in sustained complications including inflammation, hyperglycemia, and breakdown in blood supply to vital organs, all of which may result in impaired wound healing (Jeschke et al., 2008). While most other organ functions, like the liver and adipose tissue, gain some type of normalcy as early as within months, the functional disability from impaired wound closure can persist even years after the initial injury (Branski et al., 2009a; Teot et al., 2012). Although advances in acute burn management have increased survivability following initial injury, adequate wound healing is still critically important for long-term survival and a good qualitative outcome (Branski et al., 2009a; Pruitt et al., 1968). For instance, delayed wound closure has been shown to increase the incidence of wound contamination, invasive infection, and sepsis, a leading cause of death in burn patients (Branski et al., 2009a; Pruitt et al., 1968). Additionally, inadequate wound closure significantly increases the risk factor for hypertrophic scarring (Gauglitz et al., 2011). As such, the quality of life and functional outcome for burn survivors depend on the quality of wound healing and degree of scarring.

Perhaps the most devastating aspect is that burns destroy the epidermal progenitor cell pool within the hair follicles, leaving very few viable stem cells that can regenerate the burned skin tissue (Branski et al., 2009b). Furthermore, tissue regeneration and angiogenesis are further hampered by the heightened circulating levels of inflammatory cytokines and ROS that characterize severe burns (Jeschke et al., 2011). It is thus imperative for all burn caregivers to become familiar with the pathophysiological alterations to skin wound healing and the management and treatment approaches to mitigate these pathological alterations.

7.5 Current treatments of burn wounds and chronic wounds

Successful wound care involves optimizing the patient's local and systemic factors in conjunction with an ideal wound healing environment. Initially, following a burn injury a series of standardized care is initiated that involves debridement, sufficient application of dressings, and frequent dressing changes to prevent infections (Jeschke and Herndon, 2014). Many different strategies have been adopted for accelerating wound repair across burn units around the world such as skin grafting, dermal substitutes, and revolutionary synthetic skin constructs. However, autografts from uninjured skin remain the gold standard in burns (Dziewulski et al., 2012; Jeschke et al., 2004). Functionally, autografts from uninjured skin of the patient provide the most suitable environment as they not only provide a viable biochemical environment for host progenitor cells, but also antigenicity and cross-infection are nearly nonexistent, important constituents necessary for healing (Dziewulski et al., 2012). While autografts have been successful in burn patients with 30% or less total body surface area (TBSA) burns, the procedure has become problematic in patients that exceed 40% TBSA full-thickness burns as it becomes very invasive while yielding a small sample volume (Dziewulski et al., 2012).

Accordingly, clinicians have turned to allografts (uninjured skin from a donor or from cadavers) as temporary skin in burn wound coverage (Horch et al., 2005). Although these grafts provide a biologically active dermal matrix to the injured wound bed, these alternatives also have a few drawbacks such as immunological rejection by the host and increased susceptibility to cross-infections (Dziewulski et al., 2012).

The advent of bioengineering has brought about the discovery of a novel class of dermal substitutes that help in wound closure in those cases where traditional methods fail or are unavailable (Jones et al., 2002). Integra® (Integra LifeSciences, Plainsboro, New Jersey, USA), an acellular bilaminar structure that facilitates wound closure with permanent dermal replacement, is widely used (Branski et al., 2007; Dziewulski et al., 2012; Jeschke et al., 2004). The silicone layers act as a temporary epidermis to prevent fluid loss and a barrier to pathogens (Dziewulski et al., 2012). The bottom layer of Integra® is composed of a cross-linked matrix of bovine collagen and chondroitin-6-sulphate, which is crucial in preventing hypertrophic scarring by acting as a temporary dermis (Branski et al., 2007; Jeschke et al., 2004; Yannas and Burke, 1980). As the wound progresses to heal, the matrix becomes vascularized and infiltrated by host cells, eventually formulating a new dermis in the wound bed. A number of clinical trials conducted on the efficacy of Integra® in wound closure have shown promise, specifically in its ability to reduce hypertrophic scarring (Branski et al., 2007; Jeschke et al., 2004). Nevertheless, more studies are needed to fully elucidate the precise mechanisms it exerts in the wound. Unfortunately, with all the promise that dermal substitutes like Integra® have to offer, they are not ideal in the clinic as they lack to recapitulate some of the key structural properties of human skin, lack any cellular component, and are too expensive to use in large quantities. Regardless of the strategy adopted in accelerating wound repair postinjury, when choosing a method it is critical to adopt a standardized protocol that is simple, safe, and causes minimal disruption to the wound bed to avoid further complications.

Diabetes mellitus is another disease condition characterized by chronic wounds and inadequate wound closure that affect patient outcome. The "diabetic foot" is a term coined to describe foot ulcers, a common problem resulting from diabetes (Boulton, 2004; Singh et al., 2005). These foot ulcers result in a significant epidermis loss and, depending on the severity, can extend into the dermis and deeper tissue layers like muscle and bone (Boulton, 2004). With the rising rates of diabetes and obesity, foot ulcers are becoming a major health problem with significant economic costs (Tsourdi et al., 2013). The standard management for diabetic foot syndrome includes a series of treatments: debridement, sufficient application of dressings, frequent dressing changes, blood glucose control, and pressure off-loading (Tsourdi et al., 2013). The initiation of these treatment measures is to clean or remove the infected tissue as well as to keep a moist wound environment with adequate blood supply to stimulate healing (Tsourdi et al., 2013). If the diabetic foot becomes necrotic or infected, patients may need surgery to amputate the leg in order to prevent systemic infection (Margolis et al., 2005). New treatment options for diabetic patients who do not respond to conventional methods have been suggested (Tsourdi et al., 2013). One of those promising therapeutic options involves the use of stem cell therapy that can help stimulate wound closure (Dabiri et al., 2013). It is speculated that stem cell therapy will help replace damaged cells as well as stimulate other cells to secrete factors important in wound repair (Gauglitz and Jeschke, 2011). For instance, Badiavas and Falanga (2003) have shown that the application of bone marrow-derived stem cells resulted in complete wound closure in patients who were unresponsive to standard treatment, bioengineered skin, and autologous skin grafting. Thus stem cell therapy in conditions like burns and diabetes represents an exciting new treatment potential that is sure to rise in popularity in the coming decades.

7.6 Stem cell therapy and sources of stem cells

The area of stem cell therapy holds great promise for many diseases as it aims to restore damaged or nonfunctional tissue back to its functional state. This is accomplished by implanting specific types of stem cells in the damaged tissue that can provide necessary cues for regeneration and healing (Bartholomew et al., 2002; Dabiri et al., 2013). Stroke, traumatic brain injury, Alzheimer's disease, burns, and diabetes are potential stem cell therapy indications (Badiavas and Falanga, 2003; Bjorklund, 2005; Longhi et al., 2005; Loughran et al., 2013). Here we review progress in the field of skin regeneration and the use of stem cell therapy in wound healing (refer to Tables 7.1 and 7.2).

Although dermal substitutes offer excellent physical support to the wound bed they do not possess any biological activity that can accelerate skin regeneration. A pioneering strategy in skin regeneration is to improve the biological activity of these dermal substitutes by incorporating stem cells and other critical mediators of wound healing. The idea behind this concept is that these progenitor cells may facilitate recovery by providing trophic support to repair cells, counteract the damaging effects within the wound bed, and replace injured and lost cells due to the injury. Efforts to expand on the biological activity of skin substitutes have already begun with the incorporation of various stem cells and proteins critical in wound closure. But before we delve into the technical aspects of how this is being accomplished, a brief discussion on stem cells and the source of these stem cells used in skin regeneration is warranted.

Stem cells are group of primitive, undifferentiated cells capable of transforming into all types of cells in the human body (Cha and Falanga, 2007; Hu et al., 2014). These cells offer a renewable source that can be exploited for such therapies as skin and tissue regeneration (Cha and Falanga, 2007; Hu et al., 2014). This inherent feature of stem cells to self-renew and differentiate into different cell lineages has been coined "stemness." Unfortunately, a serious challenge to regenerative medicine is that not all stem cells can satisfy the "stemness" characteristic, thereby only partially fulfilling some features like differentiating into a limited type of cell lineage without self-renewal. Therefore the source of the stem cell and its destination niche plays a vital role in the definition of the acceptable degree of "stemness" attributed to that cell type. For skin healing and regeneration, there are mainly three types of stem cells (embryonic, bone marrow-derived, and adipose tissue-derived) that are utilized with some advantages and limitations to each.

Table 7.1 **Stem cell therapy used in mice**

Wound healing phase	Stem cell source	Outcome
Inflammatory phase	BM (Wu et al., 2014; Uysal et al., 2014)	Improved
	AT (Jiang et al., 2013; Kinoshita et al., 2015; Uysal et al., 2014)	Improved
Proliferation phase	BM (Ramanauskaite et al., 2010)	Improved
	AT (Gomathysankar et al., 2014)	Improved
Remodeling phase	AT (Meruane et al., 2012; Navone et al., 2014; Zamora et al., 2013)	Improved

BM, bone marrow; *AT*, adipose tissue.
Information extrapolated from a Pubmed search of journals published within the last 5 years. To group the papers into the three categories of excisional full-thickness wound healing phases, we utilized the following definitions based on the publication by Bielefeld et al. (2013). **Inflammatory phase**: as characterized by the infiltration of the wound bed by macrophages and other immune cells as well as the production of proinflammatory cytokines. **Proliferation phase**: as characterized by reepithelialization of the epidermis, neovascularization, and secretion of extracellular matrix (ECM) proteins. **Remodeling phase**: as characterized by the reorganization of the wound tissue through the degradation and replacement of immature ECM such as extra domain A fibronectin, collagen type III with collagen I, the organization of collagen I fibers into bundles, and the apoptosis of a variety of cell types at the wound site.

Table 7.2 **Stem cell therapy used in humans**

Wound healing phase	Stem cell source	Outcome
Unknown	BM (Jain et al., 2011; Kirana et al., 2012; Ravari et al., 2011)	Improved
	BM (Ding et al., 2013)	Worsened
	AT (Bura et al., 2014)	

BM, bone marrow; *AT*, adipose tissue.
Information extrapolated from a Pubmed search of journals published within the last 5 years. **Note**: we were unable to group the above stem cell clinical trials conducted in humans into the three phases of wound healing, as these trials looked at overall healing and lacked specifics to characterize which phase the therapy improved.

7.6.1 Embryonic-derived stem cells

Embryonic stem cells (ESCs) are probably characterized as the gold standard of stem cells because of their inherent potential to differentiate into multiple phenotypes given the correct stimulus, including the highly sought after neuronal cell lineage (Guan et al., 2001; Kim et al., 2007). These stem cells are derived from the inner cell mass of a blastocyst (an early-stage embryo) and have proven to be promising in disease conditions like Parkinson's, where a few clinical trials have shown favorable effects in improving patient motor activity (Bjorklund, 2005; Winkler et al., 2005). Despite these advantages, ESCs are far from the norm in regenerative medicine. Due to their embryonic origin, there has been a huge ethical backlash against the procurement and use of ESCs, making them difficult to obtain in sufficient quantity. Presently, there are no approved treatments using ESCs. Thus alternative adult donor cell sources that maintain the plasticity necessary for regenerative medicine applications and that avoid the ethical concerns surrounding ESCs have become an attractive option.

7.6.2 Bone marrow-derived mesenchymal stem cells

The bone marrow is a vital tissue source for a number of stem cells, including hematopoietic stem cells (HSCs) and mesenchymal stem cells (MSCs). HSCs are those progenitor cells responsible for repopulating the bone marrow and blood. However, MSCs are nonhematopoietic stem-like cells capable of differentiating into at least three cell lineages (osteoblasts, adipocytes, and chondrocytes) of nonmesenchymal origin (Das et al., 2013; Herzog et al., 2003; Huttmann et al., 2003). There is no single marker to identify MSCs; instead a combination of phenotypic, functional, and a lack of hematopoietic properties are used to characterize these cells. Furthermore, both of these cell types possess a high degree of plasticity that facilitates their ability to replace and increase the progenitor stem cell pool in hematopoietic and nonhematopoietic tissues (Herzog et al., 2003; Huttmann et al., 2003; Lagasse et al., 2000; Pittenger et al., 1999). In fact, a number of studies on MSCs have shown promise in their ability to promote skin regeneration and vascularization (Deng et al., 2005; Wu et al., 2007). It has been shown that MSCs promote wound repair and correct chronic wounds by stimulating the healing process to progress beyond the inflammatory phase (Tsourdi et al., 2013; Wu et al., 2007). Chronic wounds are characterized by an increased inflammatory response, and MSCs act in two ways to resolve this: (1) by attenuating the proinflammatory cytokine storm and (2) enhancing the production of antiinflammatory cytokines in the wound bed (Badiavas and Falanga, 2003; Kode et al., 2009; Ma et al., 2014; Tsourdi et al., 2013).

Other skin regenerative properties that have been linked to MSCs include their ability to migrate into the wound site and differentiate into cells required for the repair and renewal of injured tissue (Wu et al., 2007). In fact, transplanted MSCs in three patients with nonhealing wounds showed promising effects on wound closure (Badiavas and Falanga, 2003). Similar beneficial effects of MSC therapy were also noted by Lu et al. (2011) in which they observed improved healing and reduced pain in diabetic patients transplanted with MSCs.

7.6.3 Adipose-derived mesenchymal stem cells

While bone marrow-derived stem cells offer more plasticity necessary for regenerative medicine applications, harvesting them from patients is often very invasive and painful. Harvesting of adipose tissue is less invasive and is a remarkable source for MSCs, while avoiding many of the ethical, painful, and immunological concerns associated with other sources for stem cell (Chang et al., 2013; Gugerell et al., 2014; Zuk et al., 2002). Perhaps the most remarkable aspect of adipose tissue-derived MSCs is the sheer number that can be harvested with relative ease from fat biopsies or debrided fat from patients following surgery compared to other tissues (Strioga et al., 2012). A number of studies in diverse conditions have shown that adipose-derived MSCs can be therapeutic. For instance, Lee et al. (2012) have shown that intramuscular adipose-derived MSC injections in critical limb ischemia patients decreased pain and improved claudication walking distance in these patients. Additionally, Rigotti et al. (2007) found that patients undergoing chemotherapy treated with adipose-derived MSCs showed less side effects to the radiation as well as an improved clinical outcome. All of these studies and other clinical trials (Garcia-Olmo et al., 2005; Mizuno, 2010) combined with the unique features of adipose-derived progenitor cells make them an exciting alternative for stem cell therapy.

Unfortunately, most MSC transplantations are accomplished by direct injection into the wound bed, which has proven to be problematic in that such techniques not only require copious amounts of cells but also impede MSC function and survival. This has challenged researchers to come up with bioactive materials that can sustain and provide ideal microenvironments for optimal cell migration and proliferation of MSCs (Arinzeh et al., 2005). Therefore the use of bioactive scaffolds has been suggested in order to preserve and stimulate the survival, migration, and differentiation of MSCs during transplantation.

For decades, the main therapeutic treatment for burn patients has been autografts and allografts for wound repair. As discussed earlier, these treatment options have been very problematic, as they are invasive and limited in supply. Advances in regenerative medicine using stem cells have raised the possibility of offering a limitless supply of physiologically competent skin for wound healing in these patients (Gauglitz et al., 2011). In fact, many aspects of stem cell therapy are already being practiced clinically. For instance, bone marrow-derived MSCs used in a patient with deep skin burns have shown improved wound healing (Rasulov et al., 2005). In addition, similar results were found when Bey et al. (2010) treated a patient with severe radiation burns using bone marrow-derived MSCs. Unfortunately, the wide clinical use of stem cell therapy and clinical trials encompassing large numbers of burn patients are currently not possible due to ethical and legal issues with this process.

7.7 Current scaffolds for applying stem cells

For a long time, ECM and biomaterial scaffolds have been reduced to passive roles in regard to wound repair and regenerative medicine. However, it is now clear that these scaffolds have the potential to revolutionize wound therapeutics, as they provide transplanted stem cells with attachment sites similar to those normally found in the ECM. The availability of attachment sites has been shown to improve cell viability and the differentiation of stem cells (Awad et al., 2004; Bhang et al., 2007; Pashuck and Stevens, 2012). For scaffolds to be functionally effective, they must possess mechanical properties to withstand stress and physiological loads during tissue regeneration as well as be highly porous to allow for a sufficient diffusion of nutrients, the removal of metabolic waste, and neovascularization (Celiz et al., 2014; Pashuck and Stevens, 2012). Pore size is critical to scaffold effectiveness and should be in the 200–900 μm range to prevent cell occlusion and provide sufficient room for tissue growth (Celiz et al., 2014). This is essential not only for maintaining an acceptable viability but also for providing a niche for the differentiation of transplanted cells. A wide range of distinct scaffolds has been developed that not only provide a suitable environment for stem cell proliferation and survival but also biomaterials that have the capability to have specific chemokines or cues incorporated into them that can facilitate stem cell migration and differentiation in the wound bed (Pashuck and Stevens, 2012).

However, we will present mainly biomaterial scaffolds that have been widely utilized in stem cell therapy and skin regenerative applications.

7.7.1 Protein-based biomaterials

Collagen is the main ECM protein component in skin and is critical for wound healing. Many tissue-engineered scaffolds are therefore based on collagen, also because collagen is accessible in large quantities from tissues such as skin, tendon, and bone. A number of studies have also shown that scaffolds with collagen as a base provide the most optimal environment for the proliferation and differentiation of stem cells (Battista et al., 2005; Chan et al., 2007; Michelini et al., 2006). For instance, one study showed a significant difference in MSC proliferation on tissue culture dishes coated with collagen over those plated on poly-D-lysine (Qian and Saltzman, 2004). Additionally, cultured ESCs were able to differentiate into cells with neuronal phenotypes as well as endothelial cells when cultured in collagen scaffolds (Chen et al., 2003; Michelini et al., 2006). ESCs in collagen scaffolds were also able to generate new blood vessels as well as hepatocytes when provided with the appropriate chemical cue (Baharvand et al., 2006; Gerecht-Nir et al., 2004). Therefore these studies illustrate the suitability of collagen-based scaffolds in stem cell therapy.

7.7.2 Polysaccharide-based biomaterials

In keeping with using components from the ECM for establishing scaffold materials, polysaccharides are also used that not only help maintain ECM integrity but also play a major role in cell–cell recognition (Kumbar et al., 2011). Although a substantial number of polysaccharide-based scaffolds exist, our focus will be on alginate, chitosan, hyaluronan, and pullulan that have been widely used in skin regeneration in the clinic.

7.7.2.1 Alginate

Alginate, which is a polysaccharide obtained from the cell walls of brown algae, has been extensively used as a biomaterial base in a number of scaffolds. Alginate's inherent property is to act as an ionic cross-linker that allows for the encapsulation of cells seeded, which is critical for stem cell viability and protection (Awad et al., 2004). Furthermore, alginate hydrogels can be prepared by various cross-linking methods, for instance with calcium, to give them exceptional toughness and structural similarity to extracellular matrices of living tissues. In fact, a number of studies involving diverse stem cell populations have shown that alginate-based scaffolds were superior in maintaining cell survival when compared to other available biomaterials (Awad et al., 2004; Hannouche et al., 2007; Jin et al., 2007; Wayne et al., 2005). One study showed that bone marrow- and adipose tissue-derived MSCs had increased survival in alginate scaffolds. These same findings were also confirmed in neural and embryonic stem cells where improved proliferation, survival, and differentiation were shown when these were cultured in alginate-based scaffolds (Ashton et al., 2007; Gerecht-Nir et al., 2004; Maguire et al., 2006).

7.7.2.2 Chitosan and hyaluronan

These polysaccharide constituents are used in scaffold and tissue engineering (Chen et al., 2007; Jayakumar et al., 2010). Chitosan, unlike other polymers derived from costly mammalian proteins, evokes a minimal immune response as well as encapsulating seeded cells from degradation (Jayakumar et al., 2010). One study involving MSCs seeded in a chitosan scaffold transplanted on the patella of sheep promoted differentiation into chondrocyte-like cells (Mrugala et al., 2008). Additionally, studies involving the use of hyaluronan-engineered scaffolds that were seeded with MSCs demonstrated improved cartilage repair both in vitro and in vivo (Angele et al., 2008; Mehlhorn et al., 2007). Both of these polysaccharide scaffolds have also become attractive in bone tissue engineering due to their osteoconductivity and tensile properties that facilitate bone growth (Guo et al., 2011; Venkatesan and Kim, 2010).

7.7.2.3 Pullulan

As the area of scaffold engineering grows, more and more polysaccharide materials are being discovered. Pullulan is a commercially available polysaccharide purified from the fermentation medium of the fungus-like yeast *Aureobasidium pullulans* (Wong et al., 2011b). One unique inherent property of pullulan is its ability to quench free radicals (Wong et al., 2011b), an important property that may prove to be useful for stem cell delivery and viability in cutaneous wounds. As the majority of skin wounds following injury are characterized by environments rich in ROS that are highly toxic to both endogenous cells and exogenously delivered therapeutic cells, pullulan's antioxidant properties may prove to be an effective remedy. Preliminary trials indicate the positive effects of pullulan-based scaffolds (Wong et al., 2011a,b). For instance, MSC survival postdelivery and wound repair were both improved when a Pullulan-based scaffold was used, highlighting the ROS-quenching activity of this scaffold (Wong et al., 2011a). Thus if the above findings continue to hold true, and as more clinical trials are done, pullulan-based scaffolds may soon over take collagen in popularity.

In summary, the field of scaffold engineering is ever-expanding and encompasses many more biomaterials than discussed here. Our goal was to only introduce the commonly used biomaterials that are available for stem cell transplantation and regenerative medicine. Although not discussed extensively, many of the listed scaffold materials and their unique respective properties can be combined together to formulate far more effective scaffolds for tissue engineering.

7.8 Methods of applying stem cells

Despite the great potential of stem cell therapy in regenerative medicine, its therapeutic effect is limited and depends significantly on the method of delivery. There are several techniques that are either in use or in development in stem cell therapy. These techniques include injection-based stem cell delivery, scaffold-based delivery, and spray-aerosol-mediated delivery.

7.8.1 Injection-based stem cell delivery

The most commonly used method of delivery clinically, injection-based stem cell therapy involves the injection of stem cell suspensions directly into sites of injury (Zhang et al., 2008). However, the main limitations are cell survival as well as cell viability that are significantly affected by the shear pressure involved by the injection (Barbash et al., 2003; Laflamme and Murry, 2005). Furthermore, this method of delivery results in disorganized and delocalized cells in the wound area (Barbash et al., 2003). This lack of localization to the area hampers stem cell attachment and increases cell death (Zhang et al., 2008). An even more problematic issue with this method is that these cells can migrate and engraft in other organs or tissues of the body than the intended target, thereby creating detrimental cancer-prone environments in other organs. Progress has been made in finding alternative methods of delivery that can overcome the limitations and inefficiencies of injection-based stem cell delivery.

7.8.2 Scaffold-based stem cell delivery

A cornerstone in the field of bioengineering that has revolutionized stem cell delivery is the creation of scaffold biomaterials, in which stem cells can be cultured on and subsequently delivered specifically to the site of injury. As discussed earlier, numerous scaffolds have been designed using various materials such as collagen, chitosan, hyaluronan, and pullulan, among others, to mimic native tissue microenvironments (Jayakumar et al., 2010; Meinel et al., 2004). It should also have sufficient porosity with well-interconnected pores for the transport of nutrients and waste (Celiz et al., 2014). It is hypothesized that seeding the cells on these scaffolds will help improve stem cell viability, differentiation, and integration into the wound side. The limitations of three-dimensional scaffolds due to insufficient nutrient diffusion and vascularization after implantation can be overcome by improving on the porosity of these scaffolds (Celiz et al., 2014; Meinel et al., 2004). Thus there is a need to develop or find a biomaterial scaffold that will improve cell viability and aid targeted cell delivery. The use of a synthetic and porous polyethyleneglycol-polyurethane scaffold improved the engraftment of bone marrow-derived stem cells by protecting them from oxidative stress (Geesala et al., 2016).

7.8.3 Spray-based stem cell delivery

The concept of stem cell delivery was further advanced when the technology of the fibrin stem cell spray was developed (Zimmerlin et al., 2013). In essence, harvested stem cells are first mixed with fibrinogen that forms fibrin in situ by the coadministration of thrombin. The fibrin protein of the spray helps prevent the degradation of the stem cells (Falanga et al., 2007) and assists with the adherence of the stem cells to the wound site (Zimmerlin et al., 2013). In addition, the fibrin spray is unique in its ability to distribute the cells adequately across a large surface area when compared to other methods of stem cell delivery (Kaminski et al., 2011). Studies have shown that this approach to stem cell therapy is safe, effective, and as successful as topical administration in stimulating wound repair (Falanga et al., 2007; Wu et al., 2011).

7.9 Novel approaches in stem cell therapy

A number of microfluidic approaches have been introduced to define organ-mimetic tissue equivalents (Esch et al., 2011; Huh et al., 2012) as well as for the continuous formation, assembly, and in vivo application of cell-populated microfibers (Kang et al., 2011; Onoe et al., 2013). The function of skin cells and wound healing behavior have also been studied using microfluidic platforms (Morimoto et al., 2013; Sun et al., 2012), and bioprinting efforts are beginning to make their way toward the repair of burn wounds and cartilages (Albanna et al., 2012; Cohen et al., 2006). Microdevices promise to revolutionize the high-throughput formation and subsequent characterization of miniature skin tissue models for toxicological tests, potentially creating experimental capabilities that may become suitable replacements of animal experiments for cosmetic products, as required by a new European Union directive. Skin printing using different microfluidic approaches has emerged as a promising tool in those cases where conventional methods of scaffold- or transplantation-based therapies of stem cell delivery have failed. The three-dimensional seeding of stem cells in an architecture that is close to the skin holds great promise. Regenerative medicine will be revolutionized by the concept of three-dimensional tissue printing. In essence, this novel approach combines stem cell transplantation and scaffold-mediated therapy on a more grand scale with a quick turnaround. Although the exact technological working of the skin printer is beyond the scope of this chapter, it utilizes a microfluidic approach where stem cells are seeded on prefabricated biodegradable scaffolds made of candidate materials (Leng et al., 2012). It is hoped that the stem cells will populate the scaffold and create an appropriate ECM, often with the aid of perfusion, growth factors, and/or physical stimuli such as mechanical or electrical conditioning (Leng et al., 2012)

7.10 Future perspectives of stem cell therapy for wounds

Wound healing is a multifaceted process that requires the coordinated interaction of the ECM, growth factors, and cells. Stem cell therapy plays an important role in mediating each phase of the wound healing process: inflammatory, proliferative, and remodeling. A combination of the right stem cells and growth factors is a determinant for a better outcome. Although stem cell therapy has shown promising results, there are still many problems that need to be resolved before stem cells can be widely used clinically. The current data does not provide sufficient proof that the new skin cells and appendages are regenerated by stem cells. Tracking the lineage of transplanted stem cells is important. Moreover, it is essential to show that transplanted stem cells lead to functional appendages. It is also important to evaluate any malignancy potential of stem cells in therapies. The plasticity of stem cells may be a double-edged sword, and stem cells may eventually give rise to malignant cells in a nonoptimal niche. Furthermore, stem cells, like any other cells, may go through the process of aging, which might affect the therapy. More research is necessary to address different aspects of stem cells for wound therapy. Another challenge ahead of stem cell therapy is the seeding approaches for utilizing stem cells. Further preclinical and clinical studies are required to determine the effectiveness of different approaches of applying cells.

It is obvious that the complexity of stem cells, their biological behavior, their microenvironment, the three-dimensional interaction of different cells, and their clinical applications will govern the final outcome of stem cell therapy for wound healing. With a great promise, more characterization is warranted.

7.11 Conclusion

We hope that this chapter set the groundwork for answering many of the questions about stem cells and their role in regenerative medicine. Our discussion on the choice of stem cell source, scaffolds used to culture these cells, and the methods of delivery only scratched the surface of the wealth of information available. Thus while we are far from truly uncovering the signals that guide stem cells to mimic any cell phenotype and function, each new discovery in stem cell research brings us one step closer to truly exploiting the potential they possess for regenerative medicine and potential treatment options for the clinic.

References

Ågren, M.S., Eaglstein, W.H., Ferguson, M.W., Harding, K.G., Moore, K., Saarialho-Kere, U.K., Schultz, G.S., 2000. Causes and effects of the chronic inflammation in venous leg ulcers. Acta Derm. Venereol. Suppl. (Stockh.) 210, 3–17.

Albanna, M.Z., Murphy, S., Zhao, W., El-Amin, I.B., Tan, J., Jackson, J.D., Atala, A., Yoo, J.J., 2012. In situ bioprinting of autologous skin cells accelerates skin regeneration. J. Tissue Eng. Regener. Med. 6, 94–95.

Amini-Nik, S., Cambridge, E., Yu, W., Guo, A., Whetstone, H., Nadesan, P., Poon, R., Hinz, B., Alman, B.A., 2014. Beta-catenin-regulated myeloid cell adhesion and migration determine wound healing. J. Clin. Invest. 124, 2599–2610.

Amini-Nik, S., Glancy, D., Boimer, C., Whetstone, H., Keller, C., Alman, B.A., 2011. Pax7 expressing cells contribute to dermal wound repair, regulating scar size through a beta-catenin mediated process. Stem Cells 29, 1371–1379.

Angele, P., Johnstone, B., Kujat, R., Zellner, J., Nerlich, M., Goldberg, V., Yoo, J., 2008. Stem cell based tissue engineering for meniscus repair. J. Biomed. Mater. Res. A 85, 445–455.

Arinzeh, T.L., Tran, T., Mcalary, J., Daculsi, G., 2005. A comparative study of biphasic calcium phosphate ceramics for human mesenchymal stem-cell-induced bone formation. Biomaterials 26, 3631–3638.

Armstrong, D.G., Jude, E.B., 2002. The role of matrix metalloproteinases in wound healing. J. Am. Podiatr. Med. Assoc. 92, 12–18.

Ashton, R.S., Banerjee, A., Punyani, S., Schaffer, D.V., Kane, R.S., 2007. Scaffolds based on degradable alginate hydrogels and poly(lactide-co-glycolide) microspheres for stem cell culture. Biomaterials 28, 5518–5525.

Association, A.D., 2014. National Diabetes Statistics Report. (Online).

Awad, H.A., Wickham, M.Q., Leddy, H.A., Gimble, J.M., Guilak, F., 2004. Chondrogenic differentiation of adipose-derived adult stem cells in agarose, alginate, and gelatin scaffolds. Biomaterials 25, 3211–3222.

Badiavas, E.V., Falanga, V., 2003. Treatment of chronic wounds with bone marrow-derived cells. Arch. Dermatol. 139, 510–516.

Baharvand, H., Hashemi, S.M., Kazemi Ashtiani, S., Farrokhi, A., 2006. Differentiation of human embryonic stem cells into hepatocytes in 2D and 3D culture systems in vitro. Int. J. Dev. Biol. 50, 645–652.

Barbash, I.M., Chouraqui, P., Baron, J., Feinberg, M.S., Etzion, S., Tessone, A., Miller, L., Guetta, E., Zipori, D., Kedes, L.H., Kloner, R.A., Leor, J., 2003. Systemic delivery of bone marrow-derived mesenchymal stem cells to the infarcted myocardium: feasibility, cell migration, and body distribution. Circulation 108, 863–868.

Bartholomew, A., Sturgeon, C., Siatskas, M., Ferrer, K., Mcintosh, K., Patil, S., Hardy, W., Devine, S., Ucker, D., Deans, R., Moseley, A., Hoffman, R., 2002. Mesenchymal stem cells suppress lymphocyte proliferation in vitro and prolong skin graft survival in vivo. Exp. Hematol. 30, 42–48.

Battista, S., Guarnieri, D., Borselli, C., Zeppetelli, S., Borzacchiello, A., Mayol, L., Gerbasio, D., Keene, D.R., Ambrosio, L., Netti, P.A., 2005. The effect of matrix composition of 3D constructs on embryonic stem cell differentiation. Biomaterials 26, 6194–6207.

Bey, E., Prat, M., Duhamel, P., Benderitter, M., Brachet, M., Trompier, F., Battaglini, P., Ernou, I., Boutin, L., Gourven, M., Tissedre, F., Crea, S., Mansour, C.A., De Revel, T., Carsin, H., Gourmelon, P., Lataillade, J.J., 2010. Emerging therapy for improving wound repair of severe radiation burns using local bone marrow-derived stem cell administrations. Wound Repair Regen. 18, 50–58.

Bhang, S.H., Lim, J.S., Choi, C.Y., Kwon, Y.K., Kim, B.S., 2007. The behavior of neural stem cells on biodegradable synthetic polymers. J. Biomater. Sci. Polym. Ed. 18, 223–239.

Bielefeld, K.A., Amini-Nik, S., Alman, B.A., 2013. Cutaneous wound healing: recruiting developmental pathways for regeneration. Cell Mol. Life Sci. 70, 2059–2081.

Björklund, A., 2005. Cell therapy for Parkinson's disease: problems and prospects. Novartis Found Symp. 265, 174–186, discussion 187, 204–211.

Boulton, A.J., 2004. The diabetic foot: from art to science. The 18th Camillo Golgi lecture. Diabetologia 47, 1343–1353.

Branski, L.K., Al-Mousawi, A., Rivero, H., Jeschke, M.G., Sanford, A.P., Herndon, D.N., 2009a. Emerging infections in burns. Surg. Infect. (Larchmt.) 10, 389–397.

Branski, L.K., Gauglitz, G.G., Herndon, D.N., Jeschke, M.G., 2009b. A review of gene and stem cell therapy in cutaneous wound healing. Burns 35, 171–180.

Branski, L.K., Herndon, D.N., Pereira, C., Mlcak, R.P., Celis, M.M., Lee, J.O., Sanford, A.P., Norbury, W.B., Zhang, X.J., Jeschke, M.G., 2007. Longitudinal assessment of Integra in primary burn management: a randomized pediatric clinical trial. Crit. Care Med. 35, 2615–2623.

Bura, A., Planat-Benard, V., Bourin, P., Silvestre, J.S., Gross, F., Grolleau, J.L., Saint-Lebese, B., Peyrafitte, J.A., Fleury, S., Gadelorge, M., Taurand, M., Dupuis-Coronas, S., Leobon, B., Casteilla, L., 2014. Phase I trial: the use of autologous cultured adipose-derived stroma/stem cells to treat patients with non-revascularizable critical limb ischemia. Cytotherapy 16, 245–257.

Celiz, A.D., Smith, J.G., Langer, R., Anderson, D.G., Winkler, D.A., Barrett, D.A., Davies, M.C., Young, L.E., Denning, C., Alexander, M.R., 2014. Materials for stem cell factories of the future. Nat. Mater. 13, 570–579.

Cha, J., Falanga, V., 2007. Stem cells in cutaneous wound healing. Clin. Dermatol. 25, 73–78.

Chan, B.P., Hui, T.Y., Yeung, C.W., Li, J., Mo, I., Chan, G.C., 2007. Self-assembled collagen-human mesenchymal stem cell microspheres for regenerative medicine. Biomaterials 28, 4652–4666.

Chang, P., Qu, Y., Liu, Y., Cui, S., Zhu, D., Wang, H., Jin, X., 2013. Multi-therapeutic effects of human adipose-derived mesenchymal stem cells on radiation-induced intestinal injury. Cell Death Dis. 4, e685.

Chen, P.Y., Huang, L.L., Hsieh, H.J., 2007. Hyaluronan preserves the proliferation and differentiation potentials of long-term cultured murine adipose-derived stromal cells. Biochem. Biophys. Res. Commun. 360, 1–6.

Chen, S.S., Revoltella, R.P., Papini, S., Michelini, M., Fitzgerald, W., Zimmerberg, J., Margolis, L., 2003. Multilineage differentiation of rhesus monkey embryonic stem cells in three-dimensional culture systems. Stem Cells 21, 281–295.

Cheon, S.S., Cheah, A.Y., Turley, S., Nadesan, P., Poon, R., Clevers, H., Alman, B.A., 2002. Beta-catenin stabilization dysregulates mesenchymal cell proliferation, motility, and invasiveness and causes aggressive fibromatosis and hyperplastic cutaneous wounds. Proc. Natl. Acad. Sci. U.S.A. 99, 6973–6978.

Cohen, D.L., Malone, E., Lipson, H., Bonassar, L.J., 2006. Direct freeform fabrication of seeded hydrogels in arbitrary geometries. Tissue Eng. 12, 1325–1335.

Dabiri, G., Heiner, D., Falanga, V., 2013. The emerging use of bone marrow-derived mesenchymal stem cells in the treatment of human chronic wounds. Expert Opin. Emerg. Drugs 18, 405–419.

Das, M., Sundell, I.B., Koka, P.S., 2013. Adult mesenchymal stem cells and their potency in the cell-based therapy. J. Stem Cells 8, 1–16.

Deng, W., Han, Q., Liao, L., Li, C., Ge, W., Zhao, Z., You, S., Deng, H., Murad, F., Zhao, R.C., 2005. Engrafted bone marrow-derived flk-(1+) mesenchymal stem cells regenerate skin tissue. Tissue Eng. 11, 110–119.

Ding, J., Ma, Z., Shankowsky, H.A., Medina, A., Tredget, E.E., 2013. Deep dermal fibroblast profibrotic characteristics are enhanced by bone marrow-derived mesenchymal stem cells. Wound Repair Regen. 21, 448–455.

Dziewulski, P., Leon-Villapalos, J., Barret, J.P., 2012. Adult Burn Care Management. Springer, Austria.

Esch, M.B., King, T.L., Shuler, M.L., 2011. The role of body-on-a-chip devices in drug and toxicity studies. In: Yarmush, M.L., Duncan, J.S., Gray, M.L. (Eds.), Annual Review of Biomedical Engineering, vol. 13.

Falanga, V., Iwamoto, S., Chartier, M., Yufit, T., Butmarc, J., Kouttab, N., Shrayer, D., Carson, P., 2007. Autologous bone marrow-derived cultured mesenchymal stem cells delivered in a fibrin spray accelerate healing in murine and human cutaneous wounds. Tissue Eng. 13, 1299–1312.

Garcia-Olmo, D., Garcia-Arranz, M., Herreros, D., Pascual, I., Peiro, C., Rodriguez-Montes, J.A., 2005. A phase I clinical trial of the treatment of Crohn's fistula by adipose mesenchymal stem cell transplantation. Dis. Colon Rectum. 48, 1416–1423.

Gauglitz, G.G., Jeschke, M.G., 2011. Combined gene and stem cell therapy for cutaneous wound healing. Mol. Pharm. 8, 1471–1479.

Gauglitz, G.G., Korting, H.C., Pavicic, T., Ruzicka, T., Jeschke, M.G., 2011. Hypertrophic scarring and keloids: pathomechanisms and current and emerging treatment strategies. Mol. Med. 17, 113–125.

Geesala, R., Bar, N., Dhoke, N.R., Basak, P., Das, A., 2016. Porous polymer scaffold for on-site delivery of stem cells - protects from oxidative stress and potentiates wound tissue repair. Biomaterials 77, 1–13.

Gerecht-Nir, S., Cohen, S., Ziskind, A., Itskovitz-Eldor, J., 2004. Three-dimensional porous alginate scaffolds provide a conducive environment for generation of well-vascularized embryoid bodies from human embryonic stem cells. Biotechnol. Bioeng. 88, 313–320.

Gomathysankar, S., Halim, A.S., Yaacob, N.S., 2014. Proliferation of keratinocytes induced by adipose-derived stem cells on a chitosan scaffold and its role in wound healing, a review. Arch. Plast. Surg. 41, 452–457.

Guan, K., Chang, H., Rolletschek, A., Wobus, A.M., 2001. Embryonic stem cell-derived neurogenesis. Retinoic acid induction and lineage selection of neuronal cells. Cell Tissue Res. 305, 171–176.

Gugerell, A., Kober, J., Schmid, M., Nickl, S., Kamolz, L.P., Keck, M., 2014. Botulinum toxin A and lidocaine have an impact on adipose-derived stem cells, fibroblasts, and mature adipocytes in vitro. J. Plast. Reconstr. Aesthet. Surg. 67 (9), 1276–1281.

Guo, B., Finne-Wistrand, A., Albertsson, A.C., 2011. Facile synthesis of degradable and electrically conductive polysaccharide hydrogels. Biomacromolecules 12, 2601–2609.

Gurtner, G.C., Werner, S., Barrandon, Y., Longaker, M.T., 2008. Wound repair and regeneration. Nature 453, 314–321.

Hannouche, D., Terai, H., Fuchs, J.R., Terada, S., Zand, S., Nasseri, B.A., Petite, H., Sedel, L., Vacanti, J.P., 2007. Engineering of implantable cartilaginous structures from bone marrow-derived mesenchymal stem cells. Tissue Eng. 13, 87–99.

Herzog, E.L., Chai, L., Krause, D.S., 2003. Plasticity of marrow-derived stem cells. Blood 102, 3483–3493.

Horch, R.E., Jeschke, M.G., Spilker, G., Herndon, D.N., Kopp, J., 2005. Treatment of second degree facial burns with allografts–preliminary results. Burns 31, 597–602.

Hu, M.S., Rennert, R.C., Mcardle, A., Chung, M.T., Walmsley, G.G., Longaker, M.T., Lorenz, H.P., 2014. The role of stem cells during scarless skin wound healing. Adv. Wound Care (New Rochelle) 3, 304–314.

Huh, D., Torisawa, Y.-S., Hamilton, G.A., Kim, H.J., Ingber, D.E., 2012. Microengineered physiological biomimicry: organs-on-chips. Lab Chip 12, 2156–2164.

Hunt, T.K., Hopf, H., Hussain, Z., 2000. Physiology of wound healing. Adv. Skin Wound Care 13, 6–11.

Huttmann, A., Li, C.L., Duhrsen, U., 2003. Bone marrow-derived stem cells and "plasticity". Ann. Hematol. 82, 599–604.

Jain, P., Perakath, B., Jesudason, M.R., Nayak, S., 2011. The effect of autologous bone marrow-derived cells on healing chronic lower extremity wounds: results of a randomized controlled study. Ostomy Wound Manage. 57, 38–44.

Jayakumar, R., Prabaharan, M., Nair, S.V., Tamura, H., 2010. Novel chitin and chitosan nanofibers in biomedical applications. Biotechnol. Adv. 28, 142–150.

Jeschke, M.G., Chinkes, D.L., Finnerty, C.C., Kulp, G., Suman, O.E., Norbury, W.B., Branski, L.K., Gauglitz, G.G., Mlcak, R.P., Herndon, D.N., 2008. Pathophysiologic response to severe burn injury. Ann. Surg. 248, 387–401.

Jeschke, M.G., Gauglitz, G.G., Kulp, G.A., Finnerty, C.C., Williams, F.N., Kraft, R., Suman, O.E., Mlcak, R.P., Herndon, D.N., 2011. Long-term persistance of the pathophysiologic response to severe burn injury. PLoS One 6, e21245.

Jeschke, M.G., Herndon, D.N., 2014. Burns in children: standard and new treatments. Lancet 383, 1168–1178.

Jeschke, M.G., Rose, C., Angele, P., Fuchtmeier, B., Nerlich, M.N., Bolder, U., 2004. Development of new reconstructive techniques: use of Integra in combination with fibrin glue and negative-pressure therapy for reconstruction of acute and chronic wounds. Plast. Reconstr. Surg. 113, 525–530.

Jiang, D., Qi, Y., Walker, N.G., Sindrilaru, A., Hainzl, A., Wlaschek, M., Macneil, S., Scharffetter-Kochanek, K., 2013. The effect of adipose tissue derived MSCs delivered by a chemically defined carrier on full-thickness cutaneous wound healing. Biomaterials 34, 2501–2515.

Jin, X., Sun, Y., Zhang, K., Wang, J., Shi, T., Ju, X., Lou, S., 2007. Ectopic neocartilage formation from predifferentiated human adipose derived stem cells induced by adenoviral-mediated transfer of hTGF beta2. Biomaterials 28, 2994–3003.

Jones, I., Currie, L., Martin, R., 2002. A guide to biological skin substitutes. Br. J. Plast. Surg. 55, 185–193.

Kaminski, A., Klopsch, C., Mark, P., Yerebakan, C., Donndorf, P., Gabel, R., Eisert, F., Hasken, S., Kreitz, S., Glass, A., Jockenhovel, S., Ma, N., Kundt, G., Liebold, A., Steinhoff, G., 2011. Autologous valve replacement-CD133+ stem cell-plus-fibrin composite-based sprayed cell seeding for intraoperative heart valve tissue engineering. Tissue Eng. Part C Methods 17, 299–309.

Kang, E., Jeong, G.S., Choi, Y.Y., Lee, K.H., Khademhosseini, A., Lee, S.-H., 2011. Digitally tunable physicochemical coding of material composition and topography in continuous microfibres. Nat. Mater. 10, 877–883.

Kim, D.S., Kim, J.Y., Kang, M., Cho, M.S., Kim, D.W., 2007. Derivation of functional dopamine neurons from embryonic stem cells. Cell Transplant. 16, 117–123.

Kinoshita, K., Kuno, S., Ishimine, H., Aoi, N., Mineda, K., Kato, H., Doi, K., Kanayama, K., Feng, J., Mashiko, T., Kurisaki, A., Yoshimura, K., 2015. Therapeutic potential of adipose-derived SSEA-3-positive muse cells for treating diabetic skin ulcers. Stem Cells Transl. Med. 4 (2), 146–155.

Kirana, S., Stratmann, B., Prante, C., Prohaska, W., Koerperich, H., Lammers, D., Gastens, M.H., Quast, T., Negrean, M., Stirban, O.A., Nandrean, S.G., Gotting, C., Minartz, P., Kleesiek, K., Tschoepe, D., 2012. Autologous stem cell therapy in the treatment of limb ischaemia induced chronic tissue ulcers of diabetic foot patients. Int. J. Clin. Pract. 66, 384–393.

Kode, J.A., Mukherjee, S., Joglekar, M.V., Hardikar, A.A., 2009. Mesenchymal stem cells: immunobiology and role in immunomodulation and tissue regeneration. Cytotherapy 11, 377–391.

Koh, T.J., Dipietro, L.A., 2011. Inflammation and wound healing: the role of the macrophage. Expert Rev. Mol. Med. 13, e23.

Kumbar, S.G., Toti, U.S., Deng, M., James, R., Laurencin, C.T., Aravamudhan, A., Harmon, M., Ramos, D.M., 2011. Novel mechanically competent polysaccharide scaffolds for bone tissue engineering. Biomed. Mater. 6, 065005.

Laflamme, M.A., Murry, C.E., 2005. Regenerating the heart. Nat. Biotechnol. 23, 845–856.

Lagasse, E., Connors, H., Al-Dhalimy, M., Reitsma, M., Dohse, M., Osborne, L., Wang, X., Finegold, M., Weissman, I.L., Grompe, M., 2000. Purified hematopoietic stem cells can differentiate into hepatocytes in vivo. Nat. Med. 6, 1229–1234.

Lazarus, G.S., Cooper, D.M., Knighton, D.R., Percoraro, R.E., Rodeheaver, G., Robson, M.C., 1994. Definitions and guidelines for assessment of wounds and evaluation of healing. Wound Repair Regen. 2, 165–170.

Lee, H.C., An, S.G., Lee, H.W., Park, J.S., Cha, K.S., Hong, T.J., Park, J.H., Lee, S.Y., Kim, S.P., Kim, Y.D., Chung, S.W., Bae, Y.C., Shin, Y.B., Kim, J.I., Jung, J.S., 2012. Safety and effect of adipose tissue-derived stem cell implantation in patients with critical limb ischemia: a pilot study. Circ. J. 76, 1750–1760.

Leng, L., Mcallister, A., Zhang, B., Radisic, M., Guenther, A., 2012. Mosaic hydrogels: one-step formation of multiscale soft materials. Adv. Mater. 24, 3650–3658.

Lerman, O.Z., Galiano, R.D., Armour, M., Levine, J.P., Gurtner, G.C., 2003. Cellular dysfunction in the diabetic fibroblast: impairment in migration, vascular endothelial growth factor production, and response to hypoxia. Am. J. Pathol. 162, 303–312.

Longhi, L., Zanier, E.R., Royo, N., Stocchetti, N., Mcintosh, T.K., 2005. Stem cell transplantation as a therapeutic strategy for traumatic brain injury. Transpl. Immunol. 15, 143–148.

Loots, M.A., Lamme, E.N., Zeegelaar, J., Mekkes, J.R., Bos, J.D., Middelkoop, E., 1998. Differences in cellular infiltrate and extracellular matrix of chronic diabetic and venous ulcers versus acute wounds. J. Invest. Dermatol. 111, 850–857.

Loughran, J.H., Chugh, A.R., Ismail, I., Bolli, R., 2013. Stem cell therapy: promising treatment in heart failure? Curr. Heart Fail. Rep. 10, 73–80.

Lu, D., Chen, B., Liang, Z., Deng, W., Jiang, Y., Li, S., Xu, J., Wu, Q., Zhang, Z., Xie, B., Chen, S., 2011. Comparison of bone marrow mesenchymal stem cells with bone marrow-derived mononuclear cells for treatment of diabetic critical limb ischemia and foot ulcer: a double-blind, randomized, controlled trial. Diabetes Res. Clin. Pract. 92, 26–36.

Ma, S., Xie, N., Li, W., Yuan, B., Shi, Y., Wang, Y., 2014. Immunobiology of mesenchymal stem cells. Cell Death Differ. 21, 216–225.

Maguire, T., Novik, E., Schloss, R., Yarmush, M., 2006. Alginate-PLL microencapsulation: effect on the differentiation of embryonic stem cells into hepatocytes. Biotechnol. Bioeng. 93, 581–591.

Margolis, D.J., Allen-Taylor, L., Hoffstad, O., Berlin, J.A., 2005. Diabetic neuropathic foot ulcers and amputation. Wound Repair Regen 13, 230–236.

Martin, P., 1997. Wound healing–aiming for perfect skin regeneration. Science 276, 75–81.

Mast, B.A., Schultz, G.S., 1996. Interactions of cytokines, growth factors, and proteases in acute and chronic wounds. Wound Repair Regen. 4, 411–420.

Mehlhorn, A.T., Niemeyer, P., Kaschte, K., Muller, L., Finkenzeller, G., Hartl, D., Sudkamp, N.P., Schmal, H., 2007. Differential effects of BMP-2 and TGF-beta1 on chondrogenic differentiation of adipose derived stem cells. Cell Prolif. 40, 809–823.

Meinel, L., Karageorgiou, V., Fajardo, R., Snyder, B., Shinde-Patil, V., Zichner, L., Kaplan, D., Langer, R., Vunjak-Novakovic, G., 2004. Bone tissue engineering using human mesenchymal stem cells: effects of scaffold material and medium flow. Ann. Biomed. Eng. 32, 112–122.

Meruane, M.A., Rojas, M., Marcelain, K., 2012. The use of adipose tissue-derived stem cells within a dermal substitute improves skin regeneration by increasing neoangiogenesis and collagen synthesis. Plast. Reconstr. Surg. 130, 53–63.

Michelini, M., Franceschini, V., Sihui Chen, S., Papini, S., Rosellini, A., Ciani, F., Margolis, L., Revoltella, R.P., 2006. Primate embryonic stem cells create their own niche while differentiating in three-dimensional culture systems. Cell Prolif. 39, 217–229.

Mizuno, H., 2010. Adipose-derived stem and stromal cells for cell-based therapy: current status of preclinical studies and clinical trials. Curr. Opin. Mol. Ther. 12, 442–449.

Morimoto, Y., Tanaka, R., Takeuchi, S., 2013. Construction of 3D, layered skin, microsized tissues by using cell beads for cellular function analysis. Adv. Healthcare Mater. 2, 261–265.

Moseley, R., Hilton, J.R., Waddington, R.J., Harding, K.G., Stephens, P., Thomas, D.W., 2004. Comparison of oxidative stress biomarker profiles between acute and chronic wound environments. Wound Repair Regen. 12, 419–429.

Mrugala, D., Bony, C., Neves, N., Caillot, L., Fabre, S., Moukoko, D., Jorgensen, C., Noel, D., 2008. Phenotypic and functional characterisation of ovine mesenchymal stem cells: application to a cartilage defect model. Ann. Rheum. Dis. 67, 288–295.

Navone, S.E., Pascucci, L., Dossena, M., Ferri, A., Invernici, G., Acerbi, F., Cristini, S., Bedini, G., Tosetti, V., Ceserani, V., Bonomi, A., Pessina, A., Freddi, G., Alessandrino, A., Ceccarelli, P., Campanella, R., Marfia, G., Alessandri, G., Parati, E.A., 2014. Decellularized silk fibroin scaffold primed with adipose mesenchymal stromal cells improves wound healing in diabetic mice. Stem Cell Res. Ther. 5, 7.

Onoe, H., Okitsu, T., Itou, A., Kato-Negishi, M., Gojo, R., Kiriya, D., Sato, K., Miura, S., Iwanaga, S., Kuribayashi-Shigetomi, K., Matsunaga, Y.T., Shimoyama, Y., Takeuchi, S., 2013. Metre-long cell-laden microfibres exhibit tissue morphologies and functions. Nat. Mater. 12, 584–590.

Pashuck, E.T., Stevens, M.M., 2012. Designing regenerative biomaterial therapies for the clinic. Sci. Transl. Med. 4, 160–164.

Peck, M.D., 2012. Epidemiology and Prevention of Burns Throughout the World. Springer, Austria.

Pittenger, M.F., Mackay, A.M., Beck, S.C., Jaiswal, R.K., Douglas, R., Mosca, J.D., Moorman, M.A., Simonetti, D.W., Craig, S., Marshak, D.R., 1999. Multilineage potential of adult human mesenchymal stem cells. Science 284, 143–147.

Pruitt Jr., B.A., O'neill Jr., J.A., Moncrief, J.A., Lindberg, R.B., 1968. Successful control of burn-wound sepsis. JAMA 203, 1054–1056.

Qian, L., Saltzman, W.M., 2004. Improving the expansion and neuronal differentiation of mesenchymal stem cells through culture surface modification. Biomaterials 25, 1331–1337.

Ramanauskaite, G., Kaseta, V., Vaitkuviene, A., Biziuleviciene, G., 2010. Skin regeneration with bone marrow-derived cell populations. Int. Immunopharmacol. 10, 1548–1551.

Rasulov, M.F., Vasilchenkov, A.V., Onishchenko, N.A., Krasheninnikov, M.E., Kravchenko, V.I., Gorshenin, T.L., Pidtsan, R.E., Potapov, I.V., 2005. First experience of the use bone marrow mesenchymal stem cells for the treatment of a patient with deep skin burns. Bull. Exp. Biol. Med. 139, 141–144.

Ravari, H., Hamidi-almadari, D., Salimifar, M., Bonakdaran, S., Parizadeh, M.R., Koliakos, G., 2011. Treatment of non-healing wounds with autologous bone marrow cells, platelets, fibrin glue and collagen matrix. Cytotherapy 13, 705–711.

Rigotti, G., Marchi, A., Galie, M., Baroni, G., Benati, D., Krampera, M., Pasini, A., Sbarbati, A., 2007. Clinical treatment of radiotherapy tissue damage by lipoaspirate transplant: a healing process mediated by adipose-derived adult stem cells. Plast. Reconstr. Surg. 119, 1409–1422, discussion 1423–1424.

Sarkar, K., Rey, S., Zhang, X., Sebastian, R., Marti, G.P., Fox-Talbot, K., Cardona, A.V., Du, J., Tan, Y.S., Liu, L., Lay, F., Gonzalez, F.J., Harmon, J.W., Semenza, G.L., 2012. Tie2-dependent knockout of HIF-1 impairs burn wound vascularization and homing of bone marrow-derived angiogenic cells. Cardiovasc. Res. 93, 162–169.

Seibold, J.R., Uitto, J., Dorwart, B.B., Prockop, D.J., 1985. Collagen synthesis and collagenase activity in dermal fibroblasts from patients with diabetes and digital sclerosis. J. Lab. Clin. Med. 105, 664–667.

Singer, A.J., Clark, R.A., 1999. Cutaneous wound healing. N. Engl. J. Med. 341, 738–746.

Singh, N., Armstrong, D.G., Lipsky, B.A., 2005. Preventing foot ulcers in patients with diabetes. JAMA 293, 217–228.

Stocum, D.L., Cameron, J.A., 2011. Looking proximally and distally: 100 years of limb regeneration and beyond. Dev. Dyn. 240, 943–968.

Strioga, M., Viswanathan, S., Darinskas, A., Slaby, O., Michalek, J., 2012. Same or not the same? Comparison of adipose tissue-derived versus bone marrow-derived mesenchymal stem and stromal cells. Stem Cells Dev. 21, 2724–2752.

Sun, Y.-S., Peng, S.-W., Cheng, J.-Y., 2012. In vitro electrical-stimulated wound-healing chip for studying electric field-assisted wound-healing process. Biomicrofluidics 6.

Teot, L., Otman, S., Brancati, A., Mittermayr, R., 2012. Burn Wound Healing: Pathophysiology. Springer Vienna, Vienna.

Tsourdi, E., Barthel, A., Rietzsch, H., Reichel, A., Bornstein, S.R., 2013. Current aspects in the pathophysiology and treatment of chronic wounds in diabetes mellitus. Biomed. Res. Int. 2013, 385641.

Uysal, C.A., Tobita, M., Hyakusoku, H., Mizuno, H., 2014. The effect of bone-marrow-derived stem cells and adipose-derived stem cells on wound contraction and epithelization. Adv. Wound Care (New Rochelle) 3, 405–413.

Venkatesan, J., Kim, S.K., 2010. Chitosan composites for bone tissue engineering–an overview. Mar. Drugs 8, 2252–2266.

Wayne, J.S., Mcdowell, C.L., Shields, K.J., Tuan, R.S., 2005. In vivo response of polylactic acid-alginate scaffolds and bone marrow-derived cells for cartilage tissue engineering. Tissue Eng. 11, 953–963.

Wenk, J., Brenneisen, P., Wlaschek, M., Poswig, A., Briviba, K., Oberley, T.D., Scharffetter-Kochanek, K., 1999. Stable overexpression of manganese superoxide dismutase in mitochondria identifies hydrogen peroxide as a major oxidant in the AP-1-mediated induction of matrix-degrading metalloprotease-1. J. Biol. Chem. 274, 25869–25876.

Werner, S., Grose, R., 2003. Regulation of wound healing by growth factors and cytokines. Physiol. Rev. 83, 835–870.

Wetzler, C., Kampfer, H., Stallmeyer, B., Pfeilschifter, J., Frank, S., 2000. Large and sustained induction of chemokines during impaired wound healing in the genetically diabetic mouse: prolonged persistence of neutrophils and macrophages during the late phase of repair. J. Invest. Dermatol. 115, 245–253.

Winkler, C., Kirik, D., Bjorklund, A., 2005. Cell transplantation in Parkinson's disease: how can we make it work? Trends Neurosci. 28, 86–92.

Wlaschek, M., Scharffetter-Kochanek, K., 2005. Oxidative stress in chronic venous leg ulcers. Wound Repair Regen. 13, 452–461.

Wong, V.W., Rustad, K.C., Galvez, M.G., Neofytou, E., Glotzbach, J.P., Januszyk, M., Major, M.R., Sorkin, M., Longaker, M.T., Rajadas, J., Gurtner, G.C., 2011a. Engineered pullulan-collagen composite dermal hydrogels improve early cutaneous wound healing. Tissue Eng. Part A 17, 631–644.

Wong, V.W., Rustad, K.C., Glotzbach, J.P., Sorkin, M., Inayathullah, M., Major, M.R., Longaker, M.T., Rajadas, J., Gurtner, G.C., 2011b. Pullulan hydrogels improve mesenchymal stem cell delivery into high-oxidative-stress wounds. Macromol. Biosci. 11, 1458–1466.

Wu, X., Wang, G., Tang, C., Zhang, D., Li, Z., Du, D., Zhang, Z., 2011. Mesenchymal stem cell seeding promotes reendothelialization of the endovascular stent. J. Biomed. Mater. Res. A 98, 442–449.

Wu, Y., Huang, S., Enhe, J., Ma, K., Yang, S., Sun, T., fu, X., 2014. Bone marrow-derived mesenchymal stem cell attenuates skin fibrosis development in mice. Int. Wound J. 11, 701–710.

Wu, Y., Wang, J., Scott, P.G., Tredget, E.E., 2007. Bone marrow-derived stem cells in wound healing: a review. Wound Repair Regen. 15 (Suppl. 1), S18–S26.

Yannas, I.V., Burke, J.F., 1980. Design of an artificial skin. I. Basic design principles. J. Biomed. Mater. Res. 14, 65–81.

Zamora, D.O., Natesan, S., Becerra, S., Wrice, N., Chung, E., Suggs, L.J., Christy, R.J., 2013. Enhanced wound vascularization using a dsASCs seeded FPEG scaffold. Angiogenesis 16, 745–757.

Zhang, G., Hu, Q., Braunlin, E.A., Suggs, L.J., Zhang, J., 2008. Enhancing efficacy of stem cell transplantation to the heart with a PEGylated fibrin biomatrix. Tissue Eng. Part A 14, 1025–1036.

Zimmerlin, L., Rubin, J.P., Pfeifer, M.E., Moore, L.R., Donnenberg, V.S., Donnenberg, A.D., 2013. Human adipose stromal vascular cell delivery in a fibrin spray. Cytotherapy 15, 102–108.

Zuk, P.A., Zhu, M., Ashjian, P., De Ugarte, D.A., Huang, J.I., Mizuno, H., Alfonso, Z.C., Fraser, J.K., Benhaim, P., Hedrick, M.H., 2002. Human adipose tissue is a source of multipotent stem cells. Mol. Biol. Cell 13, 4279–4295.

Zykova, S.N., Jenssen, T.G., Berdal, M., Olsen, R., Myklebust, R., Seljelid, R., 2000. Altered cytokine and nitric oxide secretion in vitro by macrophages from diabetic type II-like db/db mice. Diabetes 49, 1451–1458.

Living cell products as wound healing biomaterials: current and future modalities

M. Tenenhaus[1], *H.O. Rennekampff*[2], *G. Mulder*[1]
[1]University of California at San Diego Medical Center, San Diego, CA, United States;
[2]Universitätsklinikum der RWTH Aachen, Aachen, Germany

8.1 Introduction

Living cell products are those products that are composed of varying types and numbers of cells, which may be part of a biologic, semisynthetic, or synthetic matrix delivery system. This category of materials is applied directly to a prepared wound bed for the purpose of assisting or expediting wound closure. Ideally, living cell products would be rapidly accepted in the wound bed, mimicking the activity of cells present during the normal repair process. The anticipated result is one in which the cells are accepted without rejection. The use of living cell therapy is thus considered to be a clinical approach to assist or replace the activity of endogenous cells in the tissue repair process.

A variety of products have been developed, which can be categorized as living cell biomaterials [1]. Their applications range from the treatment of acute wounds and surgically prepared graft sites to application in chronic wounds in an ambulatory setting. Studies with living cell products are constantly evolving with the hope of eventually developing a therapy, which may allow for immediate and successful wound closure for all types of wounds.

Despite the development of the many different materials, outcomes vary significantly, with no specific product being of benefit to all types of wounds. Optimal outcomes are not always as expected and can be affected by bioburden, excessive and prolonged inflammatory processes in the wound bed, poor vascular flow to the tissue, cellular senescence, and reflect characteristics such as etiology and location of wound, patient medical status, age, and numerous other factors [2–4]. The expense of many biomaterials may not always justify their use when compared to more standard approaches to wound care. Thus there is a continuous need for the development of new therapies that may accelerate wound closure, decrease complications associated with delayed healing, and result in a closed wound with decreased scarring and a higher quality of tissue, approximating that of normal, uninjured skin.

This chapter reviews the history of living cell products, including science and research performed to date, currently available living cell products, and promises of research and future technology. Limited sections on relevant stem cell applications and the use of in situ biofabrication are also included.

Wound Healing Biomaterials - Volume 1. http://dx.doi.org/10.1016/B978-1-78242-455-0.00008-2

8.2 History and new developments of living cell products for the treatment of problematic and chronic wounds

The study of living human cells is neither recent nor novel. Documentation on scientific findings and experimentation on cell growth goes back many decades [5–7]. The works most relevant to the development of tissue-engineered products for use in wound healing, in particular problematic and chronic wounds, are the pioneering studies by Eugene Bell, which led to the development of the first clinically applied bioengineered materials for use on burn wounds and chronic wounds [8]. The studies by Bell and others stimulated interest in biomatrices and biological devices, which could be applied in surgical and outpatient settings for the treatment of chronic wounds as well as complicated disease states such as epidermolysis bullosa [9].

Initial living cell products focused on the use of cultured fibroblasts and keratinocytes but expanded to studies of a variety of cell types, including endothelial cells and nonskin-derived cells like adipose cells and preadipocytes, embryonic, amniotic, mesenchymal, bone marrow-derived, and other stem cell lines [10–12]. Stem cells are considered to be particularly appealing due to their ability to differentiate into other cell lines, depending on the environment in which they were placed. Pluripotent cells are being investigated as a readily accessible source of living cells with multiple functions. These initiatives may improve the function of living cell products in the wounds further.

The delivery of living cells in a matrix that is cell friendly and accepted in the wound bed poses an additional barrier to the clinical use of cell lines. A matrix or scaffold, which facilitates cell activity while permitting cell growth and migration of other cells in the matrix, is needed to support the introduced cells and their activity. Matrices for current products include biologic, synthetic, and semisynthetic materials, which may have varying components including collagen, proteoglycans, glycosaminoglycans, cellulose, or even allo and xenograft bases. The structure and dimensions of the primary scaffold will in turn directly influence cell activity, mobility, and viability.

Although the development of living cell products has rapidly evolved, we are still in the beginning phases of developing an end product that closely and continuously mimics the normal repair process in wounds where repair is abnormal or delayed. Living cell products, skin equivalents, and cultured cell lines are all susceptible to rapid neutralization and denaturing in a hostile wound bed with a significant bacterial burden and/or high inflammatory cell proteinase levels. An adequate moisture level of the wound bed is also important for maintained cell activity [13–15]. Thus the frequency and ideal number of applications of these products required needs to be studied in more detail. Current needs also include the combination of a durable matrix, which will support cell activity for extended periods of time, a longer shelf life without the need for specialized refrigeration, and "smart biologics," which can sense the specific needs of a wound and delivery of cells as needed at an affordable cost to make these treatments available globally. Finally, adjunctive modalities, which can determine cytokine and growth factor deficiencies and levels of bacteria and proteinases would assist in determining the best treatment approach.

8.3 Current living cell products on the global market

8.3.1 Classification

As with approaches to the engineering of other tissue grafts, strategies for the fabrication of cell-based skin substitutes can be allocated to a variety of categories, as outlined in Table 8.1. These categories include temporary and permanent skin substitutes, epidermal and dermal substitutes, synthetic and biological skin substitutes, and cell-seeded formulations. The common denominator among all these products is the presence of living cells. As one might expect, a wide variety of overlap can be seen in the development and promotion of these constructs, reflecting individual performance objectives. The spectrum of possibilities is outlined in Table 8.2.

8.3.2 Temporary skin substitutes

Temporary cell-based skin substitutes provide transient physiologic wound coverage and are designed to support requisite processes of the different wound healing stages for a limited timespan. The most common indication for this type of skin substitute is the partial-thickness wound, employed in an effort to facilitate epithelialization. These

Table 8.1 Types of skin substitutes

Composition	Epidermal substitute	Dermal substitute	Epidermal–dermal substitute
Processed tissue	• Alloskin • Xenogeneic skin • Amnion • Onsite skin preparation	• Alloskin • Xenogeneic skin	• Alloskin • Xenogeneic skin
Cultured substitute	• CEA • Suspended autologous keratinocytes • Allogeneic keratinocytes		
Cell cultures in combination with biological matrix	• CEA in fibrin • CEA on hyaluronic acid • CEA on collagen	• Fibroblasts on collagen • Fibroblasts on hyaluronic acid	• Keratinocytes and fibroblasts with collagen-glucosaminoglycan matrix • Keratinocytes and fibroblasts with collagen matrix
Cell cultures in combination with synthetic matrix	• Keratinocytes on foil	• Fibroblasts on polyglactin • Fibroblasts on nylon	

CEA, cultured epithelial autograft.

Table 8.2 **Characteristics of skin substitutes**

Indication	Function	Skin substitute
Temporary application	Enhance epithelialization	Epidermal substitutes
Temporary application	Improve wound bed	Dermal substitutes Epidermal–dermal substitutes Allogeneic products
Permanent engraftment	Epidermal regeneration	Autologous keratinocyte products
Permanent engraftment	Dermal reconstruction	Autologous and possibly allogeneic fibroblast products
Permanent engraftment	Epidermal and dermal reconstruction	Autologous epidermal–dermal constructs

products may similarly be employed in the treatment of chronic wounds to improve the wound bed for subsequent healing.

Temporary skin substitutes can further be categorized as either processed biological tissue or tissue-engineered skin substitutes. Cell-based, tissue-engineered dressings are generally composed of cells, eg, keratinocytes, fibroblasts, and a carrier material. The carrier, either a fluid or a solid material or membrane, is used to deliver the cells to the wound. These materials can be of biologic origin or developed from synthetic materials. In addition to immunologic considerations, the origin of the particular biologic matrix materials may impact ethnic, cultural, and religious sensitivities. Permanent living skin substitutes are designed to replace missing skin elements like epidermis, dermis, or both, with the latter as an alternative to full-thickness skin grafts or flaps. Permanent grafts will incorporate into the host.

To protect and facilitate the early coverage of a wide variety of wound types, allogeneic and xenogeneic skin have been and continue to be employed as a temporary biologic dressing on excised or debrided wounds. An increased rate of survival was reported for burn victims covered with allogeneic cadaver skin on large excisional wounds [16,17]. When allograft skin has adhered to the wound and is subsequently removed, a well-vascularized wound bed usually remains, which is then suitable for autografting. Faster healing was observed in partial-thickness wounds covered with allogeneic skin [18]. Potentially severe problems such as viral infections have been linked to the use of allogeneic skin [19,20]. In addition, the quality of the individual grafts is variable, reflecting not only technical limitations but also tissue availability, especially when procured from elderly donors with very thin skin. Biologic properties of the allogenic skin can also vary dramatically, depending on the specific methodology employed in preservation. While fresh, irradiated GammaGraft® (Promethean LifeSciences, Pittsburgh, PA, USA), and cryopreserved skin, eg, AlloSkin™ (AlloSource, Centennial, CO, USA) and TheraSkin® (Soluble Systems, Newport News, VA, USA), have been found to maintain the highest biological properties as defined by adherence to the wound bed, glycerol-preserved allogeneic skin, eg, Euroskin (Euro Skin Bank, Beverwijk, The Netherlands) should be considered to be nonviable [21]. The potential for viral infection is the lowest in glycerol-preserved skin [22].

Xenogeneic skin is intended to be used across species. Frog skin and porcine skin are both used for the treatment and management of wounds and burns in humans, with porcine grafts most commonly employed. Porcine skin has generally been utilized as either a lyophilized or frozen product [23]. Porcine skin can be further processed to incorporate additional functions or properties such as the addition of silver. The silver-impregnated formulation EZ Derm® (Mölnlycke Health Care, Göteborg, Sweden) [24] is thought to provide additional antimicrobial protection. Xenogeneic skin may also have an inherent risk of viral transmission from, for example, porcine skin [25,26]. Allogeneic and xenogeneic skin are rejected within several weeks of placement on the wound. It has been reported that the application of a xenogeneic product may lead to antiporcine antibodies [27]. The clinical importance of this finding is unknown. Depending on the status of the patient and the quality, depth, and extent of the wound, the temporary skin substitutes will ultimately have to be replaced or a definitive wound closure performed using autologous tissue. The viability of the remaining cells in current xenogeneic products depends on the processing methods and may be low or absent. It is interesting to note that the immunologically privileged, biointegrated allograft may leave behind a well-integrated dermal layer, often successfully used to bond and accept cultured epidermal autograft preparations.

Amnion is one part of the fetal membrane and consists of a single layer of epithelium covering a stromal layer. Amnion has been used since the beginning of the last century and is a common biologic material in many countries. In fact it was not unusual in the era prior to HIV and AIDS to procure freshly donated amnion, washed and separated from the chorion, for the treatment of burn patients. Numerous publications have demonstrated that amnion can be further processed and refined (see review Ref. [28]). We have had some success with a dehydrated and sterilized human amnion/chorion allograft (EpiFix®; MiMedx, Marietta, GA, USA) [131]. Additional amniotic commercial products include Neox® (Amniox, Atlanta, GA, USA); Biovance® (Alliqua Biomedical, Langhorn, PA, USA); Amnioexcel® (Derma Sciences, Princeton, NJ, USA); and Grafix® (Osiris, Columbia, MD, USA). As amnion may have a risk of viral transmission, donor screening is necessary to minimize this risk. These products are cryopreserved, glycerol-preserved, provided in silver nitrate, or gamma-irradiated. Depending on the type of processing and preservation viability of the remaining cells in the placenta, derived products may be low or absent. In most cases, amnion is used for the treatment of partial-thickness wounds to expedite epithelialization.

8.3.3 Fibroblast-enriched skin substitutes

This type of product can be divided into allogeneic and autologous substitutes. Most products are composed of allogeneic fibroblasts and a carrier material, eg, collagen or synthetic material. Allogeneic fibroblasts can be obtained from discarded skin grafts as well as surgically resected skin from esthetic procedures or from neonatal foreskin. Fibroblasts will synthesize a variety of growth factors, cytokines, and matrix molecules, which can then be released into the wound bed. Currently fibroblast-enriched substitutes are mainly used for wounds, which show a senescent phenotype such as that seen in chronic wounds.

Transcyte® (Advanced Biohealing, San Diego, CA, USA) represents a cultured temporary skin substitute. Based on Biobrane®, human neonatal allogeneic fibroblasts are seeded on the nylon mesh. The cells proliferate rapidly within the mesh, and after several weeks a densely cellular tissue is formed [29]. Cultured human fibroblasts were shown to be minimal or nonimmunogenic, allowing for off-the-shelf use without a rejection response [123,124]. This tissue substitute contains high levels of secreted matrix molecules as well as multiple growth factors. Important factors stimulating epithelialization, such as keratinocyte growth factor and transforming growth factor (TGF)-α, were found in high levels [29]. The commercial Transcyte product is deep-frozen and nonviable. The conformability of Transcyte allows its use on wounds with an irregular shape and configuration. Two initial clinical trials demonstrated the efficacy of Transcyte for the temporary closure of excised wounds [30,31]. In addition, a controlled prospective study has demonstrated that Transcyte is effective for the treatment of partial-thickness wounds, leading to a 7-day-earlier wound closure in addition to reduced pain compared with conventional therapy [32]. Long-term results at 3, 6, and 12 months postburn showed reduced hypertrophic scarring as measured by the Vancouver scar scale. A prospective trial [126] compared the fibroblast-seeded Biobrane material (Transcyte) with the nonseeded Biobrane carrier material alone. The authors recorded 2-day-earlier epithelialization of partial-thickness burns in children and reduced autografting for the Transcyte group.

Dermagraft® (Organogenesis, Canton, MA, USA) is a living dermal replacement tissue that consists of human neonatal fibroblasts that are cultured on a biodegradable polyglactin mesh (Vicryl®) [33,34]. Fibroblasts secret multiple matrix proteins and growth factors during culturing. Dermagraft has been used in combination with meshed split-thickness skin grafts in humans and cultured keratinocytes in animal experiments [35,36]. Because fibroblasts cultured in a three-dimensional (3D) mesh appear to be nonantigenic in the allogeneic situation, Dermagraft can be considered to be a permanent replacement. Dermagraft has shown efficacy in the treatment of diabetic foot ulcers [29]. In order to improve results, Dermagraft must be reapplied in regular intervals. In a study by Pollak et al. [37], patients with diabetic foot ulcers either received Dermagraft or a conventional treatment. The study results demonstrated that 51% of the Dermagraft treatment group exhibited complete wound healing at 12 weeks, as opposed to 32% in the control group. In the pivotal trial in 314 diabetic patients with foot ulcers, the corresponding figures were 30% and 18% [128].

Hyalograft™ (Fidia Advanced Biopolymers, Abano Terme, Italy) is composed of autologous fibroblasts cultured on biodegradable esterified hyaluronic acid. Human fibroblasts are isolated from a patient's biopsy and cultured for 14 days. The cells are then seeded on the 3D nonwoven scaffold. One week later the fibroblast-enriched grafts are ready for delivery to the patient. Various observational studies and clinical trials have been performed with this material in chronic wounds and acute full-thickness wounds [38,39].

8.3.4 Permanent epidermal skin substitutes

To improve the availability of autologous wound coverage, in vitro culture techniques to grow autologous epidermis have been utilized [40,41]. The Rheinwald and Green method for in vitro passaging of single cell suspensions of keratinocytes on an irradiated mouse fibroblast line coupled with mitogens to form multilayered epithelial sheets is well described [42]. This technique represents a standard for permanent epidermal replacement and for commercially available cultured epithelial sheets (Epibase®, Laboratoires Genévrier, Antibes, France; Epicel®, Genzyme, Cambridge, MA, USA; Keratinozytensheets, DIZG, Berlin, Germany). Unfortunately, development, culturing, and processing time requirements remain high. The process starts with a biopsy size of 1–5 cm^2, and 3 weeks are required to cultivate a sufficient number of autologous epithelial sheets to cover an adult with a 70% total burn surface area, which corresponds to about 12,000 cm^2. Furthermore, cultured epithelial grafts are very expensive and difficult to handle. Confluent and stratified sheets (4–6 cell layers thick) are transferred to a backing material prior to grafting. Numerous reports on the clinical use of cultured epidermal sheets have appeared with variable results [43]. Early and late graft losses, infections, and friability of healed skin have been reported. Early take rates of cultured epithelium are reported from 40–80%, depending on the individual technique [44,45]. Actual take rates at discharge were reported at even lower rates. Nonadherence and the long-term tendency to form blisters following mechanical stress may be caused by disturbed adhesion properties of the keratinocytes as well as the abnormal structure of anchoring fibrils [46]. Basic science demonstrates that keratinocytes in differentiated cultured epithelial autografts resemble more senescent than activated cells [47]. This may at least in part explain a certain lack of adhesiveness on a fibronectin-rich wound bed, while take rates on a dermal bed are reported to be significantly higher.

Despite many drawbacks of cultured epithelial sheets, cultured grafts are utilized in the treatment of extensive skin loss in the absence of sufficient autologous donor skin. In addition, the use of cultured epithelial sheet grafts was reported for the supplementary treatment of superficial partial-thickness wounds of the face. As opposed to the full-thickness wound application, reduced scarring was noted when applied to the more superficially injured partial-thickness facial wound. Advances in the generation and release of cultured epithelial sheets could significantly enhance the use of these products.

An autologous cell substitute composed of cultured keratinocytes from the outer root sheath of scalp hair follicles (EpiDex; Modex Therapeutiques, Lausanne, Switzerland) has been successfully used to improve healing in chronic wounds [48].

An innovative carrier of esterified hyaluronic acid was developed at the University of Padova, Padova, Italy. The biopolymer has regular small holes allowing for the ingrowth of cells and subsequent outgrowth onto the wound bed. Autologous keratinocytes are isolated from a small biopsy, expanded in culture for 2 weeks, and then seeded on the carrier. The epidermal sheets are ready for transplantation 1 week after seeding [49,50].

As epithelialization is dependent on the nondifferentiated, migratory phenotype of keratinocytes, it was hypothesized that single cells, eg, suspended keratinocytes or single keratinocytes on a membrane, would be sufficient. This would also avoid the potentially damaging effect of enzymatic separation of cell sheets from culture vessels. Various approaches have been taken to deliver keratinocytes as single cells to a wound in fibrin glue, suspension, spray, or by polymeric films or spherical microcarriers at cell densities from 3×10^3 to 5×10^4 cells per cm^2 [51,52]. Autologous keratinocytes suspended in fibrin glue (TISSEEL; Baxter, Vienna, Austria) are sprayed onto the wound [132] or delivered via a membrane (MySkin®; Celltran, Sheffield, UK). Fiona Wood and her group from Perth, Australia, have demonstrated epithelialization of debrided wounds using cultured as well as noncultured keratinocytes suspended in liquid media (ReCell®, Avita Medical, Melbourn, UK). The cell suspension is prepared with the use of a commercially available kit onsite in the operating room. A small biopsy is taken from the patient and treated with trypsin, and the freed keratinocytes are resuspended in saline. A spray nozzle is used to deliver keratinocytes to the wound bed. Minimal cell loss was demonstrated and an even distribution of cells noted [53]. This method has also achieved success in reseeding melanocytes to the epidermis, which has clinical applications such as the restoration of skin color after burn wound coverage [54].

8.3.5 Temporary (allogeneic) epidermal substitutes

Allogeneic cultured keratinocyte grafts have been used with some success to expedite epithelialization in partial-thickness burn wounds [55]. Such grafts have also been used to improve the healing of chronic wounds [56]. Cells will not incorporate into the wound, as keratinocytes show strong antigenicity. While the availability of these off-the-shelf products seems intriguing, cost and practicability have hampered their use. With the availability of other biologic and synthetic dressings showing good efficacy in partial-thickness as well as deeper wounds, clinical interest in allogeneic keratinocyte products is decreasing.

Celaderm™ (Celadon Science LLC, Wellesley, MA, USA) is composed from allogeneic foreskin keratinocytes, which appear to be metabolically active but quiescent. The product comes cryopreserved and is applied without complicated thawing or rinsing procedures.

8.3.6 Permanent-cultured epidermal–dermal (composite) substitutes

Clinical results of transplanted epithelium have led to a consensus that a dermal substitute is needed to enhance the function of epithelial grafts. Therefore, efforts were taken to combine dermal substitutes and cultured keratinocytes in vitro prior to grafting. Handling the properties of the "composite" grafts improved dramatically. An additional benefit is that dislodging enzymes are unnecessary. Several variants on this technique have been described. In general, these include dermis-derived lattices, collagen-derived matrices, and cultured substrates [57].

The most refined technique was developed by Steve Boyce from the University of Cincinnati. A dermal lattice composed of collagen and glycosaminoglycans (collagen-GAG) was inoculated with autologous fibroblasts and seeded with autologous keratinocytes in vitro [58,59]. Inoculation with fibroblasts was necessary to promote epithelial growth on the upper surface. The autologous composite graft can then be transplanted onto the debrided burn wounds. Long-term follow-up clinical studies have demonstrated good results [60,61]. A major problem with dermal substrates for the delivery of keratinocytes as composite grafts is their availability for the grafting of extensive wounds.

8.3.7 Temporary cultured epidermal–dermal composites

Similarly constructed as the aforementioned permanent composite grafts, a group of products was designed to function temporarily on the wound to improve the quality of the graft bed and accelerate healing. It has been shown that such cocultured constructs exhibit synergistic expression of growth factors and cytokines as compared to a merely additive level. Similar to other allogeneic products, these products will not be incorporated into the wound [62]. These products are constructed to mimic skin with a bilayered design. Apligraf® (Organogenesis) is an allogeneic bilayered cultured skin equivalent that has both an upper epidermal and a lower dermal layer and contains human skin cells [63,64]. The dermal layer is formed by human foreskin fibroblasts, which are seeded into a purified bovine type I atelo-collagen. In this gel, fibroblasts provide additional structural and matrix proteins. The epidermal layer is formed by cultured human foreskin-derived keratinocytes, which are then differentiated to resemble the architecture of the human epidermis. Clinical trials have shown the efficacy of this material. Apligraf was compared in a large, randomized study by Veves et al. [65] with saline-moistened gauze in 208 patients with noninfected, nonischemic chronic plantar diabetic foot ulcers. At the 12-week follow-up, 56% patients treated with the human skin equivalent but only 38% control patients were healed. In another randomized controlled trial, the human skin equivalent was compared to compression therapy alone, demonstrating superiority in the percentage of healed venous leg ulcers and time to healing [122]. The cost and short shelf life are drawbacks of this material.

OrCel® (Forticell Bioscience, New York, NY) is similar to Apligraf and is also a bilayered construct composed of human neonatal allogeneic keratinocytes and fibroblasts. The scaffold cross-linked type I collagen sponge is coated on one side with atelo-collagen, which allows the seeding of keratinocytes on the nonporous side and fibroblasts on the opposite porous side. Indications for OrCel have been wounds on the hands of epidermolysis bullosa patients and skin graft donor sites in burn patients [129].

8.3.8 Temporary (allogeneic) epidermal–dermal substitutes

A temporary combined cell graft, including growth-arrested allogeneic keratinocytes and fibroblasts in a liquid carrier (Healthpoint Biotherapeutics, Fort Worth, TX, USA), was tested in a clinical phase 2b trial in venous leg ulcers [125]. It was demonstrated that cell delivery at a 2-week interval had a small but significant effect over control. The primary endpoint of this study was a reduction in ulcer size. The use of growth-arrested cells is another issue that needs further discussion [130]. The product (HP802-247) is currently evaluated in further clinical trials.

8.3.9 Conclusions

The adjunctive use of cultured epidermal autografts has shown significant applicability in the treatment of major thermal injuries. Clinical results show that keratinocytes from a small biopsy can provide lifelong coverage of extensive areas. This technology is commercially available and while expensive can be lifesaving. Unfortunately, durability, scarring, contracture, production time, and expense limit its use.

There is evidence that a small number of relatively undifferentiated keratinocytes, which may include keratinocyte stem cells, may be sufficient for achieving subsequent epidermal outgrowth. Implications from the keratinocyte stem cell models indeed suggest that these cells can give rise to a complete epidermis [66,67]. Identifying and transplanting these privileged cells as a suspension may shorten the time requirements for definitive coverage of excised extensive burn wounds.

Dermal substrates have allowed the successful use of thinner skin grafts, eased donor healing, and in some areas improved quality of healing. An interesting new field in organ transplantation has emerged from knowledge gained in the study of the survival of fetal allograft in pregnancy [68,69]. The generation of a gene-manipulated nonimmunogenic keratinocyte or skin substitute would allow the transplantation of an off the shelf product. A publication from a Canadian burn center has focused on this topic [70]. They successfully infected dermal fibroblasts with the immunosuppressant indoleamine 2,3-dioxygenase using an adenoviral vector.

8.4 Stem cells as wound healing biomaterials

The promise of employing stem cell science as an adjunct to promoting and improving the timeliness, character, and quality of wound healing is broadly publicized in evolving lay and scientific literature. While studies are encouraging and rapidly progressing, to date pure regenerative wound healing remains elusive. In this section, we review some of the pertinent available literature and clinical experience with an eye toward the future.

Besides the larger socioeconomic realities, a great challenge to our conventional wound healing approach has been the inability to specifically and temporally modulate the cascade. Growth factors, when presented with so much interest, proved to be tempered in clinical outcomes. It became rather evident clinically that a more methodical, responsive, and autoregulated approach is required for both efficacy as well as efficiency. The successful use of bone marrow transplantation over the past 40 years for a wide variety of diseases, such as leukemia, Fanconi anemia, and immunologic deficiencies has opened the door to the concept of employing stem cells as a regenerative therapeutic measure [71,72].

The term stem cell refers to an undifferentiated cell that is capable of not only proliferating but also differentiating into a wide variety of specialized cell types and still maintaining its capacity for autorenewal [73,74].

These early attributes alone established great interest in the fields of basic as well as clinical science. The possibilities seemed limitless at first, as were the concerns raised with respect to regulation, ethics, and safety.

Early studies with stem cells were developed from embryologic sources, first isolated in 1998. In an effort to resolve significant cultural, legal, and ethical challenges as well as increase the availability of autogenous transfer strategies, mature and cell-cultured populations are increasingly being studied and employed clinically. Adult stem cells can be procured from a wide variety of tissues, hematopoietic, and clinical sources. These cells are generally obtained from blood and bone marrow as well as autologous fat [75].

Two methods are most commonly employed on a cellular level to create clinically applicable stem cells. Induced pluripotent stem cells (IPS cells) describe a process by which adult cells are reverted back to a stem cell state by artificially turning on targeted genes [119,120]. Somatic cell nuclear transfer describes a process by which genetic material derived from an adult cell is transferred into an empty egg cell. A study demonstrated that nuclear transfer cells are more similar to real embryonic stem cells than IPS cells when comparing gene expression and DNA methylation [127]. IPS cells do possess several important and potentially clinical characteristics. In theory, IPS cells have the capacity of overcoming the challenges of immune rejection, as these cells are autologously derived and reprogrammed to become pluripotent. Currently, this process takes months. Many researchers believe that the most efficacious time point for introducing regenerative strategies is during the acute phase after injury. Shortening this process will certainly improve the clinical applicability.

Much of what we now understand with regard to the role of stem cells and wound healing involves the study of the mesenchymal stem cell (MSC). The chronic wound is classically characterized as a wound in which the peripheral, once actively dividing cells are senescent or refractory to the normal stimulatory messengers as well as one in which the normal extracellular matrix has been degraded with an imbalance of inflammatory mediators. MSCs have been shown to interact at all three classic wound healing stages and have been shown to be potent immunologic mediators, responsible for increasing the production of antiinflammatory cytokines such as interleukin (IL)-10 and IL-4 as well as inhibiting the production of proinflammatory cytokines such as TNF-α and interferon-γ [76]. The ability to modulate the stalled and exaggerated inflammatory phase may well be one of the most important contributions of MSCs in chronic wound healing [77,78].

The ability to clear infection by decreasing bacterial burden is a key element in successful wound healing. MSCs have been shown to have antimicrobial activity. They can combat infection by secreting immunomodulative factors in an effort to upregulate bacterial killing and phagocytosis by immune cells as well as by the direct secretion of antimicrobial peptides such as LL-37 [79,80].

8.4.1 Signaling

Complex cellular interactions and communications can be endocrine, paracrine, or autocrine in nature. Paracrine signaling describes cell signaling that occurs in the immediate extracellular environment of near and neighboring cells. Autocrine signaling defines the production of messenger which acts on the same cell, effecting change within the cell. Endocrine signaling denotes messenger effects far from the site of production of the messenger.

Stem cells have been shown to communicate via all three of the aforementioned mechanisms [81,82]. In so doing, the stem or progenitor cell may affect and modulate the wound healing cascade at a variety of time points or junctures both in near proximity to the wound site as well as from afar. It is these unique characteristics that hold particular promise in the field of regenerative wound healing.

8.4.2 Stem cell types

MSCs are considered to be multipotent in as much as they can differentiate into a wide variety of cell types but generally do not differentiate into hematopoetic cell lines [83]. In normal adult skin, MSCs are found in the perivascular spaces, dermal papilla, and the bulge of the hair follicle, suggesting their involvement in normal wound healing and epithelialization [84,85]. These stem cells are thought to originate from the mesodermal germinal layer and have been shown to differentiate into chondrocytes, osteoblasts, and adipocytes. After injury and wounding, hematopoietic stem cells and MSCs migrate to the site of injury where they appear to not only modulate the inflammatory response but also provide progenitor cell populations as well as stimulatory and paracrine effects to help regenerate, vascularize, and close the wound [86,87]. Genetic and cell culture manipulations, however, blur this distinction. Conventional sources of MSCs include a wide variety of sources, the richest of which may be Warton's Jelly from umbilical cord blood. High concentrations of MSCs have also been derived from the developing third molar tooth bud. Both modalities provide potential banking opportunities for the future. It is interesting to note that both age and sex impact the availability of stem cells for clinical application; the older the patient source, the lower the concentration of available stem cells. Higher cell counts are generally obtained from males [88].

In an effort to objectively classify this diverse population of cells the International Society for Cellular Therapy proposed a unified definition for MSCs in 2006, based on a list of minimal specifications to be met by a cell population [89]:

- The cells adhere to plastic under standard culture conditions.
- Most of them (>95%) express CD73 (5′-ectonucleotidase), CD90 (Thy-1), and CD105 (endoglin).
- At least 98% of them do not express the hematopoetic markers CD14, CD34, and CD45.
- They can differentiate in vitro along the osteoblastic, chondrogenic, and adipogenic lineages.

As opposed to treatment with embryonic and fetal-derived stem cell lines, which may require the use of immunosuppressant therapy, MSCs in general do not seem to evoke a significant immunologic response. A variety of proteins secreted by the MSCs suppress the immune system. Whether this characteristic will help balance the sometimes inhibitory inflammatory effects on local wound healing or provide the basis for allogeneic regenerative wound healing is currently being evaluated in a number of studies.

8.4.3 Bone marrow-derived stem cells

MSCs were first isolated from bone marrow [90]. Bone marrow-derived mesenchymal stem cells (BMMSCs) represent a minority of nucleated cells in the bone marrow. They are obtained from donors by aspiration, and then isolated and expanded in vitro by subculturing. In vitro studies have demonstrated that these cells have the ability to synthesize multiple growth factors like fibroblast growth factor and vascular endothelial growth factor as well as collagen, all potentially beneficial adjuncts to wound healing [91,92]. Since approximately 2003, a variety of case reports have demonstrated the beneficial effects of BMMSCs for the treatment of challenging and often refractory wounds such as radiation-induced wounds [92]. Of note is their ability to differentiate into a variety of cell/tissue types, their stimulatory effect on fibroblasts, and their ability to improve or promote local vascularity [93–95].

8.4.4 Adipose-derived stem cells

Zuk et al. [96] isolated multilineage stem cells from human lipoaspirates known as adipose-derived stem cells (ADSCs). These cells demonstrate the capability to differentiate into all three germ layers and have been shown to have low telomerase activity and low tumorigenicity [97,98]. As opposed to bone marrow where only a small percentage of the nucleated cells are MSCs, a 500-fold increase in the number of ADSCs can be obtained from a similar extracted volume. While not the same as MSCs, ADSCs are generally quite homogenous and express similar surface antigens to those found on MSCs. This abundant and generally easy to access source of stem cells provides the clinician with an expanded source of readily available viable and autologous cells.

Mouse models with deficient adipogenesis are known to have impaired wound healing. A study by Barbara Schmidt demonstrated a role for autologous intradermal adipocytes in the normal wound healing of punch biopsy sites in mice, supporting the concept of augmenting ADSCs and autologous stem cell lines in an effort to improve wound healing and intercellular communication [99].

ADSCs have been shown to secrete in significant quantities, antiapoptotic agents and angiogenic agents, including vascular endothelial growth factor and TGF-β [100]. One study has shown that ADSCs improved the biologic effects of increased blood flow in a model of hindlimb vascular injury in mice when injected intramuscularly as compared to non-ADSC-injected animals, raising the possibility of improved wound healing in ischemic and inflamed tissue [101].

8.4.5 Skin-derived stem cell populations

Stem cells can be found in both the epidermis and the dermis. Perhaps the best studied of these populations are the hair follicle-derived multipotent cells. These are found in the bulge, a part of the outer root sheath. They are thought to play a regenerative role in the healing of traumatic defects and have been shown to migrate upward in an effort to close wounds [102–104].

In the dermis, hair follicle development is thought to be regulated by a population of mesenchymal cells termed dermal papilla cells. These cells are increasingly being studied, as they are thought to represent a reservoir of multipotent cells. Skin-derived progenitor cells have been shown to differentiate into a wide variety of cell types, including glia, neurons, smooth muscle, fat, and other cell types. They have been shown to contribute to extracellular matrix homeostasis [105,106].

8.4.6 *Intravascular administration of stem cells for wound healing*

Improving the likelihood of quality wound healing is a multifactorial challenge. The control of biologic and bacterial burden, edema, trauma, and moisture are all critical considerations. However, without adequate and competent vascular perfusion, wounds cannot heal. While many novel products and strategies have directed their focus at the wound site proper, it is imperative to address peripheral vascular disease. In 2002, a novel study of peripheral vascular disease was published in *The Lancet* in which 25 patients underwent transplantation of BMMSCs. Patients were noted to have improved and sustained ankle brachial indices and tissue oxygen tensions [107].

Another study published in 2012 reviewed 15 patients with critical limb ischemia who were treated with a series of intramuscular injections of ADSCs. In this study, based on the proangiogenic effects of MSCs, approximately 67% of patients demonstrated varying degrees of wound healing, and the majority of a subset study population demonstrated an increase in the number of vascular collaterals [108]. Neither ankle brachial indices nor tissue oxygen pressures were improved while pain and ulcer size did improve. IPS cells appeared to maintain epigenetic memory derived from the cell's origin. This implied that cardiac cells are more readily derived from other cardiac cells, as compared to fibroblast cells, for example [109].

8.4.7 *Radiation injury*

Several clinicians have described the use of ADSCs for the treatment of radiation-associated wounds. Akita et al. [110] describe several cases in which ADSCs were employed to treat refractory radiation associated with chronic wounds; some patients also received a dermal substrate in addition to the stem cell fraction. Rigotti et al. [111] employed purified lipoaspirate-derived ADSCs to treat radiation wounds. Clinical improvements were noted in most patients, as was neovascularization and histology consistent with ultrastructural repair.

8.4.8 *Regenerative potential*

In mammals, scarless regenerative healing has been demonstrated in the first trimester in utero. There are several examples of postnatal limited regenerative healing, most notably in the liver [112]. Early trials introducing skeletal myocytes to the injured heart proved to be only partially successful, despite achieving structural repair. The cells were noted electromechanical coupled to the cardiac myocytes compromising potential functional contractile efficiency [113]. A meta-analysis of bone marrow cells injected via the coronary circulation for acute myocardial infection resulted in only a very modest improvement of cardiac functioning [114]. The modest beneficial effects observed may well be attributed more to the paracrine effects of delivered cells as opposed to cellular engraftment or transdifferentiation. The contribution of improved adjunctive medical regimens must also be strongly considered in interpreting these studies.

Several animal studies have demonstrated successful regeneration and repair in muscle and tendon injury repair. Of a particular interest was a study by Mori et al. [115] in which the transplantation of ADSCs into a skeletal muscle injury model promoted repair of muscle tissues, with both histological and functional muscle improvements. They were able to identify augmented angiogenesis and myogenesis and minimized fibrosis. The effects were identified as resulting from paracrine effects as opposed to differentiation.

Yamamoto et al. [116] described the treatment of three patients suffering moderate stress incontinence after radical prostatectomy with ADSCs injected into the urethral sphincter and periurethral tissue. Early results demonstrated improved continence and blood flow to the area as well as maintenance of injected cells.

There have been well over 1000 patients treated with fat and/or stem cells for mastectomy defects. Perez-Cano et al. [117] published their experience using ADSCs to treat partial mastectomy defects in 65 patients, the majority of whom also had radiation therapy. In this paper, 54 out of 65 patients demonstrated objective evidence of improvement by magnetic resonance imaging based on blinded assessment. From a clinical perspective, 34 patients had no change, and four patients reported a worsening of symptoms. Postradiation fibrosis was evaluated at 12 months and noted to be improved in 29 patients, while three patients had worse symptoms, and 35 patients related no change. No malignant change was noted in any patient. The investigators concluded that further study is warranted and ongoing.

8.4.9 Conclusions

A definitive consensus on the use of stem cells in wound healing has not been established. Stem cells are not available as easily stored products with long shelf lives. They may be time-consuming, expensive, and problematic to apply. Improved research methodologies, a better understanding of the science, and long-term, well-defined clinical studies are needed. Some effort has been taken to identify stem cells, which can give rise to a fully developed skin. Because of legal restrictions, research strategies have focused on autologous adult-derived stem cell therapies. From published experimental studies [71,72] and ongoing work from our group it seems reasonable to assume that it may be possible to generate a complete skin-like substitute from BMMSCs.

The potential of reprogramming terminally differentiated cells is a continuing challenge [119,120]. If autologous mature cell lines can be reprogrammed to an induced true pluripotent stem cell form in vivo, as has been shown in vitro, and if these lines can be directed appropriately toward regenerative pathways without the threat of immunologic, malignant, infectious, or degenerative concerns, many of our challenging wounds might well be optimally treated.

We are beginning to gain an understanding of the mitogenic and stimulatory factors that direct stem cells to the site of injury and wound. Hypoxia inducible factor (HIF)-1 has been noted to be deficient in mature individuals. Animal experiments have demonstrated improved healing when rodents were supplemented with an improved working copy of the HIF-1 gene [121]. Whether this particular factor or genetic modulation may play a relevant clinical role in human wound healing has yet to be determined. Investigation into directing native and supplemented stem cell populations to regions in need seems to be a likely course for study.

8.5 Biofabrication

Biofabrication is usually defined as the production of complex biologic products from raw materials such as living cells, matrices, biomaterials, and molecules. This rapidly evolving technology has been stimulated by the development of 3D fabrication technologies. This field builds upon currently available wound healing strategies and methodologies.

Attempts at combining the most favorable attributes of these technologies may be further bolstered by the addition of progenitor and stem cell derivatives with the hope of improving revascularization and repopulation in an effort to optimize resultant form and function. A bioprinting device may allow for in situ biofabrication of skin substitutes directly in the patient (Fig. 8.1). Complex constructs can be grown with layered living cell types to duplicate or at least mimic autologous tissue in an effort to regenerate healthy vascularized tissues, organs, and skin. Preliminary work has already been done with bottom-up and top-down approaches to produce an organized 3D multilayer skin substitute containing both keratinocytes as well as fibroblasts [118].

8.6 Clinical guidelines are needed

It is obvious that legal restrictions hamper the worldwide distribution of tissue-engineered products. Despite the fact that many of these products have undergone clinical trials and a well-documented track record can be provided, clinical usage is forbidden. A coordinated approach to providing patients worldwide with these important technologies seems necessary. Reimbursement for treatments is not always based on the efficacy of clinical trials, patient benefit, or even overall cost benefit. The development of a successful biomaterial, which potentially expedites wound closure, does not ensure that it will be used as a standard of care, unless providers understand the benefits of the therapy. Although there may be significant variations between products related to clinical study outcomes, product cost, and treatment effectiveness, there are no acceptable guidelines to assist the clinician with an appropriate selection. The majority of studies compare the new treatment modality to the standard of care rather than to similar products. Clinicians are left with their own judgment in determining which product to select for treatment. It is important for governing bodies and especially wound healing societies to initiate a comprehensive and thorough review of all the medical literature, provide treatment guidelines, and assist agencies determining reimbursement, including guidance on coding and payment for living cell products. Biomaterials should not all be placed in the same treatment category with the same payment structure.

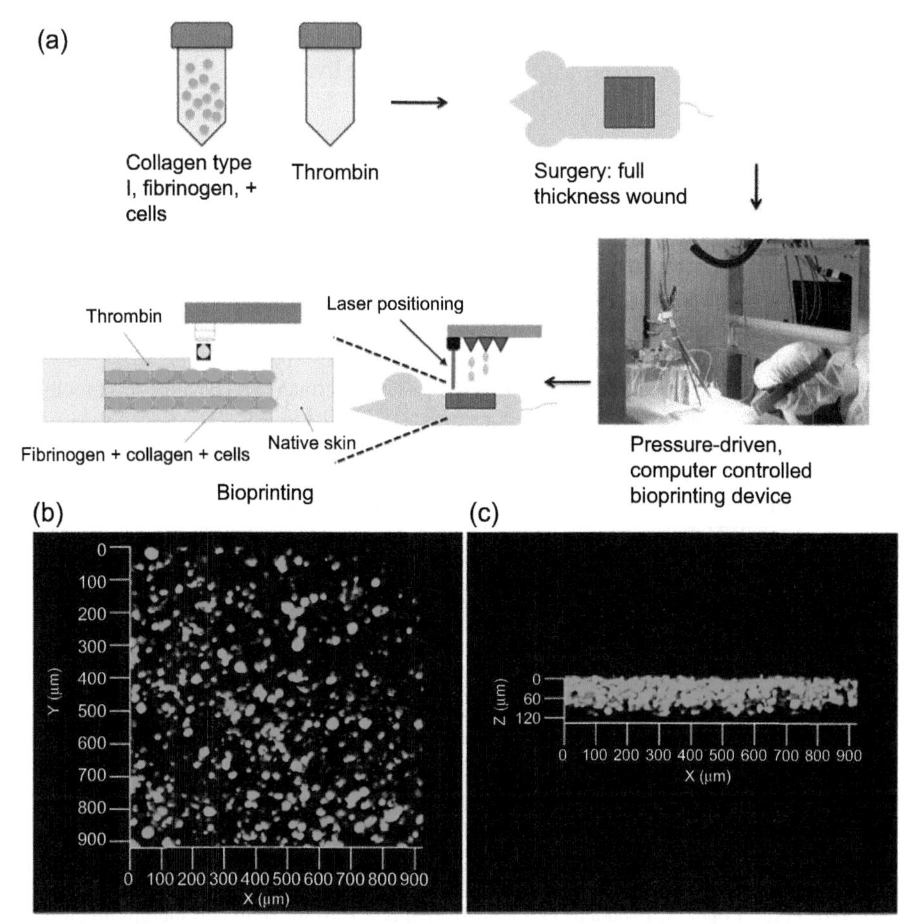

Figure 8.1 Bioprinting stem cells for the treatment of skin wounds [10]. (a) A schematic describing the approach by which amniotic fluid-derived stem (AFS) cells are bioprinted in order to increase the healing of a full-thickness skin wound. Wounds containing the deposited gels with green fluorescent protein-tagged AFS cells were harvested 24 h postprinting and analyzed with confocal microscopy. Images revealed evenly distributed cells in the gels, as viewed from above (b) or from the side (c).
From Skardal A, Mack D, Kapetanovic E, Atala A, Jackson JD, Yoo J, Soker S. Bioprinted amniotic fluid-derived cells accelerate healing of large skin wounds. Stem Cells Transl Med 2012;1:792–802.

8.7 The future

We are in the early phases of tissue regenerative technologies with almost unlimited possibilities for the development of new technologies. A list of all ongoing studies in the United States, including regenerative therapies, may be found at www.fda.gov. One can anticipate that the future will hold solutions to all of our current barriers although timelines cannot be determined. It is the hope of the authors that we will see the evolvement of cellular biomaterials into daily treatments, which will significantly affect the field of tissue repair, during the span of our lives.

References

[1] Mulder G, Wallin K, Tennenhaus M. Regenerative materials that facilitate wound healing. Clin Plastic Surg 2012;39:249–67.
[2] Goa S, DiPietro LA. Factors affecting wound healing. J Dent Res 2010;89:219–29.
[3] European Wound Management Association (EWMA). Position document: Hard-to heal wounds: a holistic approach. London: MEP Ltd; 2008.
[4] Schultz GS, Davidson JM, Kirsner RS, Bornstein P, Herman IM. Dynamic receprocity in the wound environment. Wound Repair Regen 2011;19:134–48.
[5] Falanga V, Sabolinki M. A bilayered skin construct (APLIGRAF) accelerates complete closure of hard-to-heal venous ulcers. Wound Repair Regen 1999;7:201–7.
[6] Muhart M, Mcfalls S, Kirsner RS, Elgart GW, Kerdel F, Sabolinski ML, Hardin-Young J, Eaglstein WH. Behavior of tissue-engineered skin: a comparison of living kin equivalent, autograft, and occlusive dressing in human donor sites. Arch Dermatol 1999;135:913–8.
[7] Van Winterswijk PJ, Nout E. Tissue engineering and wound healing: an overview of the past, present and future. Wounds 2007;19:277–84.
[8] Eugene Bell Center www.mbl.edu/bell.
[9] Falabella AF, Schachner LA, Valencia IC, Eaglstein WH. The use of tissue-engineered skin (Apligraf) to treat a newborn with epidermolysis bullosa. Arch Dermatol 1999;135:1219–22.
[10] Skardal A, Mack D, Kapetanovic E, Atala A, Jackson JD, Yoo J, Soker S. Bioprinted amniotic fluid-derived cells accelerate healing of large skin wounds. Stem Cells Transl Med 2012;1:792–802.
[11] Mulder GD, Lee DK, Niofar F. Autologous bone marrow-derived stem cells for chronic wounds of the lower extremity: a retrospective study. Wounds 2010;22:219–25.
[12] Menendez-Menendez Y, Alvarez-Viejo M, Ferrero-Gutierrez A, Perez-Basterrechea M, Perez Lopez S, Escudero D, Otero-Hernandez J. Adult stem cell therapy in chronic wound healing. J Stem Cell Res Ther 2014;4(1).
[13] European Wound Management Association (EWMA). Position document: Wound bed preparation in practice. London: MEP Ltd; 2004.
[14] Okan D, Wook K, Ayello EA, Sibbald G. The role of moisture balance in wound healing. Adv Skin Wound Care 2007;20:39–53.
[15] Mulder GD, Vande Berg JS. Cellular senescence and matrix metalloproteinase activity in chronic wounds. Relevance to debridement and new technologies. J Am Podiatr Med Assoc 2002;92:34–7.
[16] Wolfe RA, Roi LD, Flora JD, Feller I, Cornell RG. Mortality differences and speed of wound closure among specialized burn care facilities. JAMA 1983;250:763–6.
[17] Burke JF, Bondoc CC, Quinby WC. Primary burn excision and immediate grafting: a method shortening illness. J Trauma 1974;14:389–95.
[18] Vloemans AF, Middelkoop E, Kreis RW. A historical appraisal of the use of cryopreserved and glycerol-preserved allograft skin in the treatment of partial thickness burns. Burns 2002;28(Suppl. 1):S16–20.
[19] Kealey GP, Aguiar J, Lewis II RW, Rosenquist MD, Strauss RG, Bale Jr JF. Cadaver skin allografts and transmission of human cytomegalovirus to burn patients. J Am Coll Surg 1996;182:201–5.

[20] Kobayashi H, Kobayashi M, McCauley RL, Herndon DN, Pollard RB, Suzuki F. Cadaveric skin allograft-associated cytomegalovirus transmission in a mouse model of thermal injury. Clin Immunol 1999;92:181–7.
[21] Mackie DP. Postal survey on the use of glycerol-preserved allografts in clinical practice. Burns 2002;28(Suppl. 1):S40–4.
[22] Van Baare J, Buitenwerf J, Hoekstra MJ, du Pont JS. Virucidal effect of glycerol as used in donor skin preservation. Burns 1994;20(Suppl. 1):S77–80.
[23] Berry RB, Hackett MEJ. A comparative evaluation of lyophilized homograft, lyophilized pigskin and frozen pigskin biological dressing. Burns 1979;7:84–9.
[24] Ersek RA, Denton DR. Cross-linked silver impregnated skin for burn wound management. J Burn Care Rehabil 1988;9:476–81.
[25] Fishman JA, Patience C. Xenotransplantation: infectious risk revisited. Am J Transplant 2004;4:1383–90.
[26] Boneva RS, Folks TM. Xenotransplantation and risk of zoonotic infections. Ann Med 2004;36:504–17.
[27] Harris NS, Compton JB, Abson S, Larson DL. Comparison of fresh, frozen and lyophilized porcine skin as xenograft on burned patients. Burns 1976;2:71–5.
[28] Kesting MR, Wolff KD, Hohlweg-Majert B, Steinstraesser L. The role of allogenic amniotic membrane in burn treatment. J Burn Care Res 2008;29:907–16.
[29] Hansbrough JF, Morgan J, Greenleaf G, Underwood J. Development of a temporary living skin replacement composed of human neonatal fibroblasts cultured in Biobrane, a synthetic dressing material. Surgery 1994;115:633–44.
[30] Purdue GF, Hunt JL, Still Jr JM, Law EJ, Herndon DN, et al. A multicenter clinical trial of a biosynthetic skin replacement, dermagraft-TC, compared with cryopreserved human cadaver skin for temporary coverage of excised burn wounds. J Burn Care Rehabil 1997;18(1 Pt 1):52–7.
[31] Hansbrough JF, Mozingo DW, Kealey GP, Davis M, Gidner A, Gentzkow GD. Clinical trials of a biosynthetic temporary skin replacement, dermagraft-transitional covering, compared with cryopreserved human cadaver skin for temporary coverage of excised burn wounds. J Burn Care Rehabil 1997;18:43–51.
[32] Noordenbos J, Dore C, Hansbrough JF. Safety and efficacy of transcyte for the treatment of partial-thickness burns. J Burn Care Rehabil 1999;20:275–81.
[33] Cooper ML, Hansbrough JF, Spielvogel RL, Cohen R, Bartel RL, Naughton G. In vivo optimization of a living dermal substitute employing cultured human fibroblasts on a biodegradable polyglycolic acid or polyglactin mesh. Biomaterials 1991;12:243–8.
[34] Hansbrough JF, Morgan JL, Greenleaf GE, Bartel R. Composite grafts of human keratinocytes grown on a polyglactin mesh-cultured fibroblast dermal substitute function as a bilayer sin replacement in full thickness wound on athymic mice. J Burn Care Rehabil 1993;14:485–94.
[35] Hansbrough JF, Doré C, Hansbrough WB. Clinical trials of a living dermal tissue replacement placed beneath meshed, split-thickness skin grafts on excised burn wounds. J Burn Care Rehabil 1992;13:519–29.
[36] Rennekampff HO, Hansbrough JF, Kiessig V, Abiezzi S, Woods Jr V. Wound closure with human keratinocytes cultured on a polyurethane dressing overlaid on a cultured human dermal replacement. Surgery 1996;120:16–22.
[37] Pollak RA, Edington H, Jensen JL, Kroeker RO, Genztkow GD. A human dermal replacement for the treatment of diabetic foot ulcers. Wounds 1997;9:175–83.
[38] Galassi G, Brun P, Radice M, Cortivo R, Zanon GF, Genovese P, Abatangelo G. In vitro reconstructed dermis implanted in human wound: degradation studies of a HA-based supporting scaffold. Biomaterials 2000;21:2183–91.

[39] Caravaggi C, De Giglio R, Pritelli C, Sommaria M, Dalla Noce S, Faglia E, et al. HYAFF 11-based autologous dermal and epidermal grafts in the treatment of noninfected diabetic plantar and dorsal foot ulcers: a prospective, multicenter, controlled, randomized clinical trial. Diabetes Care 2003;26:2853–9.
[40] Green H, Kehinde O, Thomas J. Growth of cultured human epidermal cells into multiple epithelia suitable for grafting. Proc Natl Acad Sci USA 1979;76:5665–8.
[41] Gallico GG, O'Connor NE, Compton CC, Kehinde O, Green H. Permanent coverage of large burn wounds with autologous cultured human epithelium. N Engl J Med 1984;311:448–51.
[42] Rheinwald JG, Green H. Serial cultivation of strains of human keratinocytes-the formation of keratinizing colonies from single cells. Cell 1975;6:331–43.
[43] Hansbrough JF, Franco ES. Skin replacements. Clin Plast Surg 1998;25:407–23.
[44] Rue III LW, Cioffi WG, McManus WF, Pruitt Jr BA. Wound closure and outcome in extensively burned patients treated with cultured autologous keratinocytes. J Trauma 1993;34:662–7.
[45] Desai MH, Mlakar JM, McCauley RL, Abdullah KM, Rutan RL, Waymack JP, Robson MC, Herndon DN. Lack of long-term durability of cultured keratinocyte burn-wound coverage: a case report. J Burn Care Rehabil 1991;12:540–5.
[46] Woodley DT, Peterson HD, Herzog SR, Stricklin GP, Burgeson RE, Briggaman RA, Cronce DJ, O'Keefe EJ. Burn wounds resurfaced by cultured epidermal autografts show abnormal reconstitution of anchoring fibrils. JAMA 1988;259:2566–71.
[47] Rennekampff HO, Hansbrough JF, Woods Jr V, Kiessig V. Integrin and matrix molecule expression in cultured skin replacements. J Burn Care Rehabil 1996;17:213–21.
[48] Tausche AK, Skaria M, Böhlen L, Liebold K, Hafner J, Friedlein H, Meurer M, Goedkoop RJ, Wollina U, Salomon D, Hunziker T. An autologous epidermal equivalent tissue-engineered from follicular outer root sheath keratinocytes is as effective as split-thickness skin autograft in recalcitrant vascular leg ulcers. Wound Repair Regen 2003;11:248–52.
[49] Myers SR, Grady J, Soranzo C, Sanders R, Green C, Leigh IM, Navsaria HA. A hyaluronic acid membrane delivery system for cultured keratinocyte autografts: clinical "take" rates in a porcine kerato-dermal model. J Burn Care Rehabil 1997;18:214–22.
[50] Lobmann R, Pittasch D, Muhlen I, Lehnert H. Autologous human keratinocytes cultured on membranes composed of benzyl ester of hyaluronic acid for grafting in non-healing diabetic foot lesions. A pilot study. J Diabetes Complications 2003;17:199–204.
[51] Chester DL, Balderson DS, Papini RP. A review of keratinocyte delivery to the wound bed. J Burn Care Rehabil 2004;25:266–75.
[52] Rennekampff HO, Kiessig V, Griffey S, Greenlaef G, Hansbrough JF. Acellular human dermis promotes cultured keratinocyte engraftment. J Burn Care Rehabil 1997; 18:535–44.
[53] Navarro FA, Stoner ML, Park CS, Huertas JC, Lee HB, Wood FM, et al. Sprayed keratinocyte suspensions accelerate epidermal coverage in a porcine microwound model. J Burn Care Rehabil 2000;21:513–8.
[54] Navarro FA, Stoner ML, Lee HB, Park CS, Wood FM, Orgill DP. Melanocyte repopulation in full thickness wounds using a cell spray apparatus. J Burn Care Rehabil 2001;22:41–6.
[55] Alvarez-Diaz C, Cuenca-Pardo J, Sosa-Serrano A, Juárez-Aguilar E, Marsch-Moreno M, Kuri-Harcuch W. Burns treated with frozen cultured human allogeneic epidermal sheets. J Burn Care Rehabil 2000;21:291–9.
[56] Bolivar-Flores YJ, Kuri-Harcuch W. Frozen allogeneic human epidermal cultured sheets for the cure of complicated leg ulcers. Dermatol Surg 1999;25:610–7.

[57] Hansbrough JF. Wound coverage with biologic dressings and cultured skin substitutes. Austin: RG Landes Inc.; 1992. p. 93–114.
[58] Boyce S, Christianson D, Hansbrough JF. Structure of a collagen-GAG skin substitute optimized for cultured human epidermal keratinocytes. J Biomed Mater Res 1988;22:939–57.
[59] Hansbrough JF, Boyce ST, Cooper ML, Foreman TJ. Burn wound closure with cultured autologous keratinocytes and fibroblasts attached to a collagen-glycosaminoglycan substrate. JAMA 1989;262:2125–30.
[60] Boyce S, Goretsky MJ, Greenhalgh DG, Kagan RJ, Rieman MT, Warden GD. Comparative assessment of cultured skin substitutes and native skin autograft for treatment of full thickness burns. Ann Surg 1995;222:743–52.
[61] Boyce ST, Supp AP, Wickett RR, Hoath SB, Warden GD. Assessment with the dermal torque meter of skin pliability after treatment of burns with cultured skin substitutes. J Burn Care Rehabil 2000;21:55–63.
[62] Griffiths M, Ojeh N, Livingstone R, Price R, Navsaria H. Survival of Apligraf in acute human wounds. Tissue Eng 2004;10:1180–95.
[63] Bell E, Ehrlich HP, Buttle DJ, Nakatsuji T. Living tissue formed in vitro and accepted as skin-equivalent tissue of full thickness. Science 1981;211:1052–4.
[64] Eaglstein WH, Falanga V. Tissue engineering and the development of ApliGraf, a human skin equivalent. Clin Ther 1997;19:894–905.
[65] Veves A, Falanga V, Armstrong DG, Sabolinski ML. Graftskin, a human skin equivalent, is effective in the management of noninfected neuropathic diabetic foot ulcers: a prospective randomized multicenter clinical trial. Apligraf diabetic foot ulcer study. Diabetes Care 2001;24:290–5.
[66] Rochat A, Kobayashi K, Barrandon Y. Localisation of stem cells of human hair follicles by clonal analysis. Cell 1994;76:1063–73.
[67] Jones PH, Harper S, Watt FM. Stem cell patterning and fate in human epidermis. Cell 1995;80:83–93.
[68] Munn DH, Zhou M, Attwood JT, Bondarev I, Conway SJ, Marshall B, Brown C, Mellor AL. Prevention of allogeneic fetal rejection by tryptophan catabolism. Science 1998;281:1191–3.
[69] Sedlmayr P, Blaschitz A, Wintersteiger R, Semlitsch M, Hammer A, MacKenzie CR, et al. Localization of indoleamine 2,3-dioxygenase in human fetal reproductive organs and the placenta. Mol Hum Reprod 2002;8:385–91.
[70] Ghahary A, Li Y, Tredget EE, Kilani RT, Iwashina T, Karami A, et al. Expression of indoleamine 2,3-dioxygenase in dermal fibroblasts functions as a local immunosuppressive factor. J Invest Dermatol 2004;122:953–64.
[71] Borue X, Lee S, Grove J, Herzog EL, Harris R, Diflo T, Glusac E, Hyman K, Theise ND, Krause DS. Bone marrow-derived cells contribute to epithelial engraftment during wound healing. Am J Pathol 2004;165:1767–72.
[72] Kataoka K, Medina RJ, Kageyama T, Miyazaki M, Yoshino T, Makino T, Huh NH. Participation of adult bone marrow cells in reconstruction of skin. Am J Pathol 2003;163:1227–31.
[73] Watt FM, Hogan BL. Out of Eden: stem cells and their niches. Science 2000;287:1427–30.
[74] Weissman IL. Stem cells: units of development, units of regeneration, and units in evolution. Cell 2000;100:157–68.
[75] Strioga M, Viswanathan S, Darinskas A, Slaby O, Michalek J. Same or not the same? Comparison of adipose tissue-derived versus bone marrow derived mesenchymal stem and stromal cells. Stem Cells Dev 2012;21:2724–52.

[76] Aggarwal S, Pittenger MF. Human mesenchymal stem cells modulate allogeneic immune cell responses. Blood 2005;105:1815–22.
[77] Newman RE, Yoo D, LeRoux MA, Danilkovitch-Miagkova A. Treatment of inflammatory diseases with mesenchymal stem cells. Inflamm Allergy Drug Targets 2009;8:110–23.
[78] Singer NG, Caplan AI. Mesenchymal stem cells: mechanisms of inflammation. Annu Rev Pathol 2011;6:457–78.
[79] Krasnodembskaya A, Song Y, Fang X, Gupta N, Serikov V, Lee JW, et al. Antibacterial effect of human mesenchymal stem cells is mediated in part from secretion of the antimicrobial peptide LL-37. Stem Cells 2010;28:2229–38.
[80] Mei SH, Haitsma JJ, Dos Santos CC, Deng Y, Lai PF, Slutsky AS, et al. Mesenchymal stem cells reduce inflammation while enhancing bacterial clearance and improving survival in sepsis. Am J Respir Crit Care Med 2010;182:1047–57.
[81] Gnecchi M, Zhang Z, Ni A, Dzau VJ. Paracrine mechanisms in adult stem cell signaling and therapy. Circ Res 2008;103:1204–19.
[82] Chen L, Tredget EE, Wu PY, Wu Y. Paracrine factors of mesenchymal stem cells recruit macrophages and endothelial lineage cells and enhance wound healing. PLoS One 2008;3:e1886.
[83] Barry FP, Murphy JM. Mesenchymal stem cells: clinical applications and biological characterization. Int J Cell Biol 2004;36:568–84.
[84] Higgins CA, Jahoda CA, Christiano AM. Human hair follicle neogenesis using dermal papilla cells. J Invest Dermatol 2012;132:S137.
[85] Uchugonova A, Duong J, Zhang N, Koenig K, Hoffman RM. Nestin-expressing multi-potent stem cells originate in the bulge of the hair follicle and migrate to the dermal papilla. J Invest Dermatol 2012;132:S139.
[86] Crigler L, Kazhanie A, Yoon TJ, Zakhari J, Anders J, Taylor B, et al. Isolation of a mesenchymal cell population from murine dermis that contains progenitors of multiple cell lineages. FASEB J 2007;21:2050–63.
[87] Deng W, Han Q, Liao L, Li C, Ge W, Zhao Z, et al. Engrafted bone marrow derived flk-(1+) mesenchymal stem cells regenerate skin tissue. Tissue Eng 2005;11:110–9.
[88] Dedeepiya VD, Rao YY, Jayakrishnan GA, Parthiban JK, Baskar S, Manjunath SR, Senthilkumar R, Abraham SJ. Index of CD34+ cells and mononuclear cells in the bone marrow of spinal cord injury patients of different age groups: a comparative analysis. Bone Marrow Res 2012;2012:787414.
[89] Dominici M, Le Blanc K, Mueller I, Slaper-Cortenbach I, Marini F, Krause D, et al. Minimal criteria for defining multipotent mesenchymal stromal cells. The international society for cellular therapy position statement. Cytotherapy 2006;8:315–7.
[90] Badiavas EV, Falanga V. Treatment of chronic wounds with bone marrow-derived cells. Arch Dermatol 2003;139:510–6.
[91] Falanga V, Iwamoto S, Chartier M, Yufit T, Butmarc J, Kouttab N, et al. Autologous bone marrow-derived cultured mesenchymal stem cells delivered in a fibrin spray accelerates healing in murine and human cutaneous wounds. Tissue Eng 2007;13:1299–312.
[92] Lataillade JJ, Doucet C, Bey E, Carsin H, Huet C, Clairand I, et al. New approach to radiation burn treatment by dosimetry-guided surgery combined with autologous mesenchymal stem cell therapy. Regen Med 2007;2:785–94.
[93] Yoshikawa T, Mitsuno H, Nonaka I, Sen Y, Kawanishi K, Inada Y, Takakura Y, Okuchi K, Nonomura A. Wound therapy by marrow mesenchymal cell transplantation. Plast Reconstr Surg 2008;121:860–77.
[94] Wu Y, Chen L, Scott PG, Tredget EE. Mesenchymal stem cells enhance wound healing through differentiation and angiogenesis. Stem Cells 2007;25:2648–59.

[95] Sasaki M, Abe R, Fujita Y, Ando S, Inokuma D, Shimizu H. Mesenchymal stem cells are recruited into wounded skin and contribute to wound repair by transdifferentiation into multiple skin cell type. J Immunol 2008;180:2581–7.

[96] Zuk PA, Zhu M, Mizuno H, Huang J, Futrell JW, Katz AJ, Benhaim P, Lorenz HP, Hedrick MH. Multilineage cells from human adipose tissue: implications for cell-based therapies. Tissue Eng 2001;7:211–28.

[97] Ogura F, Wakao S, Kuroda Y, Tsuchiyama K, Bagheri M, Heneidi S, Chazenbalk G, Aiba S, Dezawa M. Human adipose tissue possesses a unique population of pluripotent stem cells with nontumorigenic and low telomerase activities: potential implications in regenerative medicine. Stem Cells Dev 2014;23:717–28.

[98] Kondo K, Shintani S, Shibata R, Murakami H, Murakami R, Imaizumi M, Kitagawa Y, Murohara T. Implantation of adipose-derived regenerative cells enhances ischemia-induced angiogenesis. Arterioscler Thromb Vasc Biol 2009;29:61–6.

[99] Schmidt B. Defining the function of adipocyte lineage cells during skin wound healing. J Invest Dermatol 2012;132:S138.

[100] Rehman J, Traktuev D, Li J, Merfeld-Clauss S, Temm-Grove CJ, Bovenkerk JE, Pell CL, Johnstone BH, Considine RV, March KL. Secretion of angiogenic and antiapoptotic factors by human adipose stromal cells. Circulation 2004;109:1292–8.

[101] Kim Y, Kim H, Cho H, Bae Y, Suh K, Jung J. Direct comparison of human mesenchymal stem cells derived from adipose tissues and bone marrow in mediating neovascularization in response to vascular ischemia. Cell Physiol Biochem 2007;20:867–76.

[102] Oshima H, Rochat A, Kedzia C, Kobayashi K, Barrandon Y. Morphogenesis and renewal of hair follicles from adult multipotent stem cells. Cell 2001;104:233–45.

[103] Taylor G, Lehrer MS, Jensen PJ, Sun TT, Lavker RM. Involvement of follicular stem cells in forming not only the follicle but also the epidermis. Cell 2000;102:451–61.

[104] Levy V, Lindon C, Zheng Y, Harfe BD, Morgan BA. Epidermal stem cells arise from the hair follicle after wounding. FASEB J 2007;21:1358–66.

[105] Driskell RR, Clavel C, Rendl M, Watt FM. Hair follicle dermal papilla cells at a glance. J Cell Sci 2011;124:1179–82.

[106] Biernaskie J, Paris M, Morozova O, Fagan BM, Marra M, Pevny L, Miller FD. SKPs derive from hair follicle precursors and exhibit properties of adult dermal stem cells. Cell Stem Cell 2009;5:610–23.

[107] Tateishi-Yuyama E, Matsubara H, Murohara T, Ikeda U, Shintani S, Masaki H, Amano K, Kishimoto Y, Yoshimoto K, Akashi H, Shimada K, Iwasaka T, Imaizumi T. Therapeutic Angiogenesis using Cell Transplantation (TACT) Study Investigators. Therapeutic angiogenesis for patients with limb ischaemia by autologous transplantation of bone marrow cells: a pilot study and a randomised controlled trial. Lancet 2002;360:427–35.

[108] Lee HC, An SG, Lee HW, Park JS, Cha KS, Hong TJ, Park JH, Lee SY, Kim SP, Kim YD, Chung SW, Bae YC, Shin YB, Kim JI, Jung JS. Safety and effect of adipose tissue-derived stem cell implantation in patients with critical limb ischemia. Circ J 2012;76:1750–60.

[109] Xu H, Yi BA, Wu H, Bock C, Gu H, Lui KO, Park JH, Shao Y, Riley AK, Domian IJ, Hu E, Willette R, Lepore J, Meissner A, Wang Z, Chien KR. Highly efficient derivation of ventricular cardiomyocytes from induced pluripotent stem cells with a distinct epigenetic signature. Cell Res 2012;22:142–54.

[110] Akita S, Yoshimoto H, Akino K, Ohtsuru A, Hayashida K, Hirano A, Suzuki K, Yamashita S. Early experiences with stem cells in treating chronic wounds. Clin Plastic Surg 2012;39:281–92.

[111] Rigotti G, Marchi A, Galie M, Baroni G, Benati D, Drampera M, Pasini A, Sbarbati A. Clinical t reatment of radiotherapy tissue damage by lipoaspirate transplant: a healing process medicated by adipose derived adult stem cells. Plastic Reconstr Surg 2007;119:1409–22.

[112] Taub R. Liver regeneration: from myth to mechanism. Nat Rev Mol Cell Biol 2004;5:836–47.

[113] Tang XL, Rokosh G, Sanganalmath SK, Yuan F, Sato H, Mu J, Dai S, Li C, Chen N, Peng Y, Dawn B, Hunt G, Leri A, Kajstura J, Tiwari S, Shirk G, Anversa P, Bolli R. Intracoronary administration of cardiac progenitor cells alleviates left ventricular dysfunction in rats with a 30-day-old infarction. Circulation 2010;121:293–305.

[114] Zimmet H, Porapakkham P, Porapakkham P, Sata Y, Haas SJ, Itescu S, Forbes A, Krum H. Short- and long-term outcomes of intracoronary and endogenously mobilized bone marrow stem cells in the treatment of ST-segment elevation myocardial infarction: a meta-analysis of randomized control trials. Eur J Heart Fail 2011;14:91–105.

[115] Mori R, Kamei N, Okawa S, Nakabayashi A, Yokota K, Hiqashi Y, Ochi M. Promotion of skeletal muscle repair in a rat skeletal muscle injury model by local injection of human adipose tissue-derived regenerative cells. J Tissue Eng Regen Med 2012;9(10):1150–60. [Epub ahead of print].

[116] Yamamoto T, Gotoh M, Kato M, Majima T, Toriyama K, Kamei Y, Iwaguro H, Matsukawa Y, Funahashi Y. Periurethral injection of autologous adipose-derived regenerative cells for the treatment of male stress urinary incontinence: report of three initial cases. Intl J Urol 2012;19:652–9.

[117] Perez-Cano R, Vranckx JJ, Lasso JM, Calabrese C, Merck B, Milstein AM, Sassoon E, Delay E, Weiler-Mithoff EM. Prospective trial of adipose-derived regenerative cell (ADRC)-enriched fat grafting for partial mastectomy defects: the RESTORE-2 trial. J Cancer Surg 2012;38:382–9.

[118] Pereira RF, Barrias CC, Granja PL, Bartolo PJ. Advanced biofabrication strategies for skin regeneration and repair. Nanomedicine 2013;8:603–21.

[119] Takahashi K, Yamanaka S. Induction of pluripotent stem cells from mouse embryonic and adult fibroblast cultures by defined factors. Cell 2006;126:663–76.

[120] Takahashi K, Tanabe K, Ohnuki M, Narita M, Ichisaka T, Tomoda K, Yamanaka S. Induction of pluripotent stem cells from adult human fibroblasts by defined factors. Cell 2007;131:861–72.

[121] Cerrada I, Ruiz-Saurí A, Carrero R, Trigueros C, Dorronsoro A, Sanchez-Puelles JM, Diez-Juan A, Montero JA, Sepúlveda P. Hypoxia-inducible factor 1 alpha contributes to cardiac healing in mesenchymal stem cells-mediated cardiac repair. Stem Cells Dev 2013;22:501–11.

[122] Falanga V, Margolis D, Alvarez O, Auletta M, Maggiacomo F, Altman M, Jensen J, Sabolinski M, Hardin-Young J. Rapid healing of venous ulcers and lack of clinical rejection with an allogeneic cultured human skin equivalent. Arch Dermatol 1998; 134:293–300.

[123] Kern A, Liu K, Mansbridge J. Modification of fibroblast gamma-interferon responses by extracellular matrix. J Invest Dermatol 2001;117:112–8.

[124] Kern A, Liu K, Mansbridge J. Modulation of interferon-gamma response by dermal fibroblast extracellular matrix. Ann N Y Acad Sci 2002;961:364–7.

[125] Kirsner RS, Marston WA, Snyder RJ, Lee TD, Cargill DI, Slade HB. Spray-applied cell therapy with human allogeneic fibroblasts and keratinocytes for the treatment of chronic venous leg ulcers: a phase 2, multicentre, double-blind, randomised, placebo-controlled trial. Lancet 2012;380:977–85.

[126] Kumar RJ, Kimble RM, Boots R, Pegg SP. Treatment of partial-thickness burns: a prospective, randomized trial using Transcyte. ANZ J Surg 2004;74:622–6.

[127] Ma H, Morey R, O'Neil RC, He Y, Daughtry B, Schultz MD, Hariharan M, Nery JR, Castanon R, Sabatini K, Thiagarajan RD, Tachibana M, Kang E, Tippner-Hedges R, Ahmed R, Gutierrez NM, Van Dyken C, Polat A, Sugawara A, Sparman M, Gokhale S, Amato P, Wolf DP, Ecker JR, Laurent LC, Mitalipov S. Abnormalities in human pluripotent cells due to reprogramming mechanisms. Nature 2014;511:177–83.
[128] Marston WA, Hanft J, Norwood P, Pollak R, Dermagraft Diabetic Foot Ulcer Study Group. The efficacy and safety of dermagraft in improving the healing of chronic diabetic foot ulcers: results of a prospective randomized trial. Diabetes Care 2003;26:1701–5.
[129] Still J, Glat P, Silverstein P, Griswold J, Mozingo D. The use of a collagen sponge/living cell composite material to treat donor sites in burn patients. Burns 2003;29:837–41.
[130] Zhang GY, Li QF, Cai JL, Cao YL, Fu XB. Allogeneic fibroblasts and keratinocytes for venous leg ulcers. Lancet 2013;381:372.
[131] Zelen CM, Gould L, Serena TE, Carter MJ, Keller J, Li WW. A prospective, randomised, controlled, multi-centre comparative effectiveness study of healing using dehydrated human amnion/chorion membrane allograft, bioengineered skin substitute or standard of care for treatment of chronic lower extremity diabetic ulcers. Int Wound J 2015;12:724–32.
[132] Hartmann A, Quist J, Hamm H, Bröcker EB, Friedl P. Transplantation of autologous keratinocyte suspension in fibrin matrix to chronic venous leg ulcers: improved long-term healing after removal of the fibrin carrier. Dermatol Surg 2008;34:922–9.

Biomaterials for dermal substitutes

M.M.W. Ulrich[1,2], *M. Vlig*[1], *B.K.H.L. Boekema*[1]
[1]Association of Dutch Burn Centers, Beverwijk, The Netherlands; [2]VU University Medical Center, Amsterdam, The Netherlands

9.1 Introduction

Several dermal substitutes have been developed to improve wound healing for large full-thickness skin defects. Several of these products are used in the clinic, aiming to accelerate wound closure and to overcome massive scarring. Typical wounds are diabetic foot ulcers and chronic leg ulcers, full-thickness dermal burns, and acute open or surgical wounds. Developments in primary care and treatment techniques have resulted in an increased survival of patients with large extensive skin defects. Although new treatment techniques have greatly improved the outcome of wound healing, the result is still not optimal with respect to function and cosmetic appearance. The ideal skin substitute should mimic the normal dermal structure and composition or create a microenvironment able to guide the infiltrating cells to regenerate the dermis. In addition, this biomaterial should not elicit an inflammatory reaction.

Most of the biomaterials that have been used successfully in the clinic consist of animal-derived collagen scaffolds supplemented with other extracellular matrix (ECM) components present in the skin such as elastin and glycosaminoglycan (GAG; van Zuijlen et al., 2000; Heitland et al., 2004). The different components determine the mechanical properties of the material such as viscoelasticity and stiffness. In addition, they play an important role in the storage of growth factors, determine the phenotype of resident cells, and influence the immune response. Cellular performance and immunological reactions are affected by pore size, cross-linking of the material, manufacturing procedures, and degradation rate (Badylak and Gilbert, 2008; Boekema et al., 2014; Kwon et al., 2014). An overview of materials, methods, and key parameters for the development of dermal substitutes is given in Fig. 9.1.

9.1.1 Basic compositional and functional characteristics

The main function of biomaterials for dermal substitution is to provide a stable and biodegradable scaffold, which allows the ingrowth of cells to produce dermal tissue and restore dermal function. The material should be degradable in order to allow for the synthesis of the normal complex dermal ECM by dermal fibroblasts without evoking an immunological reaction, which could hamper the healing process. The dermal scaffold should be sufficiently stable to support the fibroblasts for sustained period allowing them to build up the steady new dermal ECM, which has comparable

Wound Healing Biomaterials - Volume 1. http://dx.doi.org/10.1016/B978-1-78242-455-0.00009-4

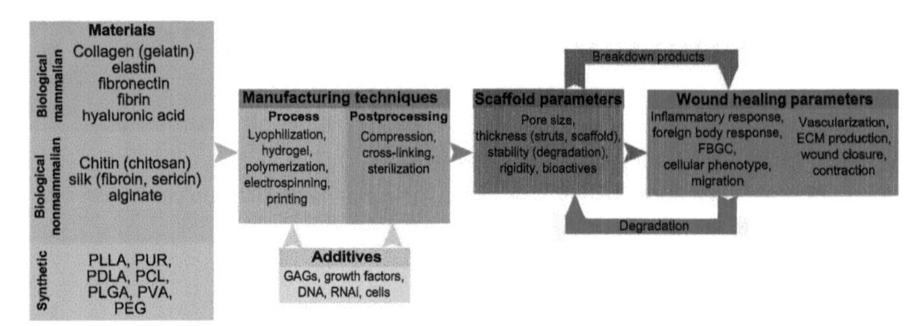

Figure 9.1 Overview of methods for biomaterial development for dermal substitutes. *DNA*, deoxyribonucleic acid; *ECM*, extracellular matrix; *FBGC*, foreign body giant cells; *GAGs*, glycosaminoglycans; *PCL*, polycaprolactone; *PDLA*, poly (D,L-lactic acid); *PEG*, poly (ethylene glycol); *PLGA*, poly (lactic-co-glycolic acid); *PLLA*, poly (L-lactic acid); *PUR*, polyurethane; *PVA*, polyvinylalcohol; *RNAi*, ribonucleic acid interference.

morphology and mechanical properties to native skin. In addition, it should not evoke an immunological reaction, which could hamper the healing process.

Several biomaterials have been evaluated for dermal substitution in preclinical studies, but only a few are currently being used in the clinic. These substitutes vary from complete decellularized dermis, which is mainly derived from human or porcine skin, to engineered constructs of (purified) biomolecules such as collagen. The latter mostly consists of a collagen backbone, which might be supplemented with other dermal ECM components such as heparin, elastin, or growth factors like fibroblast growth factor-7 (or keratinocyte growth factor) and vascular endothelial growth factor (VEGF) (Nillesen et al., 2011). These additives may improve the repopulation of the scaffold by fibroblasts and endothelial cells necessary for ECM deposition and revascularization, respectively. The initiation of these processes is the first requirement for a successful dermal substitute. The restoration of the blood circulation is one of the most important events in organ transplantation and thus also in dermal reconstruction; without restoration of the blood supply, the formation or survival of new tissue is impossible. Hence the successfulness of a biomaterial for dermal substitution is largely dependent on the angiogenic capacity of the material. Ample research has been performed to induce or increase the rate of revascularization.

Two frequently used biomaterials for skin substitution in the clinic are the products Integra® and Matriderm®. These materials are implanted onto the wound bed and are covered with a split skin autograft to provide epidermal cells. The vascularization of these materials is completely different, which has major consequences for the treatment procedure. Integra® requires a two-step procedure to allow vascularization of the substitute. This dermal substitute is supplied with a temporary silicone layer to close the wound and reduce the risk of infection during the vascularization period. Approximately 20 days after application to the wound, a split skin autograft can be transplanted onto the now vascularized dermal substitute.

Matriderm®, on the other hand, is applied in a one-step procedure where the dermal substitute and the split skin autograft are transplanted in the same operation procedure (Tong et al., 2014). Vascularization of Matriderm® proceeds much faster, probably due to the elastin lysate coating of the collagen fibers. It was shown that elastin-derived peptides attract endothelial cells and promote angiogenesis (Robinet et al., 2005; Tummalapalli and Tyagi, 1999). In contrast, the chondroitin-6-sulphate in Integra® is antiangiogenic (Hahnenberger and Jakobson, 1991).

9.1.2 Immunologic responses

The decellularization process of cadaveric donor skin may alter the composition and damage the natural structure of the components, thereby diminishing the guiding forces for the various dermal cells or stem cells. In addition, cellular remnants, eg, DNA or cytoplasmic proteins retained in the scaffold, may provoke an immunological reaction, which may harm healing due to a persistent inflammation, resulting in fibrosis (Badylak and Gilbert, 2008). Although the immunological effects in allogeneic and xenogeneic organ transplantation are commonly acknowledged, the immunological effects of biomaterials after transplantation are widely disregarded. A few studies have evaluated the mechanisms involved in the rejection of dermal substitutes. Collagen, the most widely used material in dermal substitutes, is considered only weakly antigenic, probably due to the high preservation of the amino acid sequence of the collagen triple helix among species. It was shown, however, that various dermal substitutes evoke an immunological response, which might be due to contaminating biomolecules, the addition of bioactive components, or alterations of the three-dimensional conformation as a result of, eg, chemical cross-linking (Ye et al., 2010; Lynn et al., 2004; Boekema et al., 2014).

Most studies evaluate the foreign body response with an emphasis on the formation of multinucleated foreign body giant cells (FBGCs). It is generally acknowledged that the formation of FBGCs is detrimental to the wound healing process. The immunological response to biomaterials is not necessarily harmful; the material could also be used to direct the immune response into a more favorable mode. It has been shown that biomaterials obtained from porcine intestine could skew macrophage and T cell differentiation toward the more desired phenotypes (Badylak and Gilbert, 2008; Allman et al., 2001). Macrophages are present in different subtypes, which have distinctive roles during the different phases of wound healing. The first subtype, named M1, has a proinflammatory phenotype and plays an important role in cleaning the wound environment. The M2 subtype, on the other hand, has an antiinflammatory, prorepair phenotype and has immune modulating properties (Delavary et al., 2011). Although this latter cell type is also associated with fibrosis, the switch from M1 to M2 is thought to be crucial to switch off the proinflammatory status of the wound and allow tissue repair (Sindrilaru and Scharffetter-Kochanek, 2013). However, the exact role of the different macrophage subtypes in wound healing and scar formation is still unclear.

Besides macrophages, lymphocytes are also thought to play an important role in the immune response to allo- and xenografts. Various subsets of T cells have been distinguished, mainly based on cytokine secretion. Like in macrophages, these different subsets can have distinct, sometimes opposing roles in the immune reaction. Although the role of T cells and their immunologic effects on biomaterials is indisputable, the exact mechanisms and the subtypes that are involved remain to be elucidated. It was thought that the Th1 cells are involved in transplant rejection and Th2 in transplant tolerance. Various other T cell subtypes have been characterized: Th17, Th9, Th22, and it is now thought that the balance between Th17 and regulatory T cells is the most important determinant of the immunological response to allo- or xenografts (Askar, 2014; Badylak and Gilbert, 2008).

9.2 Biomaterials for dermal substitution

Several materials are currently under investigation for their use as a dermal substitute. Collagen and chitosan are the most widely used biomaterials, and silk, polymers, and alginates are also being explored. The biomaterials used for scaffolds need to be biocompatible; cells need to adhere to and migrate into the scaffold and must be able to proliferate. Furthermore, scaffolds need to retain fluids for the distribution of nutrients throughout the scaffold. Finally, the biomaterial must be degradable and provide sufficient support for tissue regeneration, without provoking an immune response.

9.2.1 Collagen

Collagen is the most commonly used biomaterial for the fabrication of dermal substitutes; it is widely available and has ideal properties. Most collagen is isolated from the skin or tendon from cattle or pigs; other sources that are used are goats, birds, fish, and even recombinant plants (Banerjee et al., 2012; Parenteau-Bareil et al., 2011; Willard et al., 2013). Collagen is noncytotoxic, biocompatible, and degradable in a controlled manner (Clark et al., 2007; Lynn et al., 2004). Gelatin, which is formed after the partial denaturation of collagen, is also noncytotoxic and promotes cell adhesion and proliferation (Chandrasekaran et al., 2011; Nagiah et al., 2013; Huang et al., 2013), making it a good candidate as a biomaterial (Branski et al., 2007; Haslik et al., 2010; Moiemen et al., 2006). Cells are able to adhere and migrate through collagen, despite the fact that the natural binding sites in collagen, the Arg-Gly-Asp tripeptide (RGD) sequences, are inaccessible. Intracellular parts of integrins, which are linked to the cytoskeleton of cells and thereby function as signal transducers, are able to influence cell behavior, which allows cells to adhere to collagen (van der Flier and Sonnenberg, 2001). As a result, collagen scaffolds have a good biocompatibility in terms of cell viability and proliferation capacity (Willard et al., 2013; Kempf et al., 2011; Helary et al., 2010). Major drawbacks of using pure noncross-linked collagen are its poor mechanical property and tensile strength.

To be suitable as a dermal substitute, collagen scaffolds should persist long enough to allow cells to infiltrate and rebuild the ECM. However, collagen is degraded rapidly by matrix metalloproteinases produced by fibroblasts and keratinocytes. This degradation is even enhanced in an in vivo wound environment, as macrophages are also able to degrade collagen (Valentin et al., 2009). To improve the stability, collagen can be combined with ECM molecules like elastin to reduce biodegradability (de Vries et al., 1994; Emonard and Hornebeck, 1997; Lamme et al., 1996; Middelkoop et al., 1995).

Complementing scaffolds with GAGs, like hyaluronan and chondroitin sulfate, also improves cell viability and reduces degradation in vitro (Yannas and Burke, 1980), but in vivo these additions do not seem to enhance wound healing (de Vries et al., 1994). Another way to stabilize collagen scaffolds is the addition of chemical cross-links using glutaraldehyde or polyvinylalcohol (PVA). Indeed, cross-linking improves the stability and tensile strength of collagen scaffolds and reduces degradability without compromising cell viability (Charulatha and Rajaram, 2003; Hartwell et al., 2011). However, cross-linking scaffolds with chemical substances, like glutaraldehyde, are not optimal because they will be incorporated into the scaffold. When the scaffold is degraded, residues of the cross-linking agents could provoke an immune response, hampering the wound healing process. The formation of FBGCs as a result of the immunological response to chemical cross-links could lead to the loss of grafted split skin and improper or hypertrophic dermal regeneration (De Vries et al., 1993). For that reason, the carbodiimide EDC is often chosen as a cross-linking agent, because EDC itself is not incorporated into the scaffold and can be removed from the cross-linked scaffold. Cross-linking without chemical agents, for example dehydrothermal treatment (DHT), also improves the stability of collagen scaffolds (Sugiura et al., 2009; Nakada et al., 2013). However, reduced degradability could also lead to an increased inflammatory response and the formation of FBGCs due to the prolonged presence of collagen remnants (Boekema et al., 2014; Natarajan et al., 2013; Wang et al., 2013; Nakada et al., 2013; O'Brien et al., 2005; Pieper et al., 2000).

Collagen scaffolds can be used in different modalities: as a freeze-dried sponge, as a hydrogel, as nanofibers by electrospinning (ES), or as a membrane in a film. Several properties of collagen scaffolds, such as tensile strength and pore size, can be influenced by production techniques (see Section 9.3). The different manufacturing techniques are not suitable for all applications and need to be improved further. Hydrogels of collagen, for example, are fragile and difficult to handle and can be improved by increasing the collagen concentration or by compression. The different techniques for production are discussed later.

9.2.2 Chitosan

As an alternative to collagen-based scaffolds, chitosan has also gained attention for its use as a biomaterial. Chitosan is a linear polysaccharide, composed of beta-(1–4) linked D-glucosamine residues with a variable number of randomly located *N*-acetyl-glucosamine groups (Fig. 9.2), which can be obtained by partial deacetylation of chitin, the main component of the exoskeleton of invertebrates. It can also be extracted from green algae, fungi, and some yeasts, which makes it abundantly available and rather cheap. In addition, chitosan has structural similarities with GAGs, which allows cell adhesion, making chitosan interesting for biomedical applications (Suh and Matthew, 2000). Furthermore, chitosan is biocompatible and can be enzymatically degraded by hydrolysis of glucosamine–glucosamine, glucosamine–*N*-acetyl-glucosamine, and *N*-acetyl-glucosamine–*N*-acetyl-glucosamine linkages (Kean and Thanou, 2010; Rabea et al., 2003).

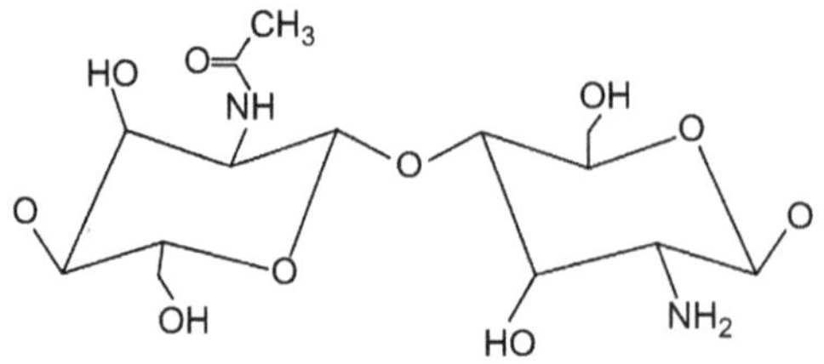

Figure 9.2 Chitosan and its derivatives.
From: Khan, F., Ahmad, S.R., 2013. Polysaccharides and their derivatives for versatile tissue engineering application. Macromol Biosci 13, 395–421.

Depending on the source and production technique, several derivatives of chitosan are available as a biomaterial (Mi et al., 2001). The source and production technique determine the molecular weight of chitosan, ranging from 30 to 1000 kDa, which influences the stability of the scaffold. The degradation rate of chitosan can be modified by altering the ratio of glucosamine to *N*-acetylglucosamine during polymerization or by changing the length of acyl side chains on *N*-acetylglucosamine (Francesko and Tzanov, 2011; Tomihata and Ikada, 1997). Neat chitosan scaffolds show good biocompatibility as hydrogels or electrospun scaffolds. Freeze-dried chitosan sponges and chitosan films do not seem to be suitable for cell adhesion and proliferation in vitro. More importantly, in vivo implantation of neat freeze-dried chitosan sponges led to rapid encapsulation and a foreign body response (Chen et al., 2013; Tchemtchoua et al., 2011). Therefore chitosan is often blended with other biomaterials like collagen, silk, or polyesters, which may improve biocompatibility and reduce biodegradation. However, there is still a need to stabilize the scaffolds. Freeze-dried chitosan–collagen scaffolds are still rapidly degraded in a rodent wound model. Reinforcement of the same scaffold with poly(lactic-co-glycolic acid) (PLGA) resulted in less contraction and improved tensile strength of the regenerated skin without altering the degradation time (Wang et al., 2012; Han et al., 2010).

Electrospun scaffolds of chitosan are easily produced but rapidly degraded. They induce a high influx of inflammatory cells, increased formation of granulation tissue, and formation of FBGCs, probably due to the presence of cross-linking agents. Furthermore, these scaffolds do not prevent contraction (Veleirinho et al., 2012; Xie et al., 2013). In addition, there is a large variation in the degradation rates of chitosan observed in several studies, varying from days to several weeks. Some studies claim that scaffold remnants are still present after 12 weeks, although it is questionable if remnants of chitosan still have a structural contribution for proper regeneration after this time. It seems more likely that the difference in degradation is due to the used source of chitosan and the used technique for producing the scaffolds (Patois et al., 2009; Tchemtchoua et al., 2011; Veleirinho et al., 2012). In addition, high degrees of acetylation can increase the inflammatory response and guide macrophages to polarize to a proinflammatory phenotype (Vasconcelos et al., 2013; Barbosa et al., 2010).

The main advantage of this material is the presence of the free amino groups and the hydroxyl groups, which can be used to link bioactive molecules (Francesko and Tzanov, 2011). In addition, the risk of transferable disease associated with the use of mammalian materials might be reduced for chitosan.

9.2.3 Silk fibroin

Silks are natural proteins that are produced by several insects and spiders. Silk fibers are composed of two proteins, a light 25 kDa and a heavy 325 kDa chain, which are held together by sericin. Silk fibers containing sericin are often referred to as virgin silk. Sericin is removed from silk fibers because of its allergenic potential (Altman et al., 2003; Zaoming et al., 1996). The remaining product, silk fibroin, is used as a biomaterial. Silk produced mostly by silk worms *Bombyx mori* (domestic silkworm) and *Antheraea pernyi* (wild silkworm) has been assessed for use as a biomaterial. Both sources support the adherence of cells due to the presence of an RGD sequence in silk proteins and therefore prove to be nontoxic (Minoura et al., 1995a,b). Although silk fibroin is resistant to degradation by collagenases, it can be degraded by other proteases (Minoura et al., 1995a,b). High percentages of silk fibroin and the use of solvents during fabrication may lead to a foreign body response (Li et al., 2003; Wang et al., 2008; Meinel et al., 2005). For tissue-engineering applications, silk is combined with other biomaterials to improve cell viability and mechanical properties of dermal substitutes. Several blends with chitosan were shown to improve cell adherence and viability in vivo (Chung and Chang, 2010). However, the use of silk-based scaffolds in full-thickness wound models still led to high contraction rates and increased granulation tissue formation. Therefore, the usefulness of silk fibroin as a biomaterial for dermal substitution remains questionable. Isolated sericin has also gained interest for use as a biomaterial despite being allergenic. Sericin is hydrophilic and possesses hydroxyl, carboxyl, and amino groups, enabling cross-linking and copolymerization, which are important properties for the manufacture of dermal substitutes (Zhang, 2002). Furthermore, sericin functions as an antioxidant and anticoagulant, and human fibroblasts are capable of adhering to it. Unfortunately, high concentrations of sericin in biomaterials proved to be toxic for cells (Tsubouchi et al., 2005; Kundu and Kundu, 2012). Although several reports claim that sericin is suitable for tissue engineering, only a few studies have made it to in vivo experiments, with questionable results (Nayak et al., 2012, 2013; Mandal et al., 2009; Siritienthong et al., 2012).

9.2.4 Polymers

Nonbiological polymers are also under investigation as part of dermal substitutes. Commonly used polymers include poly (L-lactic acid) (PLLA), polyurethane, poly (D,L-lactic acid), polycaprolactone (PCL), and PLGA. Because these nonbiological polymers lack cell-binding regions, RGD sequences need to be introduced by chemical addition using cross-links to allow for cell adhesion and migration (Kim et al., 2009). Another possibility is to combine polymers with biological sources, like collagen or gelatin, in order to enhance cell viability. Alternatively, polymer scaffolds can be prepared with the addition of growth factors to stimulate wound healing (Chandrasekaran et al., 2011; Powell and Boyce, 2009). However, despite the improved in vitro properties due to these modifications, this has not yet translated to an improved dermal substitute for clinical use (Li et al., 2009; Kim et al., 2009; Losi et al., 2013).

9.3 Manufacturing procedures

The manufacturing procedure can greatly determine the performance of a dermal substitute. Several different methods will be described with respect to the value of these techniques and the characteristics of the scaffolds. Dermal substitutes should have an appropriate pore size to allow for the sufficient diffusion of nutrients and waste products (Suzuki et al., 1990; Wang et al., 2005) and the migration of cells. Different procedures allow for the control of pore sizes and material chemistry. The role of pore size of dermal substitutes (sponges) on in vivo performance has been shown in several papers (Boekema et al., 2014; Zheng et al., 2011; O'Brien et al., 2005; Salem et al., 2002; Wang et al., 2005; Yannas et al., 1989; Zeltinger et al., 2001), which indicate that the optimal pore size appears to be in the 50–200 μm range. By studying the relation between the (micro)structure and cellular invasion, vascularization, and host response in wound models, the production processes can be fine-tuned further. Stability, ie, the resistance to degradation, is often achieved by cross-linking with glutaraldehyde, carbodiimide, or by DHT cross-linking in a vacuum oven. Finally, the sterilization method can have profound effects on the morphology, mechanical properties, and cytotoxicity of the materials (Marreco et al., 2004; Ghosh et al., 2012; Parenteau-Bareil et al., 2011).

9.3.1 *Freeze drying*

Lyophilization or freeze drying (FD) has been used for decades to produce different porous membranes or sponges (Dagalakis et al., 1980; Yannas and Burke, 1980). It is one of the most widely used methods for the fabrication of porous scaffolds for tissue engineering. Well-known examples are Integra® and Matriderm®. FD is commonly employed to produce water-soluble polymer scaffolds such as from collagen, elastin, or fibrin (Li et al., 2001; O'Brien et al., 2004). Additives (eg, GAGs or elastin) or chemical cross-linking can be used to increase stability (reduced degradation) and/or functionality (improved cellular adhesion and proliferation). Pore sizes and interconnectivity can be controlled by using different solvents, freezing temperatures, and freezing rates. During freezing of the polymer solutions, ice crystals grow and condense the surrounding polymer solution. The growth time of ice crystals is dependent on the temperature gradient during freezing (Li et al., 2001). As a result, the majority of the polymer is excluded from the ice crystals. Due to the close proximity, the macromolecules can aggregate, forming the final structure of the scaffold. Sublimation of the ice crystals during lyophilization results in a solid, porous scaffold.

Pores perpendicular to the wound bed might improve cellular infiltration (Nakada et al., 2013) and tissue regeneration. Pores can be orientated in parallel by a patented directional freezing process (Bozkurt et al., 2007; Kroehne et al., 2008). A one-dimensional heat flow was created during freezing in the direction of the temperature gradient, which was maintained during cooling and freezing. This results in directional freezing of the ice crystals; the addition of organic acid will prevent the development of side branches. By decreasing both the temperature gradient and the cooling rate, the pore size can be increased (Kuberka et al., 2002). Scaffolds with parallel-orientated pores have allowed one-stage grafting of split-thickness skin grafts, resulting in good wound healing when tested in porcine excisional wounds (Boekema et al., 2014).

Although easy and relatively cheap to produce, sponges prepared by FD may lead to a suboptimal performance because of low mechanical strength and/or dense architecture (Lee et al., 2005; Choi et al., 2001) unless properly controlled. More dense material may reduce the infiltration of cells such as fibroblasts or endothelial cells.

9.3.2 Hydrogel

Hydrogels are produced from hydrophilic monomers, forming a highly absorbent network. The components can be either natural such as collagen and chitosan or synthetic such as PCL and PVA. Poor mechanical strength is the major drawback, especially when using natural polymers. Because of their high water content, the consistency of hydrogels remarkably resembles natural soft tissue. Hydrogels are generally biocompatible and are highly permeable for oxygen, nutrients, and other water-soluble metabolites, which are prerequisites for the incorporation of living cells; the design of hydrogels for tissue engineering is excellently reviewed in Zhu and Marchant (2011). Swelling properties and elasticity are further influenced by the type and degree of cross-linking. Growth factors can be delivered by hydrogel through click chemistry (Jiang et al., 2014) or via microspheres (Huang et al., 2009). In click chemistry, diverse building blocks can be linked covalently under mild conditions, mostly via azide–alkyne cycloaddition. The coupling is highly efficient, specific, and does not interfere with proteins or living cells.

Hydrogels based on natural polymers have a low breaking strength and are easily degraded. The degradation time can be increased by increasing the concentration of the polymer, by compression, or by adding a nondegradable polymer:

1. Hydrogels with an increased concentration of collagen were easy to handle, did not contract drastically, favored cell growth, and were quickly integrated in vivo (Helary et al., 2010, 2012; Chiu et al., 2013), but the increased viscosity might be difficult for proper production (Desimone et al., 2011). Contraction of these hydrogels should be prevented as it reduces porosity and thus integration and neovascularization (Brown and Phillips, 2007).
2. Compression of hydrogels with a gravitational load for 5 min resulted in a significantly higher density of collagen fibers, increased tensile strength, and lower gel contraction and degradation (Brown et al., 2005; Mudera et al., 2007). Fibroblast proliferation is stimulated by the increased stiffness due to an increased collagen concentration (Discher et al., 2005; Hadjipanayi et al., 2009). Compression is compatible with living fibroblasts and keratinocytes, which gives rise to the opportunity to fabricate living dermal equivalents (Hu et al., 2010; Ananta et al., 2012; Braziulis et al., 2012; Pontiggia et al., 2013).
3. By adding nondegradable polymers, hydrogels with constant swelling properties but varying degradation rates can be created. By mixing nondegradable poly(ethylene glycol)–diacrylate with degradable poly(ethylene glycol)–PLLA–diacrylate, the degradation time was increased to 7 weeks, without affecting cross-link density or swelling (Chiu et al., 2013). The breaking strength can also be improved by incorporating a synthetic degradable polymer mesh (PLGA, Vicryl) during the formation of a collagen hydrogel (Ananta et al., 2012).

4. Mechanical strength can also be improved by cryogelation. Moderate freezing and subsequent thawing (like in FD) results in material with large interconnected pores suitable for the infiltration and proliferation of cells, high porosity, high elasticity, and excellent mechanical properties (Lozinsky et al., 2003; Dainiak et al., 2006; Tripathi et al., 2009; Savina et al., 2009; Shevchenko et al., 2014). It does not require any organic solvents and allows modulation of a wide range of properties. Cryogelation leads to gels with a higher mechanical strength than cross-linking gels by chemical or irradiative techniques (Yang et al., 2011).

9.3.3 Electrospinning

In ES, a solution containing the appropriate compounds is forced through a syringe by electrical forces, resulting in nano-sized fibers. These fibers are deposited on a grounded collector, often a plate or rotating drum, forming a homogenous and random three-dimensional porous structure resembling native skin (Fig. 9.3; Kim and Kim, 2007; Noh et al., 2006). The three-dimensional architecture is important as it influences cellular behavior (Alamein et al., 2013). Materials for ES can be biological, eg, collagen, silk, or chitosan, or synthetic in nature, eg, PCL, PLGA, or PVA. ES has several advantages for scaffold production; mixtures of scaffold material could be spun into single fibers or mixtures of different fibers. In addition, fibers could be coated to strengthen them and to introduce binding sites for cells. The ES technique even allows for the incorporation of living cells between or in the formed fibers.

Fiber diameter and morphology of the electrospun scaffold can be controlled by concentration and molecular weight of the polymer (Matthews et al., 2002; Powell and Boyce, 2008). The three-dimensional architecture can further be altered by varying the spinning speed of the rotating electrode, distance between electrodes, electric field, and voltage (Alamein et al., 2013). To produce electrospun composite scaffolds, coelectrospinning and coaxial ES are mostly used. In coelectrospinning, different polymer solutions are electrospun simultaneously by applying two or more syringes and power supplies to obtain hybrid fibrous scaffolds composed of two or more components

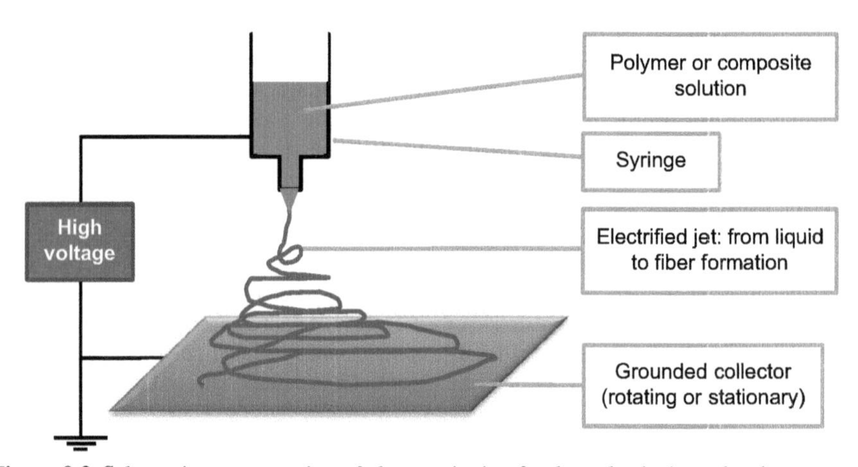

Figure 9.3 Schematic representation of electrospinning for dermal substitute development.

(Hong et al., 2011; Duan et al., 2007). In coaxial ES, two different polymer solutions can be electrospun simultaneously by using a spinneret composed of two coaxial needles to produce core-mantle structured fibers (Sun et al., 2006; Blackstone et al., 2014; Wu et al., 2010). The core material can be used to improve the mechanical properties, while the mantle material ensures biocompatibility, eg, by enabling cellular adhesion. Combining ES scaffolds with hydrogel can also alter the mechanical strength (Franco et al., 2013; Hong et al., 2011) and improve cellular infiltration (Hong et al., 2011). Multijet ES with a tube electrode enables the mass production of nanofibers (Alamein et al., 2013).

The biocompatibility of ES scaffolds in cell cultures has been shown in many papers. ES scaffolds of sheep collagen and human elastin, for example, resulted in good handling, porosity, cell migration, and proliferation (Rnjak-Kovacina et al., 2012). ES scaffolds can be used to produce full-skin equivalents with dermal fibroblasts and a differentiated epidermis (Kempf et al., 2011; Powell and Boyce, 2008). Although much progress has been made in the production of ES scaffolds, their application in rats or mice has been described only in a few cases (Kempf et al., 2011; Duan et al., 2013; Franco et al., 2013; Hong et al., 2011; Kim et al., 2012; Sundaramurthi et al., 2012). These reports show that the material is noncytotoxic, is not rejected through a foreign body response, and is degraded within weeks. However, the contribution of ES scaffolds to wound healing remains unclear because the application of ES scaffolds in mice or rats mostly resulted in highly contracted wounds. A comparison of collagen scaffolds produced by ES or FD resulted in an equal performance with respect to cell proliferation, surface hydration, cellular organization, and dermal and epidermal stratification with a continuous layer of basal keratinocytes present at the dermal–epidermal junction. After transplantation on athymic mice, scaffolds produced by ES showed a better take rate, enhanced resorption of the implanted collagen, and less wound contraction compared to the FD scaffold (Powell and Boyce, 2008). Of note, both scaffold types were seeded first with fibroblasts and keratinocytes, which already impact in vivo performance.

More research is required to obtain scaffolds by ES for use in human wound healing. Many parameters of ES can be altered to improve the strength and stability of scaffolds, but most studies still favor chemical cross-linking, which may be suboptimal. The application of cells in ES scaffolds is hampered by the low viability, the use of high electric fields, and solvents, among others.

9.3.4 Printing

Bioprinting, rapid prototyping, additive manufacturing, or solid freeform fabrication all refer to techniques to fabricate structures with high precision and reproducible geometry (Ferris et al., 2013; Ringeisen et al., 2006). For construction of a three-dimensional scaffold, a layer-by-layer approach is required. The type of technique (with or without an orifice) determines which material can be deposited. Many different materials can be printed, such as ECM molecules, cytokines, synthetic polymers, or DNA, along with viable cells. Depending on the technique applied, optimal settings, eg, the viscosity, need to be determined. At present, most studies focus on the production and in vitro performance. In vivo applications are very limited; therefore we only present a small overview of the possibilities without going into details. With printing two important aspects of tissue engineering are being addressed: the development of a vascular network and control of cell–cell contact and tissue architecture.

Inkjet printing is based on drop-on-demand of ink by a thermal or piezoelectric head. It has already been used to print endothelial cells and smooth muscle cells (Wilson and Boland, 2003; Boland et al., 2003), a three-dimensional construct containing HeLa cells (Arai et al., 2011), or a three-dimensional composite construct containing muscle cells, endothelial cells, and stem cells (Xu et al., 2013). There are several disadvantages of this technique. The high shear force in the nozzle limits the choice of hydrogels to be printed and reduces cell viability or survival (Born et al., 1992). To prevent clogging of the nozzle by the settling and aggregation of cells, low cell concentrations should be used (Xu et al., 2005; Saunders et al., 2008).

For laser-assisted bioprinting (LAB), glass slides are coated with a thin layer of laser-absorbing material (eg, gold or polyimide) and a layer of the biomaterial to be deposited (eg, fibroblasts in collagen, alginate, glycerin, or fibrin) with an adequate viscosity (Barron et al., 2004; Ferris et al., 2013). These donor slides are positioned upside down over receiver plate at a distance of a few hundred to thousand micrometers (Fig. 9.4). The laser-absorbing material can be locally evaporated by focusing laser pulses through the upper glass slide. As a result, a small amount of the biomaterial is transported toward the receiver plate through jet formation (Koch et al., 2010). The impact of the cells on the receiver plate can be dampened by, for example, a hydrogel. Living cells can survive this transfer without damage or change in phenotype (Michael et al., 2013; Hopp et al., 2005; Othon et al., 2008). Different patterns of biomaterial can be produced by moving the slides relative to each other. This procedure can be repeated layer-by-layer to generate three-dimensional patterns (Michael et al., 2013; Koch et al., 2010; Unger et al., 2011; Gruene et al., 2011). LAB is limited by production speed, throughput, and scale up (Guillotin and Guillemot, 2011; Ferris et al., 2013). Metallic residues from the laser-absorbing layer can be found in the final constructed material (Guillotin and Guillemot, 2011).

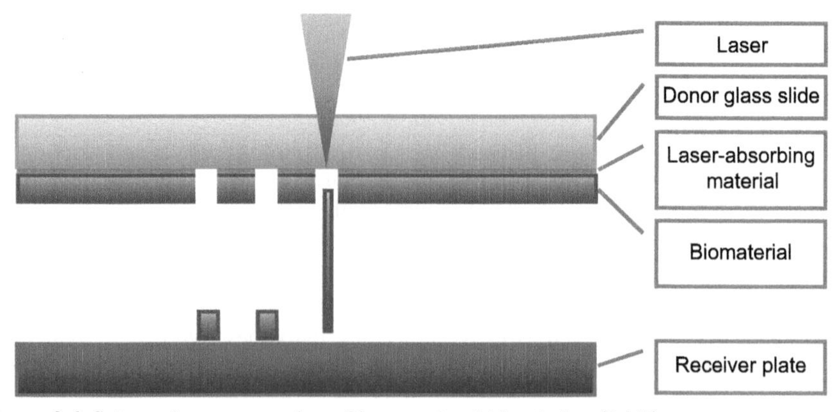

Figure 9.4 Schematic representation of laser-assisted bioprinting (LAB).

The proof of the principle of printing was shown in a study where freshly trypsinized fibroblasts (NIH3T3) and keratinocytes (HaCaT) were printed on Matriderm® by LAB. After overnight incubation, the construct was transplanted to skin fold chambers in nude mice. Eleven days later, a multilayered epidermis had formed, and fibroblasts had penetrated into the Matriderm® (Michael et al., 2013). Unfortunately, the added value of printing was not demonstrated in the study (Michael et al., 2013). In addition, the high proliferation rate of (mouse) cell lines may lead to misinterpretation of these results.

9.4 Animal studies

After in vitro testing of biomaterials for cell viability and functionality, the next step is the evaluation of the scaffolds in a suitable in vivo model. Commonly, biocompatibility is evaluated primarily by the immunological response to the scaffold implanted subcutaneously in rats or mice. It has been questioned whether results from these tests in the sterile subcutaneous compartment can predict the product behavior in a full-thickness skin defect. For example, scaffolds found to be resistant to biodegradation in the subcutaneous test proved to be unfit in a full-thickness wound model (Huang et al., 2013). Therefore we recommend that the safety of dermal substitutes be assessed in full-thickness skin wound models. In addition, the choice of animal species is paramount.

Rodents are the most commonly used animal species. The major disadvantage of using rodents for dermal research is their reliance on wound contraction to close a full-thickness wound (Davidson, 1998). This results in small scars, up to 20% of the initial wound areas (Galiano et al., 2004), with marginal new tissue formation. In contrast, wounds in humans heal primarily by granulation tissue formation. Wound contraction is an undesirable secondary process, which can result in functional impairment. Therefore wound closure or the resulting scar size in rodents must be interpreted with care. Alternatively, splinting, wound chambers, or the rabbit ear can be used to circumvent contraction in rodents (Davidson et al., 2013).

The skin anatomy of pigs is more comparable to that of humans than rodent skin, with a relatively thick epidermis with distinct rete ridges and a comparable dermal structure. Pigs are therefore suitable animals for investigating the effects of dermal substitutes. Although wound contraction is also part of the wound healing process in pig skin (Middelkoop et al., 2004; Ågren et al., 1997), the deposition of granulation tissue, epithelialization, and scar formation are more dominant in the pig than in rats (van der Veen et al., 2010; Sullivan et al., 2001).

The drawback with the experimental full-thickness wound model is that pigs normally do not develop hypertrophic scars and, in contrast to clinical studies, the long-term effects of dermal substitute application could not be demonstrated (Seo et al., 2013; Philandrianos et al., 2012). Red Duroc pigs have been proposed as a hypertrophic scar model. Although this model does show similarities with human scar formation it is not identical, and widespread use in the field of wound healing was not achieved (Domergue et al., 2014; Zhu et al., 2007). In general, the choice of animal species is dictated by the goal of the experiment.

9.5 Future perspectives

The ultimate goal of developing biomaterials for dermal substitution is to regenerate a fully functional skin with the same cosmetic appearance as normal skin. This includes the restoration of the appendages, such as hair follicles, sebaceous and sweat glands, and nerves and circulation, including the lymphatic system. Research into this area often includes the application of cells, eg, endothelial cells, mesenchymal, or epidermal (stem) cells and melanocytes.

For the cosmetic aspects, color is an important issue, and the inclusion of melanocytes to produce a correctly pigmented skin substitute is one of the topics of research (Kim et al., 2012; Bottcher-Haberzeth et al., 2013; Waaijman et al., 2010; Swope et al., 2006; Liu et al., 2011). Although the "proof of principle" was demonstrated for the transfer of melanocytes to vitiligo patients (Ghosh et al., 2012), clinical application still requires elaborate safety studies.

Rapid vascularization of the dermal substitute is a crucial requirement for the success of a dermal substitute. Different approaches toward a vascularized tissue-engineered product have been proposed (Auger et al., 2013). One of the strategies used for revascularization is the introduction of living autologous or allogeneic blood and/or lymphatic endothelial cells into biomaterials to create new tubular structures in vitro (Athanassopoulos et al., 2012; Marino et al., 2014). The development of prevascularization of skin substitutes by seeding adipose-derived stem cells is another variant of this concept (Huang et al., 2012). The use of prevascularized substitutes requires the formation of anastomoses between the substitute vessels and the existing vasculature of the wound bed of the patient. Therefore other approaches are under investigation, eg, the addition of angiogenic factors such as EGF, which stimulate the expression of VEGF and the development of new vasculature in vivo (Mineo et al., 2013). Besides the addition of bioactive, components this can also be achieved by a specific composition or three-dimensional structure of the biomaterial (Kim et al., 2012; Nillesen et al., 2011). In addition to the development of a vascular system, the development of the lymphatic system is important for proper fluid homeostasis and immune function. Both systems have their own specific endothelial cells, which do not intermix during vessel development (Marino et al., 2014).

The recovery of deep skin defects is accompanied by severe pain and itch, which can persist for years after the initial damage. This is extremely distressing to the patient and most likely caused by improper reinnervation or restoration of the sensory receptors in the recovered skin (Blais et al., 2013). Research into the regeneration of the nervous system after severe wounding is in its infancy, and knowledge of the stimulation of nerve ingrowth is limited. However, as a result of improvements in wound healing, restoring the quality of life is one of the future challenges, including the restoration of sensory functions of the skin.

Hair follicles, sweat glands, and sebaceous glands are generally lost after deep skin injuries. These structures play an important role in, eg, the production of antimicrobial substances, temperature regulation, lubrication of the skin, and sensation. In addition, appendages appear to play a role in epithelialization after wounding (Rittie et al., 2013; Cotsarelis, 2006). Hair follicles can contribute to nerve regeneration in dermal substitutes (Gagnon et al., 2011). Thus restoration of the appendages is important for proper functioning of the regenerated skin. Although current skin substitutes provide the required three-dimensional architecture for the induction of hair follicles, specific growth factors, hormones, and epithelial interactions need to be present. Specific requirements for skin substitutes to facilitate hair follicle regeneration have been addressed by Mahjour et al. (2012). Studies on the generation of skin appendages using tissue-engineered skin substitutes supplemented with fibroblasts, keratinocytes, and (mesenchymal) stem cells have been successful (Huang et al., 2012; Sriwiriyanont et al., 2013). The addition of growth factors, such as fibroblast growth factor-9 and epidermal growth factor, were found to stimulate appendage development (Gay et al., 2013; Huang et al., 2012).

To improve the regenerative capacity of biomaterials, the addition of cells, especially mesenchymal stem cells, is a promising development. Cells are typically harvested from specific tissues and expanded in tissue culture before they are used in combination with the biomaterial. This type of procedure is expensive and potentially risky for patients. Therefore the use of cultured cells is strictly regulated. The recruitment of endogenous (stem) cells to the site of injury is an attractive alternative (Suzuki et al., 2011; Lee et al., 2010). To recruit sufficient numbers of cells the homing signals need further elucidation (Vanden Berg-Foels, 2013). We also lack knowledge on the key signals responsible for the differentiation of mesenchymal stem cells.

Despite the overwhelming amount of research invested in the development of biomaterials for dermal replacement, a biomaterial allowing complete restoration of the skin is still lacking. Moreover, most of the biomaterials have been studied extensively in in vitro and in vivo, but few of them have been evaluated in randomized clinical trials. To move the field forward the translational step toward the clinic has to be taken to provide evidence-based data on the efficacy and safety of dermal substitute therapies.

Key papers:

Sindrilaru and Scharffetter-Kochanek (2013)

This paper gives an overview on the current knowledge of the role of macrophage subtypes in wound healing.

Vanden Berg-Foels (2013)

The current concept of tissue engineering is still a combination of a biomaterial with cells. Preferentially, these cells are autologous (stem) cells, which usually need to be expanded in culture to obtain a sufficient number of cells. This is a time-consuming and expensive treatment method, which is bound to very strict regulations. A different approach is to increase the recruitment of endogenous stem or differentiated cells to the site of injury in order to repair the injured tissue, omitting the abovementioned hurdles. This review gives an overview of chemoattractants for the homing of mesenchymal stem cells to the site of injury.

Badylak and Gilbert (2008)

The immunological effects (adaptive and innate) of biomaterials after transplantation have been disregarded for a long time. This paper summarizes the studies performed to evaluate the immune response of several biomaterials, including the materials used for dermal substitution, thereby drawing attention to the importance of this aspect to the successfulness of a biomaterial.

Vasconcelos et al. (2013)
This paper shows how chitosan acetylation can influence the inflammatory response through macrophage polarization.

Tsubouchi et al. (2005)
Different manufacturing procedures of chitosan biomaterials.
Comparison of chitosan scaffolds prepared by FD, ES, and film formation on cell viability in vitro and biocompatibility in vivo.

Huang et al. (2013)
Difference in degradation time of a chitosan/gelatin scaffold when implanted subcutaneously and in a full-thickness model.

Michael et al. (2013)
This paper describes the production of three-dimensional skin substitutes containing fibroblasts and keratinocytes by LAB on the collagen scaffold (Matriderm®), tested in mice.

Alamein et al. (2013)
This paper describes the development of a needle-free multijet ES device with the potential for mass production of nanofibrous ECM to allow for the production of a three-dimensional substitute for large-scale applications, such as large area skin regeneration in patients with burns. The device showed biocompatibility for the integration of stem cells.

Koch et al. (2013)
This paper gives a nice overview of the different printing techniques used in tissue engineering.

Zhu and Marchant (2011)
This article gives an overview of the different strategies used to design synthetic hydrogels to mimic ECM with bioactive properties.

References

Ågren, M.S., Mertz, P.M., Franzén, L., 1997. A comparative study of three occlusive dressings in the treatment of full-thickness wounds in pigs. J Am Acad Dermatol 36, 53–58.

Alamein, M.A., Stephens, S., Liu, Q., Skabo, S., Warnke, P.H., 2013. Mass production of nanofibrous extracellular matrix with controlled 3D morphology for large-scale soft tissue regeneration. Tissue Eng Part C Methods 19, 458–472.

Allman, A.J., Mcpherson, T.B., Badylak, S.F., Merrill, L.C., Kallakury, B., Sheehan, C., Raeder, R.H., Metzger, D.W., 2001. Xenogeneic extracellular matrix grafts elicit a TH2-restricted immune response. Transplantation 71, 1631–1640.

Altman, G.H., Diaz, F., Jakuba, C., Calabro, T., Horan, R.L., Chen, J., Lu, H., Richmond, J., Kaplan, D.L., 2003. Silk-based biomaterials. Biomaterials 24, 401–416.

Ananta, M., Brown, R.A., Mudera, V., 2012. A rapid fabricated living dermal equivalent for skin tissue engineering: an in vivo evaluation in an acute wound model. Tissue Eng Part A 18, 353–361.

Arai, K., Iwanaga, S., Toda, H., Genci, C., Nishiyama, Y., Nakamura, M., 2011. Three-dimensional inkjet biofabrication based on designed images. Biofabrication 3, 034113.

Askar, M., 2014. T helper subsets & regulatory t cells: rethinking the paradigm in the clinical context of solid organ transplantation. Int J Immunogenet 41, 185–194.

Athanassopoulos, A., Tsaknakis, G., Newey, S.E., Harris, A.L., Kean, J., Tyler, M.P., Watt, S.M., 2012. Microvessel networks pre-formed in artificial clinical grade dermal substitutes in vitro using cells from haematopoietic tissues. Burns 38, 691–701.

Auger, F.A., Gibot, L., Lacroix, D., 2013. The pivotal role of vascularization in tissue engineering. Annu Rev Biomed Eng 15, 177–200.

Badylak, S.F., Gilbert, T.W., 2008. Immune response to biologic scaffold materials. Semin Immunol 20, 109–116.

Banerjee, I., Mishra, D., Das, T., Maiti, S., Maiti, T.K., 2012. Caprine (goat) collagen: a potential biomaterial for skin tissue engineering. J Biomater Sci Polym Ed 23, 355–373.

Barbosa, J.N., Amaral, I.F., Aguas, A.P., Barbosa, M.A., 2010. Evaluation of the effect of the degree of acetylation on the inflammatory response to 3D porous chitosan scaffolds. J Biomed Mater Res A 93, 20–28.

Barron, J.A., Wu, P., Ladouceur, H.D., Ringeisen, B.R., 2004. Biological laser printing: a novel technique for creating heterogeneous 3-dimensional cell patterns. Biomed Microdevices 6, 139–147.

Blackstone, B.N., Drexler, J.W., Powell, H.M., 2014. Tunable engineered skin mechanics via coaxial electrospun fiber core diameter. Tissue Eng Part A 20 (19–20), 2746–2755.

Blais, M., Parenteau-Bareil, R., Cadau, S., Berthod, F., 2013. Concise review: tissue-engineered skin and nerve regeneration in burn treatment. Stem Cells Transl Med 2 (7), 545–551.

Boekema, B.K., Vlig, M., Olde Damink, L., Middelkoop, E., Eummelen, L., Buhren, A.V., Ulrich, M.M., 2014. Effect of pore size and cross-linking of a novel collagen-elastin dermal substitute on wound healing. J Mater Sci Mater Med 25, 423–433.

Boland, T., Mironov, V., Gutowska, A., Roth, E.A., Markwald, R.R., 2003. Cell and organ printing 2: fusion of cell aggregates in three-dimensional gels. Anat Rec A Discov Mol Cell Evol Biol 272, 497–502.

Born, C., Zhang, Z., Al-Rubeai, M., Thomas, C.R., 1992. Estimation of disruption of animal cells by laminar shear stress. Biotechnol Bioeng 40, 1004–1010.

Bottcher-Haberzeth, S., Klar, A.S., Biedermann, T., Schiestl, C., Meuli-Simmen, C., Reichmann, E., Meuli, M., 2013. "Trooping the color": restoring the original donor skin color by addition of melanocytes to bioengineered skin analogs. Pediatr Surg Int 29, 239–247.

Bozkurt, A., Brook, G.A., Moellers, S., Lassner, F., Sellhaus, B., Weis, J., Woeltje, M., Tank, J., Beckmann, C., Fuchs, P., Damink, L.O., Schugner, F., Heschel, I., Pallua, N., 2007. In vitro assessment of axonal growth using dorsal root ganglia explants in a novel three-dimensional collagen matrix. Tissue Eng 13, 2971–2979.

Branski, L.K., Herndon, D.N., Pereira, C., Mlcak, R.P., Celis, M.M., Lee, J.O., Sanford, A.P., Norbury, W.B., Zhang, X.J., Jeschke, M.G., 2007. Longitudinal assessment of integra in primary burn management: a randomized pediatric clinical trial. Crit Care Med 35, 2615–2623.

Braziulis, E., Diezi, M., Biedermann, T., Pontiggia, L., Schmucki, M., Hartmann-Fritsch, F., Luginbuhl, J., Schiestl, C., Meuli, M., Reichmann, E., 2012. Modified plastic compression of collagen hydrogels provides an ideal matrix for clinically applicable skin substitutes. Tissue Eng Part C Methods 18, 464–474.

Brown, R.A., Phillips, J.B., 2007. Cell responses to biomimetic protein scaffolds used in tissue repair and engineering. Int Rev Cytol 262, 75–150.

Brown, R.A., Wisema, M., Chuo, C.B., Cheema, U., Nazhat, S.N., 2005. Ultrarapid engineering of biomimetic materials and tissues: fabrication of nano- and microstructures by plastic compression. Adv Funct Mater 15, 1762.

Chandrasekaran, A.R., Venugopal, J., Sundarrajan, S., Ramakrishna, S., 2011. Fabrication of a nanofibrous scaffold with improved bioactivity for culture of human dermal fibroblasts for skin regeneration. Biomed Mater 6, 015001.

Charulatha, V., Rajaram, A., 2003. Influence of different crosslinking treatments on the physical properties of collagen membranes. Biomaterials 24, 759–767.

Chen, S.H., Tsao, C.T., Chang, C.H., Lai, Y.T., Wu, M.F., Chuang, C.N., Chou, H.C., Wang, C.K., Hsieh, K.H., 2013. Assessment of reinforced poly(ethylene glycol) chitosan hydrogels as dressings in a mouse skin wound defect model. Mater Sci Eng C Mater Biol Appl 33, 2584–2594.

Chiu, Y.C., Kocagoz, S., Larson, J.C., Brey, E.M., 2013. Evaluation of physical and mechanical properties of porous poly(ethylene glycol)-co-(l-lactic acid) hydrogels during degradation. PLoS One 8, e60728.

Choi, Y.S., Lee, S.B., Hong, S.R., Lee, Y.M., Song, K.W., Park, M.H., 2001. Studies on gelatin-based sponges. Part III: a comparative study of cross-linked gelatin/alginate, gelatin/hyaluronate and chitosan/hyaluronate sponges and their application as a wound dressing in full-thickness skin defect of rat. J Mater Sci Mater Med 12, 67–73.

Chung, T.W., Chang, Y.L., 2010. Silk fibroin/chitosan-hyaluronic acid versus silk fibroin scaffolds for tissue engineering: promoting cell proliferations in vitro. J Mater Sci Mater Med 21, 1343–1351.

Clark, R.A., Ghosh, K., Tonnesen, M.G., 2007. Tissue engineering for cutaneous wounds. J Invest Dermatol 127, 1018–1029.

Cotsarelis, G., 2006. Epithelial stem cells: a folliculocentric view. J Invest Dermatol 126, 1459–1468.

Dagalakis, N., Flink, J., Stasikelis, P., Burke, J.F., Yannas, I.V., 1980. Design of an artificial skin. Part III. Control of pore structure. J Biomed Mater Res 14, 511–528.

Dainiak, M.B., Kumar, A., Galaev, I.Y., Mattiasson, B., 2006. Detachment of affinity-captured bioparticles by elastic deformation of a macroporous hydrogel. Proc Natl Acad Sci USA 103, 849–854.

Davidson, J.M., 1998. Animal models for wound repair. Arch Dermatol Res 290 (Suppl.), S1–S11.

Davidson, J.M., Yu, F., Opalenik, S.R., 2013. Splinting strategies to overcome confounding wound contraction in experimental animal models. Adv Wound Care (New Rochelle) 2, 142–148.

Delavary, B.M., van der Veer, W.M., van Egmond, M., Niessen, F.B., Beelen, R.H., 2011. Macrophages in skin injury and repair. Immunobiology 216, 753–762.

Desimone, M.F., Helary, C., Quignard, S., Rietveld, I.B., Bataille, I., Copello, G.J., Mosser, G., Giraud-Guille, M.M., Livage, J., Meddahi-Pelle, A., Coradin, T., 2011. In vitro studies and preliminary in vivo evaluation of silicified concentrated collagen hydrogels. ACS Appl Mater Interfaces 3, 3831–3838.

Discher, D.E., Janmey, P., Wang, Y.L., 2005. Tissue cells feel and respond to the stiffness of their substrate. Science 310, 1139–1143.

Domergue, S., Jorgensen, C., Noel, D., 2014. Advances in research in animal models of burn-related hypertrophic scarring. J Burn Care Res 36 (5), e259–e266.

Duan, B., Wu, L., Yuan, X., Hu, Z., Li, X., Zhang, Y., Yao, K., Wang, M., 2007. Hybrid nanofibrous membranes of plga/chitosan fabricated via an electrospinning array. J Biomed Mater Res A 83, 868–878.

Duan, H., Feng, B., Guo, X., Wang, J., Zhao, L., Zhou, G., Liu, W., Cao, Y., Zhang, W.J., 2013. Engineering of epidermis skin grafts using electrospun nanofibrous gelatin/polycaprolactone membranes. Int J Nanomedicine 8, 2077–2084.

Emonard, H., Hornebeck, W., 1997. Binding of 92 kda and 72 kda progelatinases to insoluble elastin modulates their proteolytic activation. Biol Chem 378, 265–271.

Ferris, C.J., Gilmore, K.G., Wallace, G.G., In Het Panhuis, M., 2013. Biofabrication: an overview of the approaches used for printing of living cells. Appl Microbiol Biotechnol 97, 4243–4258.

van der Flier, A., Sonnenberg, A., 2001. Function and interactions of integrins. Cell Tissue Res 305, 285–298.

Francesko, A., Tzanov, T., 2011. Chitin, chitosan and derivatives for wound healing and tissue engineering. Adv Biochem Eng Biotechnol 125, 1–27.

Franco, R.A., Min, Y.K., Yang, H.M., Lee, B.T., 2013. Fabrication and biocompatibility of novel bilayer scaffold for skin tissue engineering applications. J Biomater Appl 27, 605–615.

Gagnon, V., Larouche, D., Parenteau-Bareil, R., Gingras, M., Germain, L., Berthod, F., 2011. Hair follicles guide nerve migration in vitro and in vivo in tissue-engineered skin. J Invest Dermatol 131, 1375–1378.

Galiano, R.D., Michaels, J., Dobryansky, M., Levine, J.P., Gurtner, G.C., 2004. Quantitative and reproducible murine model of excisional wound healing. Wound Repair Regen 12, 485–492.

Gay, D., Kwon, O., Zhang, Z., Spata, M., Plikus, M.V., Holler, P.D., Ito, M., Yang, Z., Treffeisen, E., Kim, C.D., Nace, A., Zhang, X., Baratono, S., Wang, F., Ornitz, D.M., Millar, S.E., Cotsarelis, G., 2013. Fgf9 from dermal γδ T cells induces hair follicle neogenesis after wounding. NatMed 19 (7), 916–923.

Ghosh, D., Kuchroo, P., Viswanathan, C., Sachan, S., Shah, B., Bhatt, D., Parasramani, S., Savant, S., 2012. Efficacy and safety of autologous cultured melanocytes delivered on poly (DL-lactic acid) film: a prospective, open-label, randomized, multicenter study. Dermatol Surg 38, 1981–1990.

Gruene, M., Unger, C., Koch, L., Deiwick, A., Chichkov, B., 2011. Dispensing pico to nanolitre of a natural hydrogel by laser-assisted bioprinting. Biomed Eng Online 10, 19.

Guillotin, B., Guillemot, F., 2011. Cell patterning technologies for organotypic tissue fabrication. Trends Biotechnol 29, 183–190.

Hadjipanayi, E., Mudera, V., Brown, R.A., 2009. Close dependence of fibroblast proliferation on collagen scaffold matrix stiffness. J Tissue Eng Regen Med 3, 77–84.

Hahnenberger, R., Jakobson, A.M., 1991. Antiangiogenic effect of sulphated and nonsulphated glycosaminoglycans and polysaccharides in the chick embryo chorioallantoic membrane. Glycoconj J 8, 350–353.

Han, C.M., Zhang, L.P., Sun, J.Z., Shi, H.F., Zhou, J., Gao, C.Y., 2010. Application of collagen-chitosan/fibrin glue asymmetric scaffolds in skin tissue engineering. J Zhejiang Univ Sci B 11, 524–530.

Hartwell, R., Leung, V., Chavez-Munoz, C., Nabai, L., Yang, H., Ko, F., Ghahary, A., 2011. A novel hydrogel-collagen composite improves functionality of an injectable extracellular matrix. Acta Biomater 7, 3060–3069.

Haslik, W., Kamolz, L.P., Manna, F., Hladik, M., Rath, T., Frey, M., 2010. Management of full-thickness skin defects in the hand and wrist region: first long-term experiences with the dermal matrix matriderm. J Plast Reconstr Aesthet Surg 63, 360–364.

Heitland, A., Piatkowski, A., Noah, E.M., Pallua, N., 2004. Update on the use of collagen/glycosaminoglycate skin substitute-six years of experiences with artificial skin in 15 german burn centers. Burns 30, 471–475.

Helary, C., Bataille, I., Abed, A., Illoul, C., Anglo, A., Louedec, L., Letourneur, D., Meddahi-Pelle, A., Giraud-Guille, M.M., 2010. Concentrated collagen hydrogels as dermal substitutes. Biomaterials 31, 481–490.

Helary, C., Zarka, M., Giraud-Guille, M.M., 2012. Fibroblasts within concentrated collagen hydrogels favour chronic skin wound healing. J Tissue Eng Regen Med 6, 225–237.

Hong, Y., Huber, A., Takanari, K., Amoroso, N.J., Hashizume, R., Badylak, S.F., Wagner, W.R., 2011. Mechanical properties and in vivo behavior of a biodegradable synthetic polymer microfiber-extracellular matrix hydrogel biohybrid scaffold. Biomaterials 32, 3387–3394.

Hopp, B., Smausz, T., Kresz, N., Barna, N., Bor, Z., Kolozsvari, L., Chrisey, D.B., Szabo, A., Nogradi, A., 2005. Survival and proliferative ability of various living cell types after laser-induced forward transfer. Tissue Eng 11, 1817–1823.

Hu, K., Shi, H., Zhu, J., Deng, D., Zhou, G., Zhang, W., Cao, Y., Liu, W., 2010. Compressed collagen gel as the scaffold for skin engineering. Biomed Microdevices 12, 627–635.

Huang, S., Lu, G., Wu, Y., Jirigala, E., Xu, Y., Ma, K., Fu, X., 2012. Mesenchymal stem cells delivered in a microsphere-based engineered skin contribute to cutaneous wound healing and sweat gland repair. J Dermatol Sci 66, 29–36.
Huang, S., Zhang, Y., Tang, L., Deng, Z., Lu, W., Feng, F., Xu, X., Jin, Y., 2009. Functional bilayered skin substitute constructed by tissue-engineered extracellular matrix and microsphere-incorporated gelatin hydrogel for wound repair. Tissue Eng Part A 15, 2617–2624.
Huang, X., Zhang, Y., Zhang, X., Xu, L., Chen, X., Wei, S., 2013. Influence of radiation crosslinked carboxymethyl-chitosan/gelatin hydrogel on cutaneous wound healing. Mater Sci Eng C Mater Biol Appl 33, 4816–4824.
Jiang, Y., Chen, J., Deng, C., Suuronen, E.J., Zhong, Z., 2014. Click hydrogels, microgels and nanogels: emerging platforms for drug delivery and tissue engineering. Biomaterials 35, 4969–4985.
Kean, T., Thanou, M., 2010. Biodegradation, biodistribution and toxicity of chitosan. Adv Drug Deliv Rev 62, 3–11.
Kempf, M., Miyamura, Y., Liu, P.Y., Chen, A.C., Nakamura, H., Shimizu, H., Tabata, Y., Kimble, R.M., Mcmillan, J.R., 2011. A denatured collagen microfiber scaffold seeded with human fibroblasts and keratinocytes for skin grafting. Biomaterials 32, 4782–4792.
Khan, F., Ahmad, S.R., 2013. Polysaccharides and their derivatives for versatile tissue engineering application. Macromol Biosci 13, 395–421.
Kim, G., Kim, W., 2007. Highly porous 3D nanofiber scaffold using an electrospinning technique. J Biomed Mater Res B Appl Biomater 81, 104–110.
Kim, H.L., Lee, J.H., Lee, M.H., Kwon, B.J., Park, J.C., 2012. Evaluation of electrospun (1,3)-(1,6)-β-D-glucans/biodegradable polymer as artificial skin for full-thickness wound healing. Tissue Eng Part A 18, 2315–2322.
Kim, K.L., Han, D.K., Park, K., Song, S.H., Kim, J.Y., Kim, J.M., Ki, H.Y., Yie, S.W., Roh, C.R., Jeon, E.S., Kim, D.K., Suh, W., 2009. Enhanced dermal wound neovascularization by targeted delivery of endothelial progenitor cells using an RGD-g-PLLA scaffold. Biomaterials 30, 3742–3748.
Koch, L., Gruene, M., Unger, C., Chichkov, B., 2013. Laser assisted cell printing. Curr Pharm Biotechnol 14, 91–97.
Koch, L., Kuhn, S., Sorg, H., Gruene, M., Schlie, S., Gaebel, R., Polchow, B., Reimers, K., Stoelting, S., Ma, N., Vogt, P.M., Steinhoff, G., Chichkov, B., 2010. Laser printing of skin cells and human stem cells. Tissue Eng Part C Methods 16, 847–854.
Kroehne, V., Heschel, I., Schugner, F., Lasrich, D., Bartsch, J.W., Jockusch, H., 2008. Use of a novel collagen matrix with oriented pore structure for muscle cell differentiation in cell culture and in grafts. J Cell Mol Med 12, 1640–1648.
Kuberka, M., Von Heimburg, D., Schoof, H., Heschel, I., Rau, G., 2002. Magnification of the pore size in biodegradable collagen sponges. Int J Artif Organs 25, 67–73.
Kundu, B., Kundu, S.C., 2012. Silk sericin/polyacrylamide in situ forming hydrogels for dermal reconstruction. Biomaterials 33, 7456–7467.
Kwon, H., Rainbow, R.S., Sun, L., Hui, C.K., Cairns, D.M., Preda, R.C., Kaplan, D.L., Zeng, L., 2014. Scaffold structure and fabrication method affect proinflammatory milieu in three-dimensional-cultured chondrocytes. J Biomed Mater Res A 103 (2), 534–544.
Lamme, E.N., de Vries, H.J., van Veen, H., Gabbiani, G., Westerhof, W., Middelkoop, E., 1996. Extracellular matrix characterization during healing of full-thickness wounds treated with a collagen/elastin dermal substitute shows improved skin regeneration in pigs. J Histochem Cytochem 44, 1311–1322.
Lee, C.H., Cook, J.L., Mendelson, A., Moioli, E.K., Yao, H., Mao, J.J., 2010. Regeneration of the articular surface of the rabbit synovial joint by cell homing: a proof of concept study. Lancet 376, 440–448.

Lee, S.B., Kim, Y.H., Chong, M.S., Hong, S.H., Lee, Y.M., 2005. Study of gelatin-containing artificial skin V: fabrication of gelatin scaffolds using a salt-leaching method. Biomaterials 26, 1961–1968.

Li, B., Davidson, J.M., Guelcher, S.A., 2009. The effect of the local delivery of platelet-derived growth factor from reactive two-component polyurethane scaffolds on the healing in rat skin excisional wounds. Biomaterials 30, 3486–3494.

Li, M., Lu, S., Wu, Z., Yan, H., Mo, J., Wang, L., 2001. Study on porous silk fibroin materials fine structure of freeze dried silk fibroin. J Appl Polym Sci 79, 2185–2191.

Li, M., Ogiso, M., Minoura, N., 2003. Enzymatic degradation behavior of porous silk fibroin sheets. Biomaterials 24, 357–365.

Liu, F., Luo, X.S., Shen, H.Y., Dong, J.S., Yang, J., 2011. Using human hair follicle-derived keratinocytes and melanocytes for constructing pigmented tissue-engineered skin. Skin ResTechnol 17 (3), 373–379.

Losi, P., Briganti, E., Errico, C., Lisella, A., Sanguinetti, E., Chiellini, F., Soldani, G., 2013. Fibrin-based scaffold incorporating vegf- and bfgf-loaded nanoparticles stimulates wound healing in diabetic mice. Acta Biomater 9, 7814–7821.

Lozinsky, V.I., Galaev, I.Y., Plieva, F.M., Savina, I.N., Jungvid, H., Mattiasson, B., 2003. Polymeric cryogels as promising materials of biotechnological interest. Trends Biotechnol 21, 445–451.

Lynn, A.K., Yannas, I.V., Bonfield, W., 2004. Antigenicity and immunogenicity of collagen. J Biomed Mater Res B Appl Biomater 71, 343–354.

Mahjour, S.B., Ghaffarpasand, F., Wang, H., 2012. Hair follicle regeneration in skin grafts: current concepts and future perspectives. Tissue Eng Part B Rev 18, 15–23.

Mandal, B.B., Priya, A.S., Kundu, S.C., 2009. Novel silk sericin/gelatin 3-d scaffolds and 2-d films: fabrication and characterization for potential tissue engineering applications. Acta Biomater 5, 3007–3020.

Marino, D., Luginbuhl, J., Scola, S., Meuli, M., Reichmann, E., 2014. Bioengineering dermo-epidermal skin grafts with blood and lymphatic capillaries. Sci Transl Med 6, 221ra14.

Marreco, P.R., da Luz Moreira, P., Genari, S.C., Moraes, A.M., 2004. Effects of different sterilization methods on the morphology, mechanical properties, and cytotoxicity of chitosan membranes used as wound dressings. J Biomed Mater Res B Appl Biomater 71, 268–277.

Matthews, J.A., Wnek, G.E., Simpson, D.G., Bowlin, G.L., 2002. Electrospinning of collagen nanofibers. Biomacromolecules 3, 232–238.

Meinel, L., Hofmann, S., Karageorgiou, V., Kirker-Head, C., Mccool, J., Gronowicz, G., Zichner, L., Langer, R., Vunjak-Novakovic, G., Kaplan, D.L., 2005. The inflammatory responses to silk films in vitro and in vivo. Biomaterials 26, 147–155.

Mi, F.L., Shyu, S.S., Wu, Y.B., Lee, S.T., Shyong, J.Y., Huang, R.N., 2001. Fabrication and characterization of a sponge-like asymmetric chitosan membrane as a wound dressing. Biomaterials 22, 165–173.

Michael, S., Sorg, H., Peck, C.T., Koch, L., Deiwick, A., Chichkov, B., Vogt, P.M., Reimers, K., 2013. Tissue engineered skin substitutes created by laser-assisted bioprinting form skin-like structures in the dorsal skin fold chamber in mice. PLoS One 8, e57741.

Middelkoop, E., de Vries, H.J., Ruuls, L., Everts, V., Wildevuur, C.H., Westerhof, W., 1995. Adherence, proliferation and collagen turnover by human fibroblasts seeded into different types of collagen sponges. Cell Tissue Res 280, 447–453.

Middelkoop, E., van den Bogaerdt, A.J., Lamme, E.N., Hoekstra, M.J., Brandsma, K., Ulrich, M.M.W., 2004. Porcine wound models for skin substitution and burn treatment. Biomaterials 25, 1559–1567.

Mineo, A., Suzuki, R., Kuroyanagi, Y., 2013. Development of an artificial dermis composed of hyaluronic acid and collagen. J Biomater Sci Polym Ed 24, 726–740.
Minoura, N., Aiba, S., Gotoh, Y., Tsukada, M., Imai, Y., 1995a. Attachment and growth of cultured fibroblast cells on silk protein matrices. J Biomed Mater Res 29, 1215–1221.
Minoura, N., Aiba, S., Higuchi, M., Gotoh, Y., Tsukada, M., Imai, Y., 1995b. Attachment and growth of fibroblast cells on silk fibroin. Biochem Biophys Res Commun 208, 511–516.
Moiemen, N.S., Vlachou, E., Staiano, J.J., Thawy, Y., Frame, J.D., 2006. Reconstructive surgery with integra dermal regeneration template: histologic study, clinical evaluation, and current practice. Plast Reconstr Surg 117, 160S–174S.
Mudera, V., Morgan, M., Cheema, U., Nazhat, S., Brown, R., 2007. Ultra-rapid engineered collagen constructs tested in an in vivo nursery site. J Tissue Eng Regen Med 1, 192–198.
Nagiah, N., Madhavi, L., Anitha, R., Anandan, C., Srinivasan, N.T., Sivagnanam, U.T., 2013. Development and characterization of coaxially electrospun gelatin coated poly (3-hydroxybutyric acid) thin films as potential scaffolds for skin regeneration. Mater Sci Eng C Mater Biol Appl 33, 4444–4452.
Nakada, A., Shigeno, K., Sato, T., Kobayashi, T., Wakatsuki, M., Uji, M., Nakamura, T., 2013. Manufacture of a weakly denatured collagen fiber scaffold with excellent biocompatibility and space maintenance ability. Biomed Mater 8, 045010.
Natarajan, V., Krithica, N., Madhan, B., Sehgal, P.K., 2013. Preparation and properties of tannic acid cross-linked collagen scaffold and its application in wound healing. J Biomed Mater Res B Appl Biomater 101, 560–567.
Nayak, S., Dey, S., Kundu, S.C., 2013. Skin equivalent tissue-engineered construct: co-cultured fibroblasts/keratinocytes on 3D matrices of sericin hope cocoons. PLoS One 8, e74779.
Nayak, S., Talukdar, S., Kundu, S.C., 2012. Potential of 2d crosslinked sericin membranes with improved biostability for skin tissue engineering. Cell Tissue Res 347, 783–794.
Nillesen, S.T., Lammers, G., Wismans, R.G., Ulrich, M.M., Middelkoop, E., Spauwen, P.H., Faraj, K.A., Schalkwijk, J., Daamen, W.F., Van Kuppevelt, T.H., 2011. Design and in vivo evaluation of a molecularly defined acellular skin construct: reduction of early contraction and increase in early blood vessel formation. Acta Biomater 7, 1063–1071.
Noh, H.K., Lee, S.W., Kim, J.M., Oh, J.E., Kim, K.H., Chung, C.P., Choi, S.C., Park, W.H., Min, B.M., 2006. Electrospinning of chitin nanofibers: degradation behavior and cellular response to normal human keratinocytes and fibroblasts. Biomaterials 27, 3934–3944.
O'brien, F.J., Harley, B.A., Yannas, I.V., Gibson, L., 2004. Influence of freezing rate on pore structure in freeze-dried collagen-gag scaffolds. Biomaterials 25, 1077–1086.
O'brien, F.J., Harley, B.A., Yannas, I.V., Gibson, L.J., 2005. The effect of pore size on cell adhesion in collagen-gag scaffolds. Biomaterials 26, 433–441.
Othon, C.M., Wu, X., Anders, J.J., Ringeisen, B.R., 2008. Single-cell printing to form three-dimensional lines of olfactory ensheathing cells. Biomed Mater 3, 034101.
Parenteau-Bareil, R., Gauvin, R., Cliche, S., Gariepy, C., Germain, L., Berthod, F., 2011. Comparative study of bovine, porcine and avian collagens for the production of a tissue engineered dermis. Acta Biomater 7, 3757–3765.
Patois, E., Osorio-Da Cruz, S., Tille, J.C., Walpoth, B., Gurny, R., Jordan, O., 2009. Novel thermosensitive chitosan hydrogels: in vivo evaluation. J Biomed Mater Res A 91, 324–330.
Philandrianos, C., Andrac-Meyer, L., Mordon, S., Feuerstein, J.M., Sabatier, F., Veran, J., Magalon, G., Casanova, D., 2012. Comparison of five dermal substitutes in full-thickness skin wound healing in a porcine model. Burns 38, 820–829.

Pieper, J.S., van Wachem, P.B., van Luyn, M.J.A., Brouwer, L.A., Hafmans, T., Veerkamp, J.H., Van Kuppevelt, T.H., 2000. Attachment of glycosaminoglycans to collagenous matrices modulates the tissue response in rats. Biomaterials 21, 1689–1699.

Pontiggia, L., Klar, A., Bottcher-Haberzeth, S., Biedermann, T., Meuli, M., Reichmann, E., 2013. Optimizing in vitro culture conditions leads to a significantly shorter production time of human dermo-epidermal skin substitutes. Pediatr Surg Int 29, 249–256.

Powell, H.M., Boyce, S.T., 2008. Fiber density of electrospun gelatin scaffolds regulates morphogenesis of dermal-epidermal skin substitutes. J Biomed Mater Res A 84, 1078–1086.

Powell, H.M., Boyce, S.T., 2009. Engineered human skin fabricated using electrospun collagen-PCL blends: morphogenesis and mechanical properties. Tissue Eng Part A 15, 2177–2187.

Rabea, E.I., Badawy, M.E.T., Stevens, C.V., Smagghe, G., Steurbaut, W., 2003. Chitosan as antimicrobial agent: applications and mode of action. Biomacromolecules 4, 1457–1465.

Ringeisen, B.R., Othon, C.M., Barron, J.A., Young, D., Spargo, B.J., 2006. Jet-based methods to print living cells. Biotechnol J 1, 930–948.

Rittie, L., Sachs, D.L., Orringer, J.S., Voorhees, J.J., Fisher, G.J., 2013. Eccrine sweat glands are major contributors to reepithelialization of human wounds. Am J Pathol 182, 163–171.

Rnjak-Kovacina, J., Wise, S.G., Li, Z., Maitz, P.K., Young, C.J., Wang, Y., Weiss, A.S., 2012. Electrospun synthetic human elastin: collagen composite scaffolds for dermal tissue engineering. Acta Biomater 8, 3714–3722.

Robinet, A., Fahem, A., Cauchard, J.H., Huet, E., Vincent, L., Lorimier, S., Antonicelli, F., Soria, C., Crepin, M., Hornebeck, W., Bellon, G., 2005. Elastin-derived peptides enhance angiogenesis by promoting endothelial cell migration and tubulogenesis through upregulation of MT1-MMP. J Cell Sci 118, 343–356.

Salem, A.K., Stevens, R., Pearson, R.G., Davies, M.C., Tendler, S.J., Roberts, C.J., Williams, P.M., Shakesheff, K.M., 2002. Interactions of 3T3 fibroblasts and endothelial cells with defined pore features. J Biomed Mater Res 61, 212–217.

Saunders, R.E., Gough, J.E., Derby, B., 2008. Delivery of human fibroblast cells by piezoelectric drop-on-demand inkjet printing. Biomaterials 29, 193–203.

Savina, I.N., Dainiak, M., Jungvid, H., Mikhalovsky, S.V., Galaev, I.Y., 2009. Biomimetic macroporous hydrogels: protein ligand distribution and cell response to the ligand architecture in the scaffold. J Biomater Sci Polym Ed 20, 1781–1795.

Seo, B.F., Lee, J.Y., Jung, S.N., 2013. Models of abnormal scarring. Biomed Res Int 2013, 423147.

Shevchenko, R.V., Eeman, M., Rowshanravan, B., Allan, I.U., Savina, I.N., Illsley, M., Salmon, M., James, S.L., Mikhalovsky, S.V., James, S.E., 2014. The in vitro characterization of a gelatin scaffold, prepared by cryogelation and assessed in vivo as a dermal replacement in wound repair. Acta Biomater 10, 3156–3166.

Sindrilaru, A., Scharffetter-Kochanek, K., 2013. Disclosure of the culprits: macrophages-versatile regulators of wound healing. Adv Wound Care (New Rochelle) 2, 357–368.

Siritienthong, T., Ratanavaraporn, J., Aramwit, P., 2012. Development of ethyl alcohol-precipitated silk sericin/polyvinyl alcohol scaffolds for accelerated healing of full-thickness wounds. Int J Pharm 439, 175–186.

Sriwiriyanont, P., Lynch, K.A., Mcfarland, K.L., Supp, D.M., Boyce, S.T., 2013. Characterization of hair follicle development in engineered skin substitutes. PLoS One 8, e65664.

Sugiura, H., Yunoki, S., Kondo, E., Ikoma, T., Tanaka, J., Yasuda, K., 2009. In vivo biological responses and bioresorption of tilapia scale collagen as a potential biomaterial. J Biomater Sci Polym Ed 20, 1353–1368.

Suh, J.K., Matthew, H.W., 2000. Application of chitosan-based polysaccharide biomaterials in cartilage tissue engineering: a review. Biomaterials 21, 2589–2598.
Sullivan, T.P., Eaglstein, W.H., Davis, S.C., Mertz, P., 2001. The pig as a model for human wound healing. Wound Repair Regen 9, 66–76.
Sun, B., Duan, B., Yuan, X.Y., 2006. Preparation of core/shell PVP/PLA ultrafine fibers by coaxial electrospinning. J Appl Polym Sci 102, 39–45.
Sundaramurthi, D., Vasanthan, K.S., Kuppan, P., Krishnan, U.M., Sethuraman, S., 2012. Electrospun nanostructured chitosan-poly(vinyl alcohol) scaffolds: a biomimetic extracellular matrix as dermal substitute. Biomed Mater 7, 045005.
Suzuki, S., Matsuda, K., Isshiki, N., Tamada, Y., Ikada, Y., 1990. Experimental study of a newly developed bilayer artificial skin. Biomaterials 11, 356–360.
Suzuki, T., Lee, C.H., Chen, M., Zhao, W., Fu, S.Y., Qi, J.J., Chotkowski, G., Eisig, S.B., Wong, A., Mao, J.J., 2011. Induced migration of dental pulp stem cells for in vivo pulp regeneration. J Dent Res 90, 1013–1018.
Swope, V.B., Supp, A.P., Schwemberger, S., Babcock, G., Boyce, S., 2006. Increased expression of integrins and decreased apoptosis correlate with increased melanocyte retention in cultured skin substitutes. Pigment Cell Res 19, 424–433.
Tchemtchoua, V.T., Atanasova, G., Aqil, A., Filee, P., Garbacki, N., Vanhooteghem, O., Deroanne, C., Noel, A., Jerome, C., Nusgens, B., Poumay, Y., Colige, A., 2011. Development of a chitosan nanofibrillar scaffold for skin repair and regeneration. Biomacromolecules 12, 3194–3204.
Tomihata, K., Ikada, Y., 1997. In vitro and in vivo degradation of films of chitin and its deacetylated derivatives. Biomaterials 18, 567–575.
Tong, E., Martin, F., Shelley, O., 2014. A novel approach to reconstruct a large full thickness abdominal wall defect: successful treatment with MatriDerm® and Split. J Wound Care 23, 355–357.
Tripathi, A., Kathuria, N., Kumar, A., 2009. Elastic and macroporous agarose-gelatin cryogels with isotropic and anisotropic porosity for tissue engineering. J Biomed Mater Res A 90, 680–694.
Tsubouchi, K., Igarashi, Y., Takasu, Y., Yamada, H., 2005. Sericin enhances attachment of cultured human skin fibroblasts. Biosci Biotechnol Biochem 69, 403–405.
Tummalapalli, C.M., Tyagi, S.C., 1999. Responses of vascular smooth muscle cell to extracellular matrix degradation. J Cell Biochem 75, 515–527.
Unger, C., Gruene, M., Koch, L., Koch, J., Chichkov, B.N., 2011. Time-resolved imaging of hydrogel printing via laser-induced forward transfer. Appl Phys A: Mat Sci Process 103, 271–277.
Valentin, J.E., Stewart-Akers, A.M., Gilbert, T.W., Badylak, S.F., 2009. Macrophage participation in the degradation and remodeling of extracellular matrix scaffolds. Tissue Eng Part A 15, 1687–1694.
Vanden Berg-Foels, W.S., 2013. In situ tissue regeneration: chemoattractants for endogenous stem cell recruitment. Tissue Eng Part B Rev 20 (1), 28–39.
Vasconcelos, D.P., Fonseca, A.C., Costa, M., Amaral, I.F., Barbosa, M.A., Aguas, A.P., Barbosa, J.N., 2013. Macrophage polarization following chitosan implantation. Biomaterials 34, 9952–9959.
van der Veen, V.C., van der Wal, M.B., van Leeuwen, M.C., Ulrich, M.M., Middelkoop, E., 2010. Biological background of dermal substitutes. Burns 36, 305–321.
Veleirinho, B., Coelho, D.S., Dias, P.F., Maraschin, M., Ribeiro-do-Valle, R.M., Lopes-da-Silva, J.A., 2012. Nanofibrous poly(3-hydroxybutyrate-co-3-hydroxyvalerate)/chitosan scaffolds for skin regeneration. Int J Biol Macromol 51, 343–350.

De Vries, H.J., Mekkes, J.R., Middelkoop, E., Hinrichs, W.L., Wildevuur, C.R., Westerhof, W., 1993. Dermal substitutes for full-thickness wounds in a one-stage grafting model. Wound Repair Regen 1, 244–252.

de Vries, H.J.C., Middelkoop, E., Mekkes, J.R., Dutrieux, R.P., Wildevuur, C.H., Westerhof, H., 1994. Dermal regeneration in native non-cross-linked collagen sponges with different extracellular matrix molecules. Wound Repair Regen 2, 37–47.

Waaijman, T., Breetveld, M., Ulrich, M., Middelkoop, E., Scheper, R.J., Gibbs, S., 2010. Use of a collagen-elastin matrix as transport carrier system to transfer proliferating epidermal cells to human dermis in vitro. Cell Transplant 19, 1339–1348.

Wang, H., Pieper, J., Peters, F., van Blitterswijk, C.A., Lamme, E.N., 2005. Synthetic scaffold morphology controls human dermal connective tissue formation. J Biomed Mater Res A 74, 523–532.

Wang, H.M., Chou, Y.T., Wen, Z.H., Wang, Z.R., Chen, C.H., Ho, M.L., 2013. Novel biodegradable porous scaffold applied to skin regeneration. PLoS One 8, e56330.

Wang, X., Li, Q., Hu, X., Ma, L., You, C., Zheng, Y., Sun, H., Han, C., Gao, C., 2012. Fabrication and characterization of poly(l-lactide-co-glycolide) knitted mesh-reinforced collagen-chitosan hybrid scaffolds for dermal tissue engineering. J Mech Behav Biomed Mater 8, 204–215.

Wang, Y., Rudym, D.D., Walsh, A., Abrahamsen, L., Kim, H.J., Kim, H.S., Kirker-Head, C., Kaplan, D.L., 2008. In vivo degradation of three-dimensional silk fibroin scaffolds. Biomaterials 29, 3415–3428.

Willard, J.J., Drexler, J.W., Das, A., Roy, S., Shilo, S., Shoseyov, O., Powell, H.M., 2013. Plant-derived human collagen scaffolds for skin tissue engineering. Tissue Eng Part A 19, 1507–1518.

Wilson Jr., W.C., Boland, T., 2003. Cell and organ printing 1: protein and cell printers. Anat Rec A Discov Mol Cell Evol Biol 272, 491–496.

Wu, L., Li, H., Li, S., Li, X., Yuan, X., Zhang, Y., 2010. Composite fibrous membranes of PLGA and chitosan prepared by coelectrospinning and coaxial electrospinning. J Biomed Mater Res A 92, 563–574.

Xie, Z., Paras, C.B., Weng, H., Punnakitikashem, P., Su, L.C., Vu, K., Tang, L., Yang, J., Nguyen, K.T., 2013. Dual growth factor releasing multi-functional nanofibers for wound healing. Acta Biomater 9, 9351–9359.

Xu, T., Jin, J., Gregory, C., Hickman, J.J., Boland, T., 2005. Inkjet printing of viable mammalian cells. Biomaterials 26, 93–99.

Xu, T., Zhao, W., Zhu, J.M., Albanna, M.Z., Yoo, J.J., Atala, A., 2013. Complex heterogeneous tissue constructs containing multiple cell types prepared by inkjet printing technology. Biomaterials 34, 130–139.

Yang, X., Tong, Y.Y., Li, Z.C., Liang, D., 2011. Aggregation-induced microgelation: a new approach to prepare gels in solution. Soft Matter 7, 978–985.

Yannas, I.V., Burke, J.F., 1980. Design of an artificial skin. I. Basic design principles. J Biomed Mater Res 14, 65–81.

Yannas, I.V., Lee, E., Orgill, D.P., Skrabut, E.M., Murphy, G.F., 1989. Synthesis and characterization of a model extracellular matrix that induces partial regeneration of adult mammalian skin. Proc Natl Acad Sci USA 86, 933–937.

Ye, Q., Harmsen, M.C., van Luyn, M.J., Bank, R.A., 2010. The relationship between collagen scaffold cross-linking agents and neutrophils in the foreign body reaction. Biomaterials 31, 9192–9201.

Zaoming, W., Codina, R., Fernandez-Caldas, E., Lockey, R.F., 1996. Partial characterization of the silk allergens in mulberry silk extract. J Investig Allergol Clin Immunol 6, 237–241.

Zeltinger, J., Sherwood, J.K., Graham, D.A., Mueller, R., Griffith, L.G., 2001. Effect of pore size and void fraction on cellular adhesion, proliferation, and matrix deposition. Tissue Eng 7, 557–572.

Zhang, Y.Q., 2002. Applications of natural silk protein sericin in biomaterials. Biotechnol Adv 20, 91–100.

Zheng, Y., Henderson, P.W., Choi, N.W., Bonassar, L.J., Spector, J.A., Stroock, A.D., 2011. Microstructured templates for directed growth and vascularization of soft tissue in vivo. Biomaterials 32, 5391–5401.

Zhu, J., Marchant, R.E., 2011. Design properties of hydrogel tissue-engineering scaffolds. Expert Rev Med Devices 8, 607–626.

Zhu, K.Q., Carrougher, G.J., Gibran, N.S., Isik, F.F., Engrav, L.H., 2007. Review of the female duroc/yorkshire pig model of human fibroproliferative scarring. Wound Repair Regen 15 (Suppl. 1), S32–S39.

van Zuijlen, P.P., van Trier, A.J., Vloemans, J.F., Groenevelt, F., Kreis, R.W., Middelkoop, E., 2000. Graft survival and effectiveness of dermal substitution in burns and reconstructive surgery in a one-stage grafting model. Plast Reconstr Surg 106, 615–623.

Engineering the tissue–wound interface: harnessing topography to direct wound healing

A.L. Clement, G.D. Pins
Worcester Polytechnic Institute, Worcester, MA, United States

10.1 Introduction

Worldwide, there are over 49 million wounds per year, including more than 14 million venous and diabetic ulcers and 5.2 million pressure ulcers (MedMarket, 2007). Since these chronic wounds are most prevalent in individuals over 60 years of age, these numbers are expected to rise as the population ages and the incidences of diabetes increase.

Although minor skin wounds will heal without significant intervention, large surface area or chronic wounds require the rapid restoration of the skin barrier function to prevent complications such as infection and desiccation, which can lead to scarring, amputation, and death. The treatment of chronic wounds presents a challenging and expensive clinical problem, with the annual cost of treatment exceeding $8 billion. One method of treatment for chronic nonhealing wounds is skin grafting; between 1998 and 1999, approximately 163,000 grafting procedures were performed on Medicare patients (Shaffer et al., 2005). In addition, in the United States, approximately 40,000 burns require hospitalization annually, resulting in over 10,000 grafts per year (American Burn Association, 2007; Milenkovic et al., 2007). While the current gold standard is a split-thickness autograft, donor site morbidity and limited availability present significant drawbacks (Sheridan and Tompkins, 1999; Blais et al., 2013). In patients with compromised wound healing, these disadvantages are even more pronounced. Several tissue-engineered skin substitutes have been developed and commercialized to address this need. Although promising, they exhibit limited mechanical stability, suboptimal wound healing, and prolonged healing times (Bar-Meir et al., 2006; Metcalfe and Ferguson, 2007; Priya et al., 2008). A common feature of current strategies is their failure to recreate the complex topography of native skin. This chapter reviews the current state of the field and the importance of the physical microenvironment in cellular regulation and presents new models for investigating the role of topography in regulating cellular function.

10.1.1 Structure and function of skin

Skin, a complex multilayered organ, creates a physical barrier between the body and the external environment. In addition to providing protection from pathogens,

Wound Healing Biomaterials - Volume 1. http://dx.doi.org/10.1016/B978-1-78242-455-0.00010-0

ultraviolet damage, and chemical exposure, it is also responsible for sensory detection and maintaining tissue homeostasis (Madison, 2003; Rook and Burns, 2010).

10.1.1.1 Epidermis

The outer most layer of the skin, the epidermis, provides a protective barrier to the external environment. The epidermis is avascular and is comprised primarily of keratinocytes (95%), which are responsible for creating the semipermeable barrier. To create the complex structure of the epidermis, keratinocytes cycle through four epidermal strata: the stratum basale, stratum spinosum, stratum granulosum, and the stratum corneum. This constant cycling makes the epidermis highly regenerative, with a complete turnover time of approximately 28 days (Rook and Burns, 2010).

The innermost epidermal layer, the stratum basale, consists of a single layer of small, cuboidal keratinocytes and contains both slow-cycling keratinocyte stem cells and highly proliferative transit-amplifying (TA) cells. The keratinocyte stem cells are responsible for the regenerative potential of skin. As they divide, keratinocyte stem cells give rise to daughter TA cells, which rapidly divide to produce the terminally differentiated cells that comprise the other three epidermal strata. Terminal differentiation is characterized by downregulation of integrin expression, which results in keratinocyte detachment from the underlying basal lamina and initiates their upward migration (Senoo, 2013; Jones and Watt, 1993). The keratinocytes in the next layer, the stratum spinosum, are larger and begin synthesizing the keratin that will eventually end up in the stratum corneum. As the keratinocytes continue to differentiate, they migrate upward into the stratum granulosum where they begin to flatten. The keratinocytes of the stratum granulosum contain both keratohyalin granules composed of profillagrin (the precursor to fillagrin) and involucrin and lamellar bodies comprised of lipids and polysaccharides. The stratum granulosum is the last layer, comprised of viable cells. As cells transition from the stratum granulosum to the stratum corneum, they enucleate and release the granule contents. The flattened, keratinized cells or corneocytes, together with the lipid-enriched extracellular matrix (ECM), of the stratum corneum provide the semipermeable barrier of skin (Rook and Burns, 2010).

10.1.1.2 Dermis

The dermis is a supporting matrix predominantly comprised of type I collagen, elastin, and glycosaminoglycans. Type I collagen, the primary ECM component of the dermis, is arranged in the dermal tissue as a planar network of cross-linked fibers arranged in a "basket woven" conformation that provides the tensile strength of skin. The interwoven elastic fibers, composed of elastin and elastin-associated microfibrils, provide skin with its elasticity. Interspersed between the collagen and elastic fibers, glycosaminoglycans and proteoglycan macromolecules help maintain tissue hydration (Rook and Burns, 2010). The dermis itself can be divided into two layers: the uppermost papillary layer and the innermost reticular layer. The reticular dermis is characterized by larger, more densely packed collagen fibers and is the principal location of elastic fibers (Janson et al., 2012; Tajima and Pinnell, 1981; Chang et al., 2002). In contrast, the papillary dermis is characterized by dermal papillae, finger-like projections that extend upward into the epidermis. The papillary dermis is highly porous and densely populated by fibroblasts. It also contains a rich microcapillary network that is responsible for sustaining the epidermis. The microvascular network in the dermis also regulates tissue homeostasis, hydration, and thermal stability. As such, when the structural and functional properties of the tissue are disrupted by trauma, revascularization of the dermis is a critical component of the skin wound healing process (Tonnesen et al., 2000).

10.1.1.3 Dermal–epidermal junction

The dermal–epidermal junction (DEJ) is the dynamic basement membrane interface between the stratified epidermis and the dermis. The DEJ not only regulates the transport of molecules between the dermis and the epidermis but also regulates keratinocyte behavior by modulating cellular functions, including proliferation, migration, and differentiation (Burgeson and Christiano, 1997).

In native skin, the DEJ is not flat but rather is defined by the architecture of the dermal papillae. This interdigitated topography significantly increases the surface area between the dermis and the epidermis, which enhances the resistance to mechanical shearing and promotes dermal–epidermal adhesion, resulting in increased structural and mechanical stability of the epidermal. The increased surface area afforded by the topography of the DEJ also promotes nutrient diffusion from the microvasculature of the dermis to the epidermis. In addition, the dermal projections define microniches with dimensions ranging from 50 to 400 μm width and 50–200 μm depths (Odland, 1950; Fawcett and Jensh, 1997b). These keratinocyte microniches serve a critical role in directing epidermal morphogenesis and regulating epidermal wound healing (Jones et al., 1995; Lavker and Sun, 1982, 1983; Connelly et al., 2010; Arwert et al., 2012). In addition, microtopographic niches create distinct cellular microenvironments that differentially direct keratinocyte phenotypes and cellular function.

Notably, there is a high degree of variation in DEJ topography based on anatomical location. Areas exposed to increased friction, including palmar and planter surfaces, are characterized by longer, more densely packed papillae as compared to areas of less friction, such as dorsal and abdominal surfaces (Odland, 1950). In addition, some skin pathologies involve changes to the DEJ topography. For example, the dermal papillae are lengthened in psoriasis and flattened in aged skin (Murphy et al., 2007; Montagna and Carlisle, 1979).

Conforming to the topography of the DEJ is a thin (30–50 nm), ECM-dense layer called the basal lamina (Briggaman, 1982; Briggaman and Wheeler, 1975). The basal lamina is comprised of collagen (types IV, VII, XV, XVIII), fibronectin, and laminin (Alberts, 2002). The basal lamina functions as a physical barrier between the fibroblasts of the dermis and the epidermis and provides biochemical cues, which direct keratinocyte proliferation and differentiation.

10.1.2 Need for wound healing models

Skin substitutes provide a valuable platform for the in vitro investigation of skin pathologies, including epidermal cancers. There is a high incidence of skin diseases reported annually, with one in three Americans suffering from some form of skin disease (Bickers et al., 2006). This number includes 3.5 million new cases of skin cancer diagnosed annually (Rogers et al., 2010). For over 100 years, in vitro cell culture has allowed researchers to investigate the physical and biochemical signals that affect cell fate and function. In vitro cell culture has also provided model systems for investigating disease pathogenesis and wound healing as well as for testing toxicity and carcinogenicity (Mather and Roberts, 1998). Most cell culture is carried out in two dimensions on treated glass or plastic substrates; however, this oversimplified environment fails to recreate the physiological conditions in native tissues. This led to the development of three-dimensional (3D) organotypic models of skin to better investigate the cellular and molecular mechanisms involved in epidermal morphogenesis, wound healing, and disease progression (Gautrot et al., 2012; Powell et al., 2010; Bellas et al., 2012; Waugh and Sherratt, 2007; Huang et al., 2010; Ojeh and Navsaria, 2013; Groeber et al., 2011; Vaccariello et al., 1999). These models can also be used for drug and cosmetics screening. Although no model fully recreates the complexity of native skin, several groups developed methods to incorporate skin appendages into in vitro models to create a more physiologically relevant model (Atac et al., 2013; Huang et al., 2010). However, there is still a significant need for a 3D in vitro model of native skin that fully recreates complex biochemical and biophysical cues in the keratinocyte microenvironment.

10.1.3 Clinical need for engineered skin substitutes

Each year, an estimated 80 skin grafts are performed per 100,000 population (Milenkovic et al., 2007). Tissue-engineered skin substitutes represent a promising alternative therapy. In 2013, tissue-engineered skin substitutes made up nearly 3% of the multibillion dollar global wound healing market (MedMarket Diligence, 2013). Although these skin substitutes have exhibited some clinical success, they are not perfect (Bar-Meir et al., 2006; Metcalfe and Ferguson, 2007; Priya et al., 2008). This highlights a critical need for more advanced off-the-shelf tissue-engineered skin substitutes that can rapidly and robustly restore the skin barrier.

10.1.4 Current strategies for wound healing models and engineered skin substitutes

A variety of treatment strategies have been employed to address the large and growing number of chronic and traumatic cutaneous wounds. Since the primary concern with skin wounds is the rapid restoration of barrier function, the most fundamental approach for their treatment is cleaning and dressing the wound. In fact, bandages and wound dressings account for over 50% of the wound healing market (MedMarket Diligence, 2013). While wound dressings may prevent desiccation and promote skin regeneration, in large surface area and chronic nonhealing wounds, dressings alone are insufficient for preventing wound contraction and restoring the epidermal barrier. Furthermore, choosing a dressing can be difficult, because each wound situation is unique. To rapidly restore skin functionality in large surface area and nonhealing wounds, bioengineered skin substitutes are being investigated (Powers et al., 2013). The ideal bioengineered skin substitute will promote rapid restoration of the barrier integrity while supporting the regeneration of normal anatomy and physiology of the skin in the healing wound. Specifically, it should create a semipermeable barrier that serves to prevent infection while maintaining tissue homeostasis with respect to temperature and hydration, integrate efficiently into the host tissue with minimal impact on cosmesis, promote angiogenesis, and prevent wound contraction. In addition, it should be cost-efficient, easy to handle, and approximate native skin mechanically and anatomically (Metcalfe and Ferguson, 2007). There are three primary strategies for bioengineered skin grafts: dermal-only substitutes, epidermal-only substitutes, and composite dermal–epidermal substitutes. Below, we will review the current state-of-the-art therapeutic approaches for each of these strategies and discuss the limitations of each approach.

10.1.4.1 Autografts

Autografting is considered to be the "gold standard" treatment for the rapid closure of large skin wounds. The procedure involves the harvest of healthy, undamaged skin from one area of the patient's body and subsequent transplantation into the wound bed. Currently, the most common type of autograft is the split-thickness skin graft, in which the harvested skin consists of the epidermis and the uppermost layer of the dermis. This treatment strategy allows for a more structurally robust graft, while somewhat limiting the donor site trauma. However, because it involves a secondary surgery site, autografting is often not an option for patients with compromised wound healing or for the treatment of large area burns and skin traumas due to the lack of available donor sites (Sheridan and Tompkins, 1999). As such, there is a need for an off-the-shelf skin substitute that provides immediate and permanent coverage to skin traumas and chronic wounds.

10.1.4.2 Dermal skin substitutes

Wound contraction and excessive fibrotic scarring are major concerns in the healing of large area, split- and full-thickness skin injuries. The ability of dermal substitutes to prevent wound contraction and to limit fibrotic tissue deposition is well characterized in full-thickness wounds (Singer and Clark, 1999). Dermal substitute strategies are primarily acellular, which makes them more cost-effective and simpler to regulate than cell-based, epidermal, and composite substitute strategies. However, dermal substitutes are commonly used in conjunction with epidermal substitutes or temporary synthetic barrier layers because they do not provide immediate barrier function. FDA-approved dermal substitutes composed of natural or synthetic biomaterials have demonstrated some efficacy for the treatment of chronic ulcers and burns.

Alloderm® (LifeCell, Branchburg, NJ), GraftJacket® (KCI, San Antonio, TX), and Primatrix® (TEI Biosciences, Waltham, MA) are scaffolds created by decellularizing human or bovine dermal matrices. The proprietary lyophilization processes preserve the ECM and basement membrane structure while eliminating the immunogenicity associated with allogeneic skin transplant (Woo et al., 2007; Wainwright, 1995). Since Alloderm® does not provide an epidermal barrier, it is often used in conjunction with a meshed split-thickness autograft. Although the autografting procedure still requires a secondary surgery, the use of a dermal support substitute improves graft take as well as cosmesis of the meshed graft and minimizes the amount of autologous skin required for grafting (Wainwright, 1995). In addition to requiring an autograft or epidermal substitute, Alloderm® is also limited by inconsistent revascularization and the availability of donor skin (Shakespeare, 2005; Shevchenko et al., 2010).

Integra® (Integra LifeSciences, Plainsboro, NJ), based on the research of Yannas et al., is a collagen-glycosaminoglycan (GAG) sponge with a temporary silicone rubber epidermal barrier (Shastri, 2006; Stern et al., 1990; Dagalakis et al., 1980; Yannas and Burke, 1980). The sponge is gradually infiltrated by the patient's own cells and revascularized over a period of 2–3 weeks. Following the integration of the neodermal and vascular components, the silicone barrier layer is removed and replaced with a meshed split-thickness autograft. Integra® has several advantages, including ease of handling and a long shelf life. It has demonstrated widespread success as a dermal substitute for preventing wound contraction in full-thickness burns and chronic ulcers (Shakespeare, 2005; Yannas and Burke, 1980). Despite its success, Integra® requires a second procedure to replace the epidermis and does not completely eliminate the need for autografting (Shastri, 2006; Supp and Boyce, 2005).

Dermagraft® (Organogenesis, Canton, MA) is a dermal substitute that is primarily used for the treatment of chronic diabetic foot ulcers. It is composed of a polyglactin mesh that is cryopreserved following seeding with neonatal fibroblasts (Marston et al., 2003; Shevchenko et al., 2010). As the scaffold degrades in the wound bed, growth factors and ECM components synthesized by the fibroblasts are released, promoting dermal regeneration and keratinocyte epithelialization. A major disadvantage of Dermagraft® is its high cost as compared with other dermal substitutes associated with the incorporation of fibroblasts into the scaffold. In addition, it requires multiple applications to facilitate wound healing (Hart et al., 2012).

10.1.4.3 Epidermal skin substitutes

Since the prompt restoration of barrier function is essential for positive healing outcomes, one strategy is the use of a cellular epidermal layer to provide immediate barrier protection. There are many strategies for developing bioengineered epidermal skin substitutes. In general the approaches differ in three aspects: cell source and cell culture techniques, cell differentiation state and level of epidermal organization, and delivery method.

Although cultured epidermal substitutes have demonstrated some clinical success, the lack of an underlying dermal support matrix leads to poor graft stability and subpar wound healing, especially in the treatment of full-thickness wounds (Compton et al., 1993; Atiyeh and Costagliola, 2007). Accordingly, epidermal substitutes are often used in conjunction with dermal substitutes. Some commercially available epidermal skin substitutes are described below.

The pioneering technology for the first cultured epithelial autografts (CEAs) suitable for grafting into a wound bed was developed in the late 1970s by James Rheinwald, Ph.D., and Howard Green, M.D., who described the process for culturing human keratinocytes in vitro into sheets of epithelial tissue (Green et al., 1979; O'Connor et al., 1981; Gallico and O'Connor, 1985; Gallico et al., 1984; Shevchenko et al., 2010). Today, the product Epicel® (Genzyme, Cambridge, MA) is made by expanding and culturing autologous keratinocytes to confluence over a period of 2–3 weeks. The keratinocytes are then enzymatically detached from the culture substrate as a contiguous sheet that can be transplanted to the wound. CEAs provide wound covering and begin to promote the regeneration of the DEJ within weeks, although full maturation requires over a year (Compton et al., 1989). Despite early reports of clinical success, the clinical excitement for CEAs was dampened by the lengthy culture times required for CEA application and the highly variable rates of graft take (Atiyeh and Costagliola, 2007; Williamson et al., 1995; Carsin et al., 2000). Studies demonstrated that CEAs perform best when applied to wounds that retain some dermal appendages or in conjunction with dermal substitutes (Atiyeh and Costagliola, 2007).

Another epithelial autograft strategy is MySkin® (Regenerys, Cambridge, UK). Although not FDA approved for use in the United States, MySkin® is approved for use in the United Kingdom for the treatment of burns and chronic wounds. MySkin® consists of a temporary silicone substrate, which is used to deliver subconfluent autologous keratinocytes to the wound bed. In contrast to Epicel®, MySkin® does not require enzymatic detachment of the keratinocytes from the culture substrate. Instead, the keratinocytes are cultured on the polymer substrate, which is then used to transfer a subconfluent population of undifferentiated keratinocytes to the wound bed (Haddow et al., 2003). The silicone layer is anchored in place for 4 days as a dressing before being replaced with an absorbent dressing (Hernon et al., 2006). Multiple weekly applications are required for wound closure. MySkin® offers several advantages over traditional CEAs. Not only is the culture time following biopsy shorter (approximately 10 days compared to 14 days for CEAs), but if the treatment needs to be rescheduled for any reason, such as infection in the wound bed, the keratinocytes can be harvested from the MySkin® substrate, replated, and transplanted within 2 days. This flexibility increases the chance of a positive treatment outcome. Additionally, the enzymatic keratinocyte detachment required for the generation of CEAs may negatively impact the ability of the basal keratinocytes to reattach and engraft in the wound bed, whereas MySkin® does not require dissociation of the cells from the substrate. Instead, the substrate itself aids in graft transfer and enhances mechanical stability. However, MySkin® still requires multiple applications and lacks an underlying dermal support (Shevchenko et al., 2010).

ReCell® (Avita Medical, Melbourn, UK) is an autologous cell harvest and delivery system for the treatment of full-thickness burns that does not require culture time. Instead a small, split-thickness biopsy is obtained during surgery and enzymatically digested. The epidermis is then separated from the dermis and the dissociated epidermal cells are resuspended in lactate solution, resulting in a cell suspension predominantly consisting of keratinocytes (>60%) and also containing fibroblasts and melanocytes (Wood et al., 2012). The cell suspension is then applied to the prepared wound bed within an hour of isolation (Gravante et al., 2007). ReCell® can be applied directly to the wound or used in conjunction with a split-thickness autograft or a dermal skin substitute. ReCell® has been shown to enhance epithelialization and promote improved cosmetic outcomes. However, some authors have reported poor healing outcomes when used alone without a split-thickness autograft. In addition, as with other autografts, the ReCell® technique can result in donor site morbidity (Tenenhaus and Rennekampff, 2012). The results of these studies suggest that ReCell® and other epidermal grafting strategies would benefit from an engineered dermal matrix.

10.1.4.4 Composite skin substitutes

The most advanced skin substitutes are dermal–epidermal composite grafts designed to replace both layers of skin with a single treatment. Because they provide both dermal support to prevent wound contraction and an epidermal layer to reestablish barrier function, these substitutes are also the most expensive. Due to their complexity and required culture time, most composite skin substitutes utilize allogeneic keratinocytes.

One of the first composite skin grafts utilized neonatal allogenic cells and consisted of keratinocytes cultured on the surface of a flat, precontracted, fibroblast-populated type I bovine collagen matrix. The keratinocytes formed a stratified epidermis that resembles native epidermal strata, and a fibroblast-populated collagen matrix served as a provisional dermal matrix. This technology was developed and commercialized by Organogenesis as Apligraf® as a treatment for chronic wounds. Apligraf® has demonstrated clinical success for the treatment of diabetic and venous leg ulcers. Compared to compression therapy alone, Apligraf® has been shown to expedite wound closure and to promote healing of wounds that failed to respond to conventional therapy (Veves et al., 2001; Sams et al., 2002; Sheridan and Tompkins, 1999; Shevchenko et al., 2010; Falanga et al., 1998). However, its high cost of $28/cm^2, short shelf life of 5 days, and instability in the wound bed remain significant limitations.

OrCel® (Forticell Bioscience, Englewood Cliffs, NJ) is a tissue-engineered skin construct similar to Apligraf®. OrCel® is fabricated by seeding neonatal allogenic fibroblasts into type I bovine collagen sponge with a nonporous collagen-gel coating. Keratinocytes are seeded on the nonporous coating and cultured to form a confluent layer. OrCel® is used in the treatment of dystrophic epidermolysis bullosa, a blistering skin disease, as well as to promote healing at the autograft donor site in burn patients. Like Apligraf®, OrCel® does not function like a permanent skin substitute but rather as a temporary (7–14 days) bioactive dressing. As the grafts are broken down in the wound bed they release important ECM proteins and growth factors that stimulate the wound healing cascade (Shevchenko et al., 2010). Compared with an acellular bioactive dressing, OrCel® treatment resulted in shorter healing times (Still et al., 2003).

10.1.5 Importance of the physical microenvironment

A common limitation of current strategies is the lack of topography at the DEJ. In native skin, the epidermis conforms to the topography of the dermal grooves and papillae. This complex interdigitated topography not only promotes mechanical integrity between the dermis and the epidermis but also creates microniches with dimensions ranging from 50–400 μm in width and 50–200 μm in depth (Odland, 1950; Fawcett and Jensh, 1997a). These microtopographic niches create cellular microenvironments that promote keratinocyte clustering and differentially drive keratinocyte functions (Butler and Orgill, 2005; Jones et al., 1995; Lavker and Sun, 1982, 1983; Odland, 1950; Vracko, 1974). Additionally, $\beta_1^{bri}p63^+$ keratinocytes have been shown to localize to the deep rete ridges and the tips of the dermal papillae of the DEJ (Jensen et al., 1999; Jones and Watt, 1993; Jones et al., 1995; Kai-Hong et al., 2007; Lavker and Sun, 1982, 1983). Keratinocytes that express high levels of β_1 integrin and the nuclear transcription factor p63 have demonstrated the highest colony-forming efficiency and have been identified as putative stem cells (Barrandon and Green, 1987; Hotchin et al., 1995; Hotchin and Watt, 1992; Jones and Watt, 1993; Jones et al., 1995; Kai-Hong et al., 2007; Pellegrini et al., 2001; Van Rossum et al., 2004; Yang et al., 1999). In native skin, β_1^{bri} keratinocytes have been shown to cluster in specific topographical locations at the DEJ based on the location of the skin. In skin from foreskin and scalp, β_1^{bri} keratinocytes cluster in the tips of the dermal papillae, whereas in skin from the palm, β_1^{bri} cells cluster in the bottoms of the rete ridges (Hotchin et al., 1995; Jones et al., 1995).

Recreating stem cell patterning in 3D in vitro models will provide a model for systematically evaluating the roles of biochemical and physical cues in modulating epidermal morphogenesis as well as the etiologies of disease pathologies. Further, incorporating native stem cell clustering in therapeutic skin substitutes may provide a more stable graft to help reduce wound healing times and increase graft persistence. In addition to promoting stem cell localization, the microtopography of the DEJ may play a role in directing keratinocyte proliferation, differentiation, and migration. In vitro studies have shown that culture substrates direct keratinocyte shape, differentiation, and proliferation (Bush and Pins, 2010; Fujisaki et al., 2008; Fujisaki and Hattori, 2002). We will now discuss several in vitro and in vivo models for evaluating the role of topography in regulating cellular function and directing epidermal morphogenesis.

10.2 In vitro approaches to assessing the role of the physical microenvironment in the regulation of cellular function and wound healing

In vitro models are valuable tools for studying cellular responses to microenvironmental cues in epidermal wound healing and pathologies. In vitro models provide several advantages to in vivo and clinical studies. Significantly, they offer a controlled environment, which allows for the isolation of different environmental factors. Several in vitro model systems are available for studying the keratinocyte microenvironment and the effect of topography on cellular function.

10.2.1 Keratinocyte signaling within the microenvironment

In their study of the keratinocyte microenvironment, Chandrasekaran et al. (2011) incorporate 3D topography to investigate how epithelial cells interact with their microenvironment and autoregulate the epidermal to mesenchymal transition. They seeded immortalized keratinocytes (HaCaT) into microfabricated, spherical microbubble polydimethylsiloxane (PDMS) wells with diameters ranging from 100 to 300 μm. The geometries of the microbubble wells promoted autocrine signaling by allowing the concentrations of secreted factors to reach bioactive levels. Although cells seeded on both flat PDMS and in wells formed spheroids, they found that only HaCaT cells cultured in microbubbles underwent a transition from spheroid to sheeting morphology, similar to the epidermal to mesenchymal transition observed in cancer metastasis (Chandrasekaran et al., 2011). In a follow-up study, the microbubble arrays were used to investigate the clonogenic potential of two different subpopulations (adherent and nonadherent) of the tumorigenic WM115 melanoma cell line (Chandrasekaran and DeLouise, 2011). One, two, three, or more cells from each population were seeded into individual microwells and monitored for proliferation over 5 days. The nonadherent cells were found to have a greater clonogenic potential. These microbubble arrays provide a novel platform for examining the roles of soluble factors on keratinocyte functions in cellular microenvironments.

10.2.2 Cell–cell and cell–matrix interactions within the keratinocyte microenvironment

To study the microenvironment of multicellular constructs, Gautrot et al. (2012) developed microepidermis, small (<10 cells) stratified tissues grown on 100 μm diameter collagen-coated disks and rings. Soft lithography was used to create micropatterns of protein-resistant poly(oligo(ethylene glycol methacrylate)) brushes, which enabled the production of high-fidelity, collagen-coated, cell-adhesive islands (Gautrot et al., 2010). They created adhesive disk- and ring-shaped islands with an outer diameter of 100 μm and inner, nonadhesive regions ranging from 0 to 60 μm in diameter. They found that changing the shape of the micropatterned collagen coating from disk- to ring-shaped led to changes in the epidermal organization (Gautrot et al., 2012). Specifically, although both ring and disk microepidermis contained a central, involucrin positive region, the involucrin positive cells were partitioned more tightly on rings with 40 μm nonadhesive centers than on solid disks. In addition, the location of proliferative Ki-67 positive nuclei and β_1 integrins was altered in rings compared to disks; in disks the expression was centralized, while in rings there was little expression in the nonadhesive centers but increased expression at the periphery of the nonadhesive centers. This model provides a powerful tool for facile, high-throughput analysis of tissue morphogenesis and homeostasis.

10.2.3 Physical topography of the keratinocyte microenvironment

Lammers et al. (2012) developed a method for producing micropatterned membranes with topographic features mimicking the structure of dermal papillae. They created a polycarbonate mold that replicated dermal papillae using conical frustrum geometries (175 μm in diameter, 100 μm in height), which they used to create a PDMS negative master. The microstructured collagen membrane was then fabricated by air-drying a collagen fibril suspension on the PDMS molds overnight. Gene expression microarray data showed differential gene expression on micropatterned membranes compared to flat controls, suggesting a role for topography in defining keratinocyte functional niches.

Other models for evaluating cellular responses to microtopographic cues are micropatterned dermal–epidermal regeneration matrices (μDERMs). This 3D in vitro model system consists of a collagen-GAG sponge laminated to a micropatterned collagen-gel layer, which allows for the study of biochemical and cellular cues as well as topographic cues (Fig. 10.1; Bush and Pins, 2012; Clement et al., 2013). μDERMs are cultured at the air–liquid interface to create full-thickness in vitro skin substitutes. Like the dermal papillae template model, μDERMs recreate the invaginations found at the native DEJ. Bush and Pins (2012) showed increased epidermal thickness in narrow channels and β_1 integrin bright keratinocyte clustering in the narrow channels and corners of the wide channels. To investigate the role of topography in conjunction with fibroblast paracrine signaling, Clement et al. (2013) seeded the dermal collagen-GAG sponge with fibroblasts 48 h prior to the seeding of μDERMs. Their data suggested the possibility of three distinct microniches residing in different channel topographies.

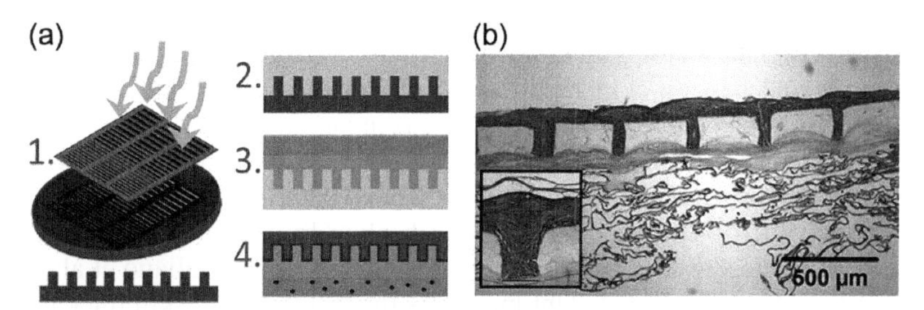

Figure 10.1 Micropatterned dermal–epidermal regeneration matrices (μDERMs).
(a) Production of μDERMs and 3D skin model system. Photolithography is used to create a silicon wafer with microtopographic features resembling the dermal–epidermal junction (DEJ) (1). Polydimethylsiloxane (PDMS) is cured on the wafer's surface, creating a negative mold (2). Type I collagen is self-assembled on the micropatterned mold, and a collagen-glycosaminoglycan sponge is laminated to the collagen matrix and cross-linked to form the μDERM (3). μDERMs are conjugated with fibronectin and sterilized. The dermal side is seeded with fibroblasts and cultured for 48 h, and then the micropattterned epidermal surface is seeded with keratinocytes (4). After 48 h, μDERMs were cultured at the air–liquid interface for up to 7 days. (b) Hematoxylin and eosin stain of μDERM after 7 days of air–liquid interface culture. High magnification of 100 μm channel in inset. Scale = 500 μm.

In particular, they noted increased keratinocyte proliferation in the narrow channels of these constructs, enhanced laminin 332 deposition in the wider channels, and putative keratinocyte stem cell clustering in the corners of wide channels and the bottoms of narrow channels.

10.3 In vivo approaches to assess cellular function in wound healing

While in vitro models provide an efficient platform for studying cellular pathways involved in epidermal morphogenesis and wound healing, they lack the complexity of in vivo wound healing. In particular, complex multitissue processes such as angiogenesis are better studied using animal models. Several in vivo models have been used to study epidermal wound healing, including excisional models, incisional models, and burn models (Wang et al., 2013; Hamoen et al., 2002; Medalie and Morgan, 1999; Boyce, 1999; Middelkoop et al., 2004; Davidson, 1998). Each model has its own advantages, depending on the specific study goals. Excisional wound healing models, in which a small piece of full-thickness skin is completely removed from the wound bed, are particularly valuable because they allow for the assessment of epithelialization, granulation tissue formation, scar formation, contraction, and angiogenesis in a single model (Galiano et al., 2004).

Small rodents (mice and rats) and pigs are the most commonly utilized animals in skin wound healing models. Small rodent models are popular because they are less expensive and allow for larger experimental numbers than large animal studies. Additionally, the availability of transgenic mice has made them a useful model for studying the influence of specific factors on wound healing (Grose and Werner, 2004). However, a major limitation of murine models is the significant differences in wound healing compared to humans. In rodents, wound contraction is the primary means of wound closure, whereas in humans the driving mechanisms are granulation and epithelialization. This has led to the development of murine excisional splinting models, which prevent wound contraction (Wang et al., 2013; Dunn et al., 2013). This results in wound healing by granulation tissue formation and more closely approximates human wound healing (Galiano et al., 2004). Another limitation of small rodent models is the difference in skin thickness and dermal architecture between rodent and human skin. In particular, rodent skin is much thinner and lacks the dermal papillae pattern characteristic of human skin (Middelkoop et al., 2004).

Alternatives to rodent models are pig models. Despite being more expensive, these models are advantageous particularly because of the greater similarity between human and porcine skin, although pig skin lacks eccrine sweat glands that are the major source for new epithelium in superficial wounds in human skin (Rittie et al., 2013). To mitigate costs, multiple treatments can be tried on the same animal, reducing variability, although systemic effects cannot be evaluated in this manner (Middelkoop et al., 2004). As with rodent models, porcine models can be created with excisional or burn wounds, depending on the goals of the study (Ponticorvo et al., 2014; Singer and McClain, 2003).

Another strategy for assessing wound healing in vivo is grafting bioengineered skin substitutes onto these animal models and then creating a wound on the implant. This method allows for the in vivo assessment of cell signaling pathways of human keratinocytes, which can differ from murine pathways. For example, Greenberg et al. (2006) employed this strategy to examine the effect of a chemical insult (sulfur mustard) on skin. The results of this study showed that bioengineered skin substitutes can be used to study human keratinocyte responses in an in vivo model and provide a useful platform for evaluating wound healing.

Although current animal models do not allow for the systematic evaluation of a defined keratinocyte microenvironment, in the future, these wound healing models may be utilized to look at the role of the cellular microenvironment in directing wound healing in vivo. By adapting the in vitro strategies previously discussed for in vivo implantation, the role of microtopography and microenvironmental cell cues in epithelialization, angiogenesis, and wound integration can be better elucidated. This could provide a platform for the development of future clinical substitutes for the treatment of chronic and traumatic wounds.

10.4 Future trends

Here, we present several 3D in vitro skin models that allow for the analysis of cellular responses to the DEJ microenvironment. These models advance the current state of research by incorporating biochemical, topographic, and 3D architectural cues of the keratinocyte microniche. In addition to utilizing these models to elucidate the role of the physical microenvironment in regulating cell fate and function, we expect that these model systems can advance in vitro models of wound healing and disease pathogenesis, particularly for pathologies such as psoriasis and basal cell carcinoma, which may have a topographic component to the etiology. Finally, these models provide a basis for designing the next generation of skin substitutes to improve clinical outcomes and the quality of life for patients suffering from large area skin traumas and chronic wounds.

10.4.1 Psoriasis models

Psoriasis is a chronic inflammatory skin disease that is characterized by hyperproliferative keratinocytes that result in the formation of red, scaly plaques (Rook and Burns, 2010). In addition to the physical symptoms, psoriasis is associated with psychological and social problems. Estimated to affect 2% of the population in the United States, psoriasis has been shown to negatively impact the health-related quality of life (a measure of physical and emotional well-being established by patient self-evaluation) as much as major chronic conditions such as cancer, diabetes, and arthritis (Rapp et al., 1999). Histopathologically, psoriatic skin exhibits abnormal keratinization as well as elongated rete ridges and correspondingly elongated dermal papillae

(Murphy et al., 2007). Currently, in vitro psoriasis models utilize keratinocytes cultured on flat substrates or on deepidermized dermis (DED) from nonpsoriatic patients; varying strategies are employed to establish the psoriatic phenotype, including the use of keratinocytes harvested from psoriatic lesions, coculture of normal keratinocytes with fibroblasts harvested from psoriatic lesions, and the addition of cytokines such as interleukin (IL)-20, IL-22, or oncostatin-M (Barker et al., 2004; Tjabringa et al., 2008; Saiag et al., 1985; Danilenko, 2008). Although these models are useful for investigating the psoriatic keratinocyte phenotype and identifying signaling pathways involved in psoriasis, they lack the topographic cues necessary to fully recreate the psoriatic microenvironment.

μDERMS provide an in vitro platform for investigating the role of topography in psoriatic pathogenesis. Computational modeling of the changes in psoriatic DEJ architecture with respect to keratinocyte proliferation suggest that hyperproliferation is strongly correlated with DEJ elongation, leading to the mechanistic hypothesis that the decreased keratinocyte turnover time leads to an expansion of the basal proliferative compartment and consequently the increase in DEJ surface area (Iizuka et al., 1997, 1999, 2004). However, the results of our studies demonstrate an increase in proliferation in deep, narrow channels. Similarly, the hyperproliferation exhibited by psoriatic keratinocytes may be stimulated by the topographical changes. This suggests that a positive feedback mechanism may be partially responsible for the development of psoriatic lesions. Accordingly, using a psoriatic μDERM model to gain a better understanding of the psoriatic microtopographic environment may lead to a mechanistic understanding of the etiology of the pathology as well as new therapeutic strategies for promoting plaque resolution.

10.4.2 Epithelial cancer models

Skin cancer is the most common type of cancer, affecting an estimated 20% of Americans over the course of their lives (Robinson, 2005; Stern, 2010). Nonmelanoma cancers, including basal cell carcinomas and cutaneous squamous cell carcinomas, comprise the majority of skin cancers and have a low mortality rate. However, melanoma, which accounts for less than 2% of skin cancers, is expected to claim nearly 10,000 lives in 2014 (American Cancer Association, 2014). Both 2D and 3D in vitro models of skin cancer have been used for drug screening and to study disease pathogenesis/carcinogenesis (Borchers et al., 1997; Commandeur et al., 2009; Obrigkeit et al., 2009; Eves et al., 2000). Like other in vitro models of skin, these lack defined topography at the DEJ. Research has shown that skin cancer cells cultured in vitro respond differently to treatment in three dimension as compared to two dimension (Vorsmann et al., 2013; Smalley et al., 2006). As such, there is a need to use 3D in vitro models to better understand the relationships between cellular microniches and the pathogenesis of epithelial cancers (Yamada and Cukierman, 2007).

Further, research into other types of cancer suggests an important role of the tumor microenvironment in tumorogenesis (Gudjonsson et al., 2003; Weaver et al., 1996; Weigelt and Bissell, 2008; Bissell et al., 2002). For example, while normal mammary cells and malignant mammary cells grown in a two dimensional monolayer culture are morphologically similar, when grown in laminin-rich ECM gels, the cancer cells formed disorganized tumor-like aggregates, whereas the normal cells formed organized acinar structures (Weigelt and Bissell, 2008). Additionally, in colon cancer, the underlying matrix topography is altered, with a loss of the characteristic crypt organization in areas with poorly differentiated tumors (Anderson et al., 2006; Rapier et al., 2010). In vitro models that recreate the native microenvironment provide a platform for studying both tumorogenesis and drug delivery targets. Therefore, we propose that the introduction of microtopography into an epidermal cancer model will not only allow us to investigate the role of microtopography in tumorogenesis but will also create a more physiologically relevant cancer model with a high utility for drug screening.

10.4.3 Wound healing models

Previously, several different methods have been utilized to create an in vitro cutaneous wound, including cryoburn (El Ghalbzouri et al., 2004; Han et al., 2005), scratch assay (Kandyba et al., 2010), and incisional wounding (Garlick and Taichman, 1994a,b). These models rely on flat substrates or DED with an undefined surface topography. Although these models provide valuable insight into epithelialization and the mechanisms of wound healing, there are currently no model systems that allow for the investigation of the role of topographical cues in wound healing. The studies presented here demonstrate that microtopography and the keratinocyte microenvironments play an integral role in regulating cellular functions such as proliferation and basement membrane protein deposition, which are critical to wound healing. By using the strategies discussed previously as a platform for developing novel in vitro wound healing models, the role of DEJ architecture in guiding wound healing can be identified and used as a basis for developing the next-generation wound healing strategies and clinical skin substitutes.

10.4.4 Creating robust bioengineered skin substitutes for traumatic and chronic wounds

The next-generation skin substitute must be a robust graft that promotes the prompt reestablishment of the epidermal barrier, supports angiogenesis, and rapidly integrates with the surrounding tissue. It is clear that the ideal skin substitute will replace both the dermal and epidermal components (Sheridan and Tompkins, 1999; Shevchenko et al., 2010; Kamel et al., 2013). Current composite skin substitutes fail to recreate the keratinocyte microenvironment created by the native DEJ. The increased surface area created by the series of dermal papillae found that the DEJ is thought to increase the structural stability of the tissue. In addition, studies have demonstrated that the incorporation of microtopography into skin substitutes leads to a change in keratinocyte morphology and gene expression, resulting in increased epidermal thickness and superior epidermal morphology (Lammers et al., 2012; Clement et al., 2013; Bush and Pins, 2012). Further, we showed that the dimensions of these topographic features differentially promote keratinocyte proliferation and basement membrane protein synthesis (Clement et al., 2013). Based on these results, we suggest that the incorporation of microtopography into the next-generation skin substitutes will improve clinical outcomes. Accordingly, future in vivo studies to assess the ability of μDERMs to serve as a template for autologous skin regeneration are essential.

10.5 Conclusions

The in vitro and in vivo experimental strategies discussed in this chapter provide a toolbox of model systems that allows for the further elucidation of the keratinocyte microniche. Together, they allow for the investigation of cell–cell, cell–matrix, and cell-soluble factor interactions in the context of the native microenvironment. These models represent an important advancement for the in vitro study of disease pathologies and wound healing mechanisms. In the future, these in vitro models may lead to the development of a next-generation clinical skin substitute that incorporates topography and matrix cues to better recreate keratinocyte microenvironments and enhance the rate of wound healing.

References

Alberts, E.A., 2002. Molecular Biology of the Cell. Garland Science, New York.

American Burn Association, 2007. Burn Incidence and Treatment in the US. 2007. Fact Sheet.

American Cancer Association, 2014. Cancer Facts & Figures 2014. American Cancer Association, Atlanta.

Anderson, R., Anderson, E., Shakir, L., Glover, S., 2006. Image analysis of extracellular matrix topography of colon cancer cells. Microscopy and Analysis 20, 5–7.

Arwert, E.N., Hoste, E., Watt, F.M., 2012. Epithelial stem cells, wound healing and cancer. Nature Reviews. Cancer 12, 170–180.

Atac, B., Wagner, I., Horland, R., Lauster, R., Marx, U., Tonevitsky, A.G., Azar, R.P., Lindner, G., 2013. Skin and hair on-a-chip: in vitro skin models versus ex vivo tissue maintenance with dynamic perfusion. Lab on a Chip 13, 3555–3561.

Atiyeh, B.S., Costagliola, M., 2007. Cultured epithelial autograft (CEA) in burn treatment: three decades later. Burns 33, 405–413.

Bar-Meir, E., Mendes, D., Winkler, E., 2006. Skin substitutes. IMAJ 8, 188–191.

Barker, C.L., Mchale, M.T., Gillies, A.K., Waller, J., Pearce, D.M., Osborne, J., Hutchinson, P.E., Smith, G.M., Pringle, J.H., 2004. The development and characterization of an in vitro model of psoriasis. The Journal of Investigative Dermatology 123, 892–901.

Barrandon, Y., Green, H., 1987. Three clonal types of keratinocyte with different capacities for multiplication. Proceedings of the National Academy of Sciences of the United States of America 84, 2302–2306.

Bellas, E., Seiberg, M., Garlick, J., Kaplan, D.L., 2012. In vitro 3D full-thickness skin-equivalent tissue model using silk and collagen biomaterials. Macromolecular Bioscience 12, 1627–1636.

Bickers, D.R., Lim, H.W., Margolis, D., Weinstock, M.A., Goodman, C., Faulkner, E., Gould, C., Gemmen, E., Dall, T., 2006. The burden of skin diseases: 2004 a joint project of the American Academy of Dermatology Association and the Society for Investigative Dermatology. Journal of the American Academy of Dermatology 55, 490–500.

Bissell, M.J., Radisky, D.C., Rizki, A., Weaver, V.M., Petersen, O.W., 2002. The organizing principle: microenvironmental influences in the normal and malignant breast. Differentiation 70, 537–546.

Blais, M., Parenteau-Bareil, R., Cadau, S., Berthod, F., 2013. Concise review: tissue-engineered skin and nerve regeneration in burn treatment. Stem Cells Translational Medicine 2, 545–551.

Borchers, A.H., Steinbauer, H., Schafer, B.S., Kramer, M., Bowden, G.T., Fusenig, N.E., 1997. Fibroblast-directed expression and localization of 92-kDa type IV collagenase along the tumor-stroma interface in an in vitro three-dimensional model of human squamous cell carcinoma. Molecular Carcinogenesis 19, 258–266.

Boyce, S.T., 1999. Methods for the serum-free culture of keratinocytes and transplantation of collagen-GAG-based skin substitutes. Methods in Molecular Medicine 18, 365–389.

Briggaman, R.A., 1982. Biochemical composition of the epidermal-dermal junction and other basement membrane. Journal of Investigative Dermatology 78, 1–6.

Briggaman, R.A., Wheeler Jr., C.E., 1975. The epidermal-dermal junction. Journal of Investigative Dermatology 65, 71–84.

Burgeson, R.E., Christiano, A.M., 1997. The dermal-epidermal junction. Current Opinion in Cell Biology 9, 651–658.

Bush, K., Pins, G., 2010. Carbodiimide conjugation of fibronectin on collagen basal lamina analogs enhances cellular binding domains and epithelialization. Tissue Eng Part A 16, 829–838.

Bush, K., Pins, G., 2012. Development of microfabricated dermal epidermal regenerative matrices to evaluate the role of cellular microenvironments on epidermal morphogenesis. Tissue Eng Part A 18, 2343–2353.

Butler, C.E., Orgill, D.P., 2005. Simultaneous in vivo regeneration of neodermis, epidermis, and basement membrane. Advances in Biochemical Engineering/Biotechnology 94, 23–41.

Carsin, H., Ainaud, P., Le Bever, H., Rives, J., Lakhel, A., Stephanazzi, J., Lambert, F., Perrot, J., 2000. Cultured epithelial autografts in extensive burn coverage of severely traumatized patients: a five year single-center experience with 30 patients. Burns 26, 379–387.

Chandrasekaran, S., Delouise, L.A., 2011. Enriching and characterizing cancer stem cell sub-populations in the WM115 melanoma cell line. Biomaterials 32, 9316–9327.

Chandrasekaran, S., Giang, U.B., King, M.R., Delouise, L.A., 2011. Microenvironment induced spheroid to sheeting transition of immortalized human keratinocytes (HaCaT) cultured in microbubbles formed in polydimethylsiloxane. Biomaterials 32, 7159–7168.

Chang, H.Y., Chi, J.T., Dudoit, S., Bondre, C., van de Rijn, M., Botstein, D., Brown, P.O., 2002. Diversity, topographic differentiation, and positional memory in human fibroblasts. Proceedings of the National Academy of Sciences of the United States of America 99, 12877–12882.

Clement, A.L., Moutinho, T., Pins, G.D., 2013. Micropatterned dermal-epidermal regeneration matrices create functional niches that enhance epidermal morphogenesis. Acta Biomaterialia 9, 9474–9484.

Commandeur, S., de Gruijl, F.R., Willemze, R., Tensen, C.P., El Ghalbzouri, A., 2009. An in vitro three-dimensional model of primary human cutaneous squamous cell carcinoma. Experimental Dermatology 18, 849–856.

Compton, C.C., Gill, J.M., Bradford, D.A., Regauer, S., Gallico, G.G., O'connor, N.E., 1989. Skin regenerated from cultured epithelial autografts on full-thickness burn wounds from 6 days to 5 years after grafting. A light, electron microscopic and immunohistochemical study. Laboratory Investigation 60, 600–612.

Compton, C.C., Hickerson, W., Nadire, K., Press, W., 1993. Acceleration of skin regeneration from cultured epithelial autografts by transplantation to homograft dermis. Journal of Burn Care and Rehabilitation 14, 653–662.

Connelly, J.T., Gautrot, J.E., Trappmann, B., Tan, D.W., Donati, G., Huck, W.T., Watt, F.M., 2010. Actin and serum response factor transduce physical cues from the microenvironment to regulate epidermal stem cell fate decisions. Nature Cell Biology 12, 711–718.

Dagalakis, N., Flink, J., Stasikelis, P., Burke, J.F., Yannas, I.V., 1980. Design of an artificial skin. Part III. Control of pore structure. Journal of Biomedical Materials Research 14, 511–528.

Danilenko, D.M., 2008. Review paper: preclinical models of psoriasis. Veterinary Pathology 45, 563–575.

Davidson, J.M., 1998. Animal models for wound repair. Archives of Dermatological Research 290 (Suppl.), S1–S11.

Dunn, L., Prosser, H.C., Tan, J.T., Vanags, L.Z., Ng, M.K., Bursill, C.A., 2013. Murine model of wound healing. Journal of Visualized Experiments: JoVE e50265.

El Ghalbzouri, A., Hensbergen, P., Gibbs, S., Kempenaar, J., Van Der Schors, R., Ponec, M., 2004. Fibroblasts facilitate re-epithelialization in wounded human skin equivalents. Laboratory Investigation 84, 102–112.

Eves, P., Layton, C., Hedley, S., Dawson, R.A., Wagner, M., Morandini, R., Ghanem, G., Mac Neil, S., 2000. Characterization of an in vitro model of human melanoma invasion based on reconstructed human skin. British Journal of Dermatology 142, 210–222.

Falanga, V., Margolis, D., Alvarez, O., Auletta, M., Maggiacomo, F., Altman, M., Jensen, J., Sabolinski, M., Hardin-Young, J., 1998. Rapid healing of venous ulcers and lack of clinical rejection with an allogeneic cultured human skin equivalent. Human skin equivalent investigators group. Archives of Dermatology 134, 293–300.

Fawcett, D.W., Jensh, R.P., 1997a. Bloom & Fawcett: Concise Histology. Chapman and Hall, New York.

Fawcett, D.W., Jensh, R.P., 1997b. Concise Histology. Chapman and Hall, New York, NY.

Fujisaki, H., Adachi, E., Hattori, S., 2008. Keratinocyte differentiation and proliferation are regulated by adhesion to the three-dimensional meshwork structure of type IV collagen. Connective Tissue Research 49, 426–436.

Fujisaki, H., Hattori, S., 2002. Keratinocyte apoptosis on type I collagen gel caused by lack of laminin 5/10/11 deposition and Akt signaling. Experimental Cell Research 280, 255–269.

Galiano, R.D., Michaels, J.T., Dobryansky, M., Levine, J.P., Gurtner, G.C., 2004. Quantitative and reproducible murine model of excisional wound healing. Wound repair and regeneration : official publication of the Wound Healing Society [and] the European Tissue Repair Society 12, 485–492.

Gallico 3rd, G.G., O'connor, N.E., 1985. Cultured epithelium as a skin substitute. Clinics in Plastic Surgery 12, 149–157.

Gallico 3rd, G.G., O'connor, N.E., Compton, C.C., Kehinde, O., Green, H., 1984. Permanent coverage of large burn wounds with autologous cultured human epithelium. The New England Journal of Medicine 311, 448–451.

Garlick, J.A., Taichman, L.B., 1994a. Effect of TGF-beta 1 on re-epithelialization of human keratinocytes in vitro: an organotypic model. The Journal of Investigative Dermatology 103, 554–559.

Garlick, J.A., Taichman, L.B., 1994b. Fate of human keratinocytes during reepithelialization in an organotypic culture model. Laboratory Investigation 70, 916–924.

Gautrot, J.E., Trappmann, B., Oceguera-Yanez, F., Connelly, J., He, X., Watt, F.M., Huck, W.T., 2010. Exploiting the superior protein resistance of polymer brushes to control single cell adhesion and polarisation at the micron scale. Biomaterials 31, 5030–5041.

Gautrot, J.E., Wang, C., Liu, X., Goldie, S.J., Trappmann, B., Huck, W.T., Watt, F.M., 2012. Mimicking normal tissue architecture and perturbation in cancer with engineered micro-epidermis. Biomaterials 33, 5221–5229.

Gravante, G., Di Fede, M.C., Araco, A., Grimaldi, M., De Angelis, B., Arpino, A., Cervelli, V., Montone, A., 2007. A randomized trial comparing ReCell system of epidermal cells delivery versus classic skin grafts for the treatment of deep partial thickness burns. Burns 33, 966–972.

Green, H., Kehinde, O., Thomas, J., 1979. Growth of cultured human epidermal cells into multiple epithelia suitable for grafting. Proceedings of the National Academy of Sciences of the United States of America 76, 5665–5668.

Greenberg, S., Kamath, P., Petrali, J., Hamilton, T., Garfield, J., Garlick, J.A., 2006. Characterization of the initial response of engineered human skin to sulfur mustard. Toxicological Sciences: An Official Journal of the Society of Toxicology 90, 549–557.

Groeber, F., Holeiter, M., Hampel, M., Hinderer, S., Schenke-Layland, K., 2011. Skin tissue engineering–in vivo and in vitro applications. Advanced Drug Delivery Reviews 63, 352–366.

Grose, R., Werner, S., 2004. Wound-healing studies in transgenic and knockout mice. Molecular Biotechnology 28, 147–166.

Gudjonsson, T., Ronnov-Jessen, L., Villadsen, R., Bissell, M.J., Petersen, O.W., 2003. To create the correct microenvironment: three-dimensional heterotypic collagen assays for human breast epithelial morphogenesis and neoplasia. Methods 30, 247–255.

Haddow, D.B., Steele, D.A., Short, R.D., Dawson, R.A., Macneil, S., 2003. Plasma-polymerized surfaces for culture of human keratinocytes and transfer of cells to an in vitro wound-bed model. Journal of Biomedical Materials Research Part A 64, 80–87.

Hamoen, K., Erdag, G., Cusick, J., Rakhorst, H., Morgan, J., 2002. Genetically modified skin substitutes. In: Morgan, J. (Ed.), Gene Therapy Protocols. Springer, New York.

Han, B., Grassl, E.D., Barocas, V.H., Coad, J.E., Bischof, J.C., 2005. A cryoinjury model using engineered tissue equivalents for cryosurgical applications. Annals of Biomedical Engineering 33, 972–982.

Hart, C.E., Loewen-Rodriguez, A., Lessem, J., 2012. Dermagraft: use in the treatment of chronic wounds. Advances in Wound Care 1, 138–141.

Hernon, C.A., Dawson, R.A., Freedlander, E., Short, R., Haddow, D.B., Brotherston, M., Macneil, S., 2006. Clinical experience using cultured epithelial autografts leads to an alternative methodology for transferring skin cells from the laboratory to the patient. Regenerative Medicine 1, 809–821.

Hotchin, N., Gandarillas, A., Watt, F., 1995. Regulation of cell surface beta 1 integrin levels during keratinocyte terminal differentiation. The Journal of Cell Biology 128, 1209–1219.

Hotchin, N., Watt, F., 1992. Transcriptional and post-translational regulation of beta 1 integrin expression during keratinocyte terminal differentiation. Journal of Biological Chemistry 267, 14852–14858.

Huang, S., Xu, Y., Wu, C., Sha, D., Fu, X., 2010. In vitro constitution and in vivo implantation of engineered skin constructs with sweat glands. Biomaterials 31, 5520–5525.

Iizuka, H., Honda, H., Ishida-Yamamoto, A., 1997. Epidermal remodeling in psoriasis (II): a quantitative analysis of the epidermal architecture. The Journal of Investigation Dermatology 109, 806–810.

Iizuka, H., Honda, H., Ishida-Yamamoto, A., 1999. Epidermal remodelling in psoriasis (III): a hexagonally-arranged cylindrical papilla model reveals the nature of psoriatic architecture. Journal of Dermatological Science 21, 105–112.

Iizuka, H., Takahashi, H., Ishida-Yamamoto, A., 2004. Psoriatic architecture constructed by epidermal remodeling. Journal of Dermatological Science 35, 93–99.

Janson, D.G., Saintigny, G., Van Adrichem, A., Mahe, C., El Ghalbzouri, A., 2012. Different gene expression patterns in human papillary and reticular fibroblasts. Journal of Investigative Dermatology 132, 2565–2572.

Jensen, U.B., Lowell, S., Watt, F.M., 1999. The spatial relationship between stem cells and their progeny in the basal layer of human epidermis: a new view based on whole-mount labelling and lineage analysis. Development 126, 2409–2418.

Jones, P., Watt, F., 1993. Separation of human epidermal stem cells from transit amplifying cells on the basis of differences in integrin function and expression. Cell 73, 713–724.

Jones, P.H., Harper, S., Watt, F.M., 1995. Stem cell patterning and fate in human epidermis. Cell 80, 83–93.

Kai-Hong, J., Jun, X., Kai-Meng, H., Ying, W., Hou-Qi, L., 2007. P63 expression pattern during rat epidermis morphogenesis and the role of p63 as a marker for epidermal stem cells. Journal of Cutaneous Pathology 34, 154–159.

Kamel, R.A., Ong, J.F., Eriksson, E., Junker, J.P., Caterson, E.J., 2013. Tissue engineering of skin. Journal of the American College of Surgeons 217, 533–555.

Kandyba, E., Hodgins, M., Martin, P., 2010. A versatile murine 3D organotypic model to evaluate aspects of wound healing and epidermal organization. Methods in Molecular Biology 585, 303–312.

Lammers, G., Roth, G., Heck, M., Zengerle, R., Tjabringa, G.S., Versteeg, E.M., Hafmans, T., Wismans, R., Reinhardt, D.P., Verwiel, E.T., Zeeuwen, P.L., Schalkwijk, J., Brock, R., Daamen, W.F., Van Kuppevelt, T.H., 2012. Construction of a microstructured collagen membrane mimicking the papillary dermis architecture and guiding keratinocyte morphology and gene expression. Macromolecular Bioscience 12, 675–691.

Lavker, R.M., Sun, T., 1982. Heterogeneity in epidermal basal keratinocytes: Morphological and functional correlations. Science 215, 1239–1241.

Lavker, R.M., Sun, T., 1983. Epidermal stem cells. Journal of Investigative Dermatology 81, 121s–127s.

Madison, K.C., 2003. Barrier function of the skin: "la raison d'etre" of the epidermis. Journal of Investigative Dermatology 121, 231–241.

Marston, W.A., Hanft, J., Norwood, P., Pollak, R., 2003. The efficacy and safety of dermagraft in improving the healing of chronic diabetic foot ulcers: results of a prospective randomized trial. Diabetes Care 26, 1701–1705.

Mather, J.P., Roberts, P.E., 1998. Introduction to Cell and Tissue Culture: Theory and Technique. Plenum Press, New York.

Medalie, D., Morgan, J., 1999. Preparation and transplantation of a composite graft of epidermal keratinocytes on acellular dermis. In: Morgan, J., Yarmush, M. (Eds.), Tissue Engineering Methods and Protocols. Humana Press.

MedMarket, 2007. Worldwide Wound Management, 2007–2016.

MedMarket Diligence, 2013. Worldwide wound management, forecast to 2021: established and emerging products, technologies and markets in the Americas, Europe, Asia/Pacific and Rest of World. Wound Management 2013.

Metcalfe, A.D., Ferguson, M.W., 2007. Tissue engineering of replacement skin: the crossroads of biomaterials, wound healing, embryonic development, stem cells and regeneration. Journal of the Royal Society Interface 4, 413–437.

Middelkoop, E., Van Den Bogaerdt, A.J., Lamme, E.N., Hoekstra, M.J., Brandsma, K., Ulrich, M.M., 2004. Porcine wound models for skin substitution and burn treatment. Biomaterials 25, 1559–1567.

Milenkovic, M., Russo, C.A., Elixhauser, A., 2007. Hospital Stays for Burns, 2004. HCUP Statistical Brief #25. Agency for Healthcare Research and Quality, Rockville, MD. http://www.hcup-us.ahrq.gov/reports/statbriefs/sb25.pdf.
Montagna, W., Carlisle, K., 1979. Structural changes in aging human skin. Journal of Investigative Dermatology 73, 47–53.
Murphy, M., Kerr, P., Grant-Kels, J.M., 2007. The histopathologic spectrum of psoriasis. Clinics in Dermatology 25, 524–528.
O'connor, N.E., Mulliken, J.B., Banks-Schlegel, S., Kehinde, O., Green, H., 1981. Grafting of burns with cultured epithelium prepared from autologous epidermal cells. Lancet 1, 75–78.
Obrigkeit, D.H., Jugert, F.K., Beermann, T., Baron, J.M., Frank, J., Merk, H.F., Bickers, D.R., Abuzahra, F., 2009. Effects of photodynamic therapy evaluated in a novel three-dimensional squamous cell carcinoma organ construct of the skin. Photochemistry and Photobiology 85, 272–278.
Odland, G., 1950. The morphology of the attachment between the dermis and the epidermis. The Anatomical Record 108, 399–413.
Ojeh, N.O., Navsaria, H.A., 2013. An in vitro skin model to study the effect of mesenchymal stem cells in wound healing and epidermal regeneration. Journal of Biomedical Materials Research Part A 102 (8), 2785–2792.
Pellegrini, G., Dellambra, E., Golisano, O., Martinelli, E., Fantozzi, I., Bondanza, S., Ponzin, D., Mckeon, F., De Luca, M., 2001. p63 identifies keratinocyte stem cells. Proceedings of the National Academy of Sciences of the United States of America 98, 3156–3161.
Ponticorvo, A., Burmeister, D.M., Yang, B., Choi, B., Christy, R.J., Durkin, A.J., 2014. Quantitative assessment of graded burn wounds in a porcine model using spatial frequency domain imaging (SFDI) and laser speckle imaging (LSI). Biomedical Optics Express 5, 3467–3481.
Powell, H.M., Mcfarland, K.L., Butler, D.L., Supp, D.M., Boyce, S.T., 2010. Uniaxial strain regulates morphogenesis, gene expression, and tissue strength in engineered skin. Tissue Engineering Part A 16, 1083–1092.
Powers, J.G., Morton, L.M., Phillips, T.J., 2013. Dressings for chronic wounds. Dermatologic Therapy 26, 197–206.
Priya, S.G., Jungvid, H., Kumar, A., 2008. Skin tissue engineering for tissue repair and regeneration. Tissue Engineering Part B 114, 105–118.
Rapier, R., Huq, J., Vishnubhotla, R., Bulic, M., Perrault, C.M., Metlushko, V., Cho, M., Tay, R.T., Glover, S.C., 2010. The extracellular matrix microtopography drives critical changes in cellular motility and Rho A activity in colon cancer cells. Cancer Cell International 10, 24.
Rapp, S.R., Feldman, S.R., Exum, M.L., Fleischer Jr., A.B., Reboussin, D.M., 1999. Psoriasis causes as much disability as other major medical diseases. Journal of the American Academy of Dermatology 41, 401–407.
Rittie, L., Sachs, D.L., Orringer, J.S., Voorhees, J.J., Fisher, G.J., 2013. Eccrine sweat glands are major contributors to reepithelialization of human wounds. The American Journal of Pathology 182, 163–171.
Robinson, J.K., 2005. Sun exposure, sun protection, and vitamin D. JAMA 294, 1541–1543.
Rogers, H.W., Weinstock, M.A., Harris, A.R., Hinckley, M.R., Feldman, S.R., Fleischer, A.B., Coldiron, B.M., 2010. Incidence estimate of nonmelanoma skin cancer in the United States, 2006. Archives of Dermatology 146, 283–287.
Rook, A., Burns, T., 2010. Rook's Textbook of Dermatology, eigth ed. Wiley-Blackwell, Chichester, West Sussex, UK; Hoboken, NJ.

Saiag, P., Coulomb, B., Lebreton, C., Bell, E., Dubertret, L., 1985. Psoriatic fibroblasts induce hyperproliferation of normal keratinocytes in a skin equivalent model in vitro. Science 230, 669–672.

Sams, H.H., Chen, J., King, L.E., 2002. Graftskin treatment of difficult to heal diabetic foot ulcers: one center's experience. Dermatologic Surgery 28, 698–703.

Senoo, M., 2013. Epidermal stem cells in homeostasis and wound repair of the skin. Advances in Wound Care 2, 273–282.

Shaffer, C.L., Feldman, S.R., Fleischer, A.B., Huether, M.J., Chen, J., 2005. The cutaneous surgery experience of multiple specialties in the medicare population. Journal of the American Academy of Dermatology 52, 1045–1048.

Shakespeare, P.G., 2005. The role of skin substitutes in the treatment of burn injuries. Clinics in Dermatology 23, 413–418.

Shastri, V.P., 2006. Future of regenerative medicine: challenges and hurdles. Artificial Organs 30, 828–834.

Sheridan, R.L., Tompkins, R.G., 1999. Skin substitutes in burns. Burns 25, 97–103.

Shevchenko, R.V., James, S.L., James, S.E., 2010. A review of tissue-engineered skin bioconstructs available for skin reconstruction. Journal of the Royal Society Interface 7, 229–258.

Singer, A.J., Clark, R.A., 1999. Cutaneous wound healing. The New England Journal of Medicine 341, 738–746.

Singer, A.J., Mcclain, S.A., 2003. Development of a porcine excisional wound model. Academic Emergency Medicine: Official Journal of the Society for Academic Emergency Medicine 10, 1029–1033.

Smalley, K.S., Lioni, M., Herlyn, M., 2006. Life isn't flat: taking cancer biology to the next dimension. In Vitro Cellular & Developmental Biology. Animal 42, 242–247.

Stern, R., Mcpherson, M., Longaker, M., 1990. Histological study of artificial skin used in the treatment of full-thickness thermal injury. Journal of Burn Care and Rehabilitation 11, 7–13.

Stern, R.S., 2010. Prevalence of a history of skin cancer in 2007: results of an incidence-based model. Archives of Dermatology 146, 279–282.

Still, J., Glat, P., Silverstein, P., Griswold, J., Mozingo, D., 2003. The use of a collagen sponge/living cell composite material to treat donor sites in burn patients. Burns 29, 837–841.

Supp, D., Boyce, S., 2005. Engineered skin substitutes: practices and potentials. Clinics in Dermatology 23, 403–412.

Tajima, S., Pinnell, S.R., 1981. Collagen synthesis by human skin fibroblasts in culture: studies of fibroblasts explanted from papillary and reticular dermis. Journal of Investigative Dermatology 77, 410–412.

Tenenhaus, M., Rennekampff, H.O., 2012. Surgical advances in burn and reconstructive plastic surgery: new and emerging technologies. Clinics in Plastic Surgery 39, 435–443.

Tjabringa, G., Bergers, M., Van Rens, D., De Boer, R., Lamme, E., Schalkwijk, J., 2008. Development and validation of human psoriatic skin equivalents. American Journal of Pathology 173, 815–823.

Tonnesen, M.G., Feng, X., Clark, R.A., 2000. Angiogenesis in wound healing. Journal of Investigative Dermatology 5, 40–46.

Vaccariello, M.A., Javaherian, A., Parenteau, N., Garlick, J.A., 1999. Use of skin equivalent technology in a wound healing model. Methods in Molecular Medicine 18, 391–405.

Van Rossum, M.M., Framssem, M.E.J., Cloin, W.A.H., Van De Bosch, G.J.M., Boezmman, J.B.M., Schalkwijk, J., Van De Kerkhof, P.C.M., Van Erp, P.E.J., 2004. Functional characterization of beta1-integrin-positive epidermal cell populations. Acta Dermato-Venereologica 84, 265–270.

Veves, A., Falanga, V., Armstrong, D.G., Sabolinski, M.L., 2001. Graftskin, a human skin equivalent, is effective in the management of noninfected neuropathic diabetic foot ulcers: a prospective randomized multicenter clinical trial. Diabetes Care 24, 290–295.

Vorsmann, H., Groeber, F., Walles, H., Busch, S., Beissert, S., Walczak, H., Kulms, D., 2013. Development of a human three-dimensional organotypic skin-melanoma spheroid model for in vitro drug testing. Cell Death and Disease 4, e719.

Vracko, R., 1974. Basal lamina scaffold-anatomy and significance for maintenance of orderly tissue structure. American Journal of Pathology 77, 314–346.

Wainwright, D.J., 1995. Use of an acellular allograft dermal matrix (AlloDerm) in the management of full-thickness burns. Burns 21, 243–248.

Wang, X., Ge, J., Tredget, E.E., Wu, Y., 2013. The mouse excisional wound splinting model, including applications for stem cell transplantation. Nature protocols 8, 302–309.

Waugh, H.V., Sherratt, J.A., 2007. Modeling the effects of treating diabetic wounds with engineered skin substitutes. Wound Repair and Regeneration 15, 556–565.

Weaver, V.M., Fischer, A.H., Peterson, O.W., Bissell, M.J., 1996. The importance of the microenvironment in breast cancer progression: recapitulation of mammary tumorigenesis using a unique human mammary epithelial cell model and a three-dimensional culture assay. Biochemistry and Cell Biology 74, 833–851.

Weigelt, B., Bissell, M.J., 2008. Unraveling the microenvironmental influences on the normal mammary gland and breast cancer. Seminars in Cancer Biology 18, 311–321.

Williamson, J.S., Snelling, C.F., Clugston, P., Macdonald, I.B., Germann, E., 1995. Cultured epithelial autograft: five years of clinical experience with twenty-eight patients. Journal of Trauma 39, 309–319.

Woo, K., Ayello, E.A., Sibbald, R.G., 2007. The edge effect: current therapeutic options to advance the wound edge. Advances in Skin and Wound Care 20, 99–117.

Wood, F.M., Giles, N., Stevenson, A., Rea, S., Fear, M., 2012. Characterisation of the cell suspension harvested from the dermal epidermal junction using a ReCell(R) kit. Burns 38, 44–51.

Yamada, K.M., Cukierman, E., 2007. Modeling tissue morphogenesis and cancer in 3D. Cell 130, 601–610.

Yang, A., Schweitzer, R., Sun, D., Kaghad, M., Walker, N., Bronson, R., Tabin, C., Sharpe, A., Caput, D., Crum, C., Mckeon, F., 1999. p63 is essential for regenerative proliferation in limb, craniofacial and epithelial development. Nature 398, 714–718.

Yannas, I.V., Burke, J.F., 1980. Design of an artificial skin. I. Basic design principles. Journal of Biomedical Materials Research 14, 65–81.

Autologous cell-rich biomaterial (LeucoPatch) in the treatment of diabetic foot ulcers

R. Lundquist
Reapplix ApS, Birkerød, Denmark

11.1 Introduction

Chronic wounds fail to proceed through the orderly and timely series of events of acute wounds to produce a durable structural, functional, and cosmetic closure. They are often stuck in the inflammatory phase, but the mechanisms for their inability to progress into healing remain unanswered (Ågren et al., 2000).

The major types of chronic wounds defined by their etiology are venous leg ulcers (VLUs), pressure ulcers (PUs), and diabetic foot ulcers (DFUs). The prevalence of VLUs in the United States has been estimated at 0.60 million, PUs at 2.50 million, and DFUs at 1.60 million (Sen et al., 2009). In addition to these major groups, there are additional chronic wounds represented by nonhealing surgical wounds, ischemic wounds, and wounds due to vasculitis, accounting for about 40% of all wounds treated in outpatient wound centers in the United States (Fife et al., 2012).

Autologous platelet-derived products promote wound healing processes (Martinez et al., 2015). A Cochrane systematic review concluded that there is insufficient evidence for using platelet-rich plasma (PRP) in the treatment of chronic wounds (Martinez-Zapata et al., 2012). In contrast, another systematic review concluded that PRP is beneficial for the treatment of DFUs (Villela and Santos, 2010). The diverse results may be attributed to differences in the methods and systems applied to produce PRP and similar products (Kushida et al., 2014; Magalon et al., 2014; Mazzucco et al., 2009).

LeucoPatch (Reapplix ApS, Birkerød, Denmark) is a unique autologous biomaterial of a robust fibrin matrix that not only contains a platelet concentrate fraction and their growth factors but also functional leukocytes. This chapter focuses on the clinical development of LeucoPatch for nonhealing DFUs.

11.2 The pathogenesis of diabetic foot ulcers

Due to the epidemic of type 2 diabetes and increased incidences of type 1 diabetes, DFUs are a growing clinical problem (Boulton et al., 2005). DFUs are divided into

Wound Healing Biomaterials - Volume 1. http://dx.doi.org/10.1016/B978-1-78242-455-0.00011-2

neuropathic, ischemic, and neuroischemic, based on the neurological and vascular pathologies. The neuroischemic DFUs are the most common (50%) followed by the neuropathic (35%) and the ischemic (15%) DFUs (Prompers et al., 2007). Ischemia in diabetic patients is usually manifested in peripheral arterial occlusive disease (PAOD), which is 2–4 times more common in diabetic as compared with nondiabetic patients (Beckman et al., 2002) and associated with nonhealing wounds.

The debilitated peripheral neurological system results from the diabetic disease and poorly controlled blood glucose levels. Studies of skin biopsies from lateral malleolus in individuals with type 1 and type 2 diabetes mellitus have shown lowered neuropeptide levels and degeneration of the peripheral sensory and autonomic nerves (Galkowska et al., 2006; Lindberger et al., 1989).

Due to the sensory neuropathy, diabetic patients are unable to walk properly, react normally to nonfitting footwear, or sense injuries. This is most often the cause of the acute injury that eventually leads to an open lesion that may be converted into a nonhealing wound. In addition, the abrogated sympathetic nerve system can result in dry skin and fissures that increase the risk of ulcer formation.

Wound healing demands an increased blood supply via the microvascular system. This is accomplished primarily by the contraction of arteriovenous shunts that direct the blood to the wound area (Braverman, 2000; Rendell et al., 1997). The shunting is followed by vasodilation and the extravasation of immune cells, primarily neutrophils, macrophages, and lymphocytes, into the wound bed. These cells are important for clearing pathogens and debris, but they also orchestrate healing. The shift from a nonhealing but pathogen-clearing macrophage phenotype (M1) to a healing one (M2) is crucial for the transition into the reparative wound healing phase (Aller et al., 2014; Mosser and Edwards, 2008). In diabetic patients, the neuropathy causes nonfunctional shunting and impaired microvascular blood flow (Sun et al., 2013). These microcirculatory vasomotor changes correlate with the severity of peripheral neuropathy in patients with type 2 diabetes. The consequence is reduced vasodilation and leukocyte infiltration of the wound area (Galkowska et al., 2005), as summarized in Fig. 11.1. This combined with lowered blood pressure in the extremities due to PAOD is often described as a state of "chronic capillary ischemia" (Jörneskog, 2012). While transcutaneous oxygen tension ($TcPO_2$) measurements predict the healing of DFUs, toe blood pressure measurements are less predictive (Kalani et al., 1999). This finding supports the importance of the microvascular blood flow in the healing of DFUs.

The microangiopathic changes together with continuous systemic inflammation impair the acute inflammatory response in diabetes; infections are not cleared properly, which delay wound healing (Pradhan et al., 2009).

11.3 Recommended treatments of diabetic foot ulcers

The severe consequences of DFUs have been realized (Boulton et al., 2005). This has led to the development of guidelines for the treatment of DFUs (Lavery et al.,

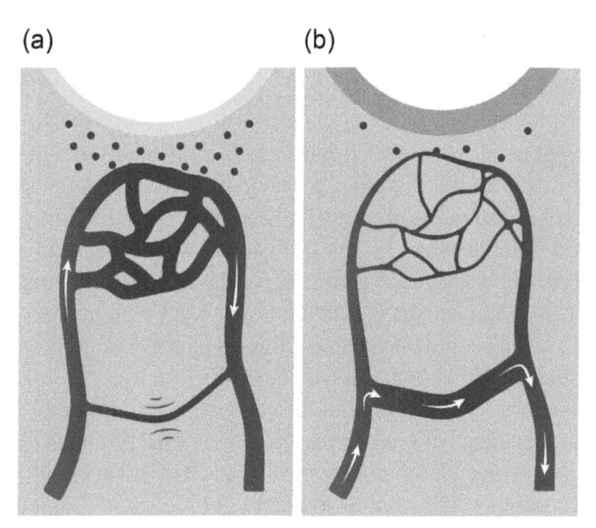

Figure 11.1 Arteriovenous (AV) shunting occurs as part of the normal healing response, which promotes the delivery of immune cells, platelets, growth factors, and oxygen (a). The lack of AV shunting due to neuropathy leads to a local deficiency of these cells and factors (b).

2016). The National Institute for Health and Clinical Excellence (NICE) recommends that chronic DFU patients be referred to multidisciplinary diabetic foot care teams (MDTs). These are often hospital-based and include a diabetologist, vascular surgeon, orthopedic surgeon, microbiologist, interventional radiologist, orthotist, specialist podiatrist, diabetes specialist nurse, and/or tissue viability nurse. Visits should include an assessment of peripheral arterial disease, neuropathy, infection, and relevant medical, personal, and social factors (Boulton et al., 2005). In practice, DFU treatment consists of vascular surgery when indicated, systemic (antibiotics) and local (antiseptics) infection control, wound bed debridement, bandaging, and off-loading of the foot/limb using a total contact cast or removable boots in addition to metabolic control (McInnes, 2012; de Olivera et al., 2015).

NICE states: "In the absence of strong evidence of clinical or cost effectiveness, healthcare professionals should use wound dressings that *best match clinical experience*, patient preference, and the site of the wound, and consider the cost of the dressings." This rather conservative view of this health authority has not prevented the use of more active treatments, including recombinant human platelet-derived growth factor (PDGF)-BB (Regranex®, Smith & Nephew, London, United Kingdom), dermal substitutes (Dermagraft®, Organogenesis, Canton, MA, United States), and skin substitutes (Apligraf®, Organogenesis). The effectiveness of these products has been documented in properly conducted randomized controlled trials (RCTs) (Wieman et al., 1998; Marston, 2004; Veves and Falanga, 2001). In addition to these FDA-approved products there are a number of products in development such as biological matrices of animal or human origin with or without allogeneic cells.

11.4 Autologous blood-derived biomaterials for wound treatment

The use of autologous blood-derived biomaterials in the treatment of chronic wounds was introduced in the mid-1980s by Dr David Knighton and his colleagues. They developed a platelet-derived wound healing factor (PDWHF) formula derived from autologous blood (Knighton et al., 1986). Platelets were first isolated from the anti-coagulated whole blood and then activated by the addition of thrombin (1 U/mL). The supernatant from the activated platelets was mixed with microcrystalline bovine collagen to generate an acellular salve containing a plethora of growth factors at super-physiological concentrations (Knighton et al., 1986).

Since the launch of PDWHF several different platelet-derived or platelet concentrate products have been developed. From the generic PRP products four distinct product categories have evolved (Dohan Ehrenfest et al., 2014). These autologous products are classified according to cell composition and fibrin content (Table 11.1):

- Pure platelet-rich plasma (P-PRP) products consist of platelets without leukocytes in plasma and can be used either as a fluid for injection into orthopedic injuries or as a gel for topical application to skin wounds. Gels form after activation by calcium and thrombin, releasing growth factors and polymerizing fibrin. The level of fibrin (2–3 mg/mL) typically matches that of plasma. Products that belong to this category are Aurix™ (Nuo Therapeutics, Gaithersburg, MD, United States) and Endoret® (prgf®) (BTI Biotechnology Institute, Vitoria, Spain).
- Leukocyte- and platelet-rich plasma (L-PRP) products are similar to P-PRPs, but they contain leukocytes in addition to the platelets. Typically leukocytes are concentrated 3–5 times as compared with whole blood counts. L-PRPs are administered as a liquid without activation or in the form of a gel after platelet activation. GPS® III Platelet Separation System (Biomet Biologics, Warsaw, IN, United States) and Harvest® SmartPrep® (Harvest Terumo BCT, Lakewood, CO, United States) are products in this category (Magalon et al., 2014).
- Pure platelet-rich fibrin (P-PRF) preparations constitute activated platelets in a polymerized fibrin matrix. The fibrin content is higher than in P-PRP and L-PRP products, and the viscosity normally prevents P-PRF products from being injected. Instead P-PRF products are applied to the wound. Examples of products in this category are Fibrinet® PRFM (Cascade Medical, Wayne, NJ, United States) and Vivostat® PRF® (Vivostat A/S, Allerød, Denmark). Vivostat PRF is formed in situ via a spraying device with a final fibrin concentration of 15–20 mg/mL fibrin (Ågren et al., 2014; Lundquist et al., 2008).

Table 11.1 Categories of autologous platelet-containing products

	Fibrin concentration	
Major cell types	**Low**	**High**
Platelets	P-PRP	P-PRF
Platelets and leukocytes	L-PRP	L-PRF

P-PRP, pure platelet-rich plasma; *L-PRP*, leukocyte- and platelet-rich plasma; *P-PRF*, pure platelet-rich fibrin; *L-PRF*, leukocyte- and platelet-rich fibrin.

From Dohan Ehrenfest DM, Andia I, Zumstein MA, Zhang CQ, Pinto NR, Bielecki T. Classification of platelet concentrates (Platelet-Rich Plasma-PRP, Platelet-Rich Fibrin-PRF) for topical and infiltrative use in orthopedic and sports medicine: current consensus, clinical implications and perspectives. Muscles Ligaments Tendons J 2014;4:3–9.

- Leukocyte- and platelet-rich fibrin (L-PRF) is similar to P-PRF preparations with respect to the high fibrin content. As opposed to P-PRFs, they contain significant number of leukocytes apart from the platelets derived from the blood. Choukroun's PRF (Del Corso and Dohan, 2009) and IntraSpin – L-PRF™ (Intra-Lock, Boca Raton, FL, United States) belong to this group of products.

The P-PRP, L-PRP, and P-PRF products are made from anticoagulated whole blood in multiple steps. L-PRF products are prepared in two steps without chemical additives. L-PRF is isolated manually from the fibrin clot that forms after instant centrifugation at a low speed of whole blood where "most platelet aggregates and leukocytes are concentrated within the end of the PRF clot, close to the border with the base of red blood cells. The way the clot is separated considerably influences the final biologic content of the PRF" (Del Corso and Dohan, 2009). Thus the resulting L-PRF product will vary in composition, depending on the person preparing the product.

11.5 Clinical evidence for platelet-derived products in wound care

Only the near-physiological concentration P-PRP gel Aurix (former AutoloGel™) has been evaluated in an RCT on DFUs. Although no statistically significant improvement of healing was demonstrated in the RCT, Aurix has received FDA approval for wound healing (Driver et al., 2006). Meta-analysis of small-sized trials on platelet products indicate a positive effect on the healing of DFUs (Martínez-Zapata et al., 2009; Villela and Santos, 2010), and a retrospective analysis of the data from wound care centers in the United States indicated positive effects of a platelet releasate in promoting healing of DFUs (Margolis et al., 2001). Despite these data neither NICE nor the Center for Medicare Services of the United States have reimbursed these products for routine use in wound management.

11.6 LeucoPatch

LeucoPatch is a purely autologous cell-rich fibrin-based biomaterial without chemical additives and belongs to the L-PRF group (Dohan Ehrenfest et al., 2014). LeucoPatch is produced by mechanical means by the patented 3CP process (centrifugation, coagulation, and compaction procedure) using a proprietary closed sterile device (Lundquist and Holm, 2010) in three steps at the bedside: (1) blood is drawn by venipuncture into the sterile vacuum device in a process identical to normal blood sampling, (2) the device is positioned in a specially designed centrifuge insert and spun in an automated two-step process at the bedside, and (3) the device is opened. This process takes 20 min. The formed patch is then applied to the wound without further processing.

Leukocytes and platelets are concentrated in LeucoPatch as compared with whole blood. In Table 11.2, the cell counts of whole blood and LeucoPatch from 10 healthy donors are presented (Lundquist et al., 2013).

Three distinct layers form in LeucoPatch: (1) polymerized and cross-linked fibrin matrix, (2) compacted platelets, and (3) leukocytic concentrates facing the wound. In Fig. 11.2, the three-layered structure of LeucoPatch is shown (Lundquist et al., 2013).

Table 11.2 Leukocyte and platelet numbers in whole blood and LeucoPatch derived from 10 healthy donors

	Whole blood in 10^6/mL	LeucoPatch in 10^6
Neutrophils	2.73 (2.05–3.96)	24.6 (8.9–48.7)
Monocytes	0.42 (0.34–0.53)	5.9 (4.6–7.6)
Lymphocytes	2.03 (1.17–2.49)	25.3 (14.4–40.1)
Platelets	218 (151–382)	3528 (2442–6221)

Mean (range).
From Lundquist R, Holmstrøm K, Clausen C, Jørgensen B, Karlsmark T. Characteristics of an autologous leukocyte and platelet-rich fibrin patch intended for the treatment of recalcitrant wounds. Wound Repair Regen 2013;21:66–76.

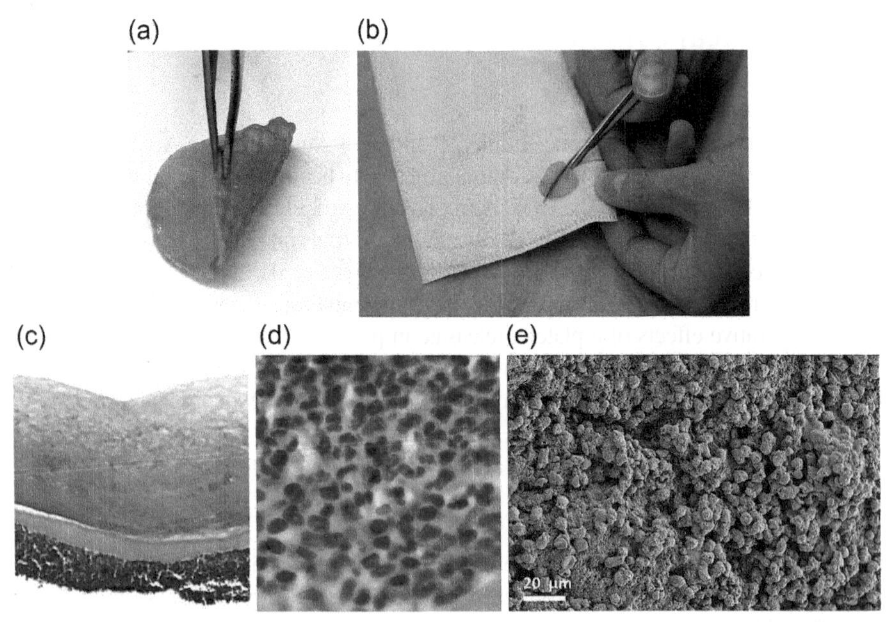

Figure 11.2 The LeucoPatch physical structure. (a) The circular LeucoPatch. (b) The consistency and toughness of the LeucoPatch enables trimming of the medical device with surgical instruments. (c) A cross-section of hematoxylin-eosin-stained LeucoPatch showing the three layers: fibrin (top); platelets (light pink, middle); and leukocytes (bottom layer). (d, e) The leukocytic layer stained with hematoxylin-eosin (d) and examined by scanning electron microscopy (e).

The levels of some growth factors in LeucoPatch were compared with standard P-PRP (Lundquist et al., 2008). PDGF-AB levels were increased by a factor of 2.6 in LeucoPatch and in VEGF by a factor of 9.9 as compared with P-PRP. The neutrophil chemokine IL-8 was 282 times higher in LeucoPatch than in P-PRP (Lundquist et al., 2013).

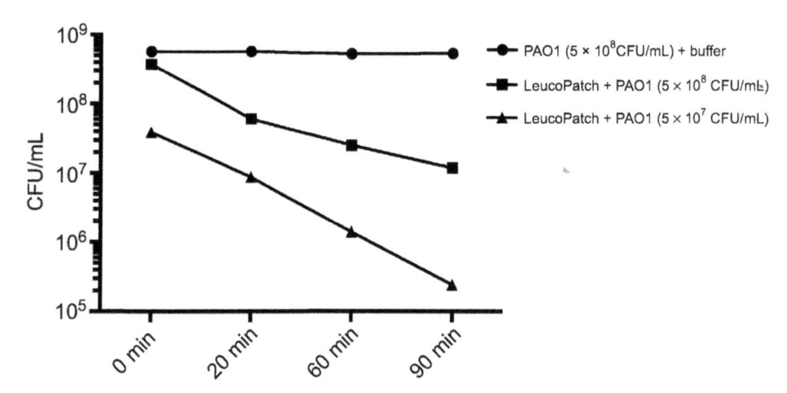

Figure 11.3 Bactericidal activity of LeucoPatch during 90 min of incubation with planktonic *Pseudomans aeruginosa* (PAO1). LeucoPatches were mixed with PAO1, and at the indicated time points samples were removed for bacterial counting. Samples were serially diluted, plated on Conradi–Drigalski medium, and the colony-forming units (CFU) were determined after an overnight incubation at 37°C.

It was demonstrated in vitro that one growth factor, PDGF-AB, is released continuously from LeucoPatch for at least 1 week. Other in vitro studies showed that LeucoPatch enhanced the proliferation and migration of human fibroblasts and keratinocytes (Lundquist et al., 2013).

Interestingly, the addition of chronic wound fluid increased the rate of release of PDGF-AB from LeucoPatch (Lundquist et al., 2013). This biochemical feature has relevance when using LeucoPatch in the treatment of chronic wounds. *Pseudomans aeruginosa* is common in chronic wounds that exhibit impaired wound healing (Renner et al., 2012). The effect of LeucoPatch on bacterial survival was studied in one *P. aeruginosa* strain (PAO1). These tests indicated that the leukocytes in LeucoPatch are viable and capable of phagocytosis and killing the bacteria (Fig. 11.3). Moreover, initial tests indicate that LeucoPatch reduces *P. aeruginosa*-derived biofilms (Thomsen et al., 2016).

11.7 Clinical investigations of LeucoPatch

LeucoPatch has been tested in a number of clinical studies.

The initial studies were carried out in 15 patients with 16 lower extremity wounds of different etiologies. The median duration of the wounds was 24 months, and the median size was 2.3 cm^2 (Jørgensen et al., 2011). Wounds were treated weekly with LeucoPatch for 6 weeks. During this period the wounds decreased by 65% in surface wound area. In addition, four wounds were completely healed. No serious adverse events were observed (Jørgensen et al., 2011).

A case series of DFUs with probe-to-bone (Wagner grade 3) is ongoing. Patients with nonischemic ($TcPO_2 \geq 30$ mm Hg) DFUs with a duration of at least 6 weeks and a positive probing-to-bone test are consecutively enrolled. LeucoPatch is applied weekly for up to 20 weeks. All patients are given oral antibiotics at least until bone coverage is achieved and thereafter at the discretion of the physician. Twenty-two patients with 26 DFUs of a median 26 week of duration have been treated for a median of eight times with LeucoPatch. Bone was covered with new tissue in 18 DFUs and 15 of these were completely epithelialized.

A multicenter prospective cohort study has been performed in 44 hard-to-heal DFUs in patients being treated at five MDT clinics in Denmark and two MDT clinics in Sweden. Wagner grade 1 or 2 DFUs with a wound duration of more than 6 weeks and a maximal area of $10\,cm^2$ were included (Löndahl et al., 2015). Patients with DFUs that decreased more than 40% in wound area during a 2-week run-in period were excluded. The primary endpoint was healing within 20 weeks. LeucoPatch was applied weekly or until the DFU was completely epithelialized, whichever occurred first. Complete epithelialization was achieved in 16 (36%) DFUs at 12 weeks and in 26 (59%) of the DFUs at 20 weeks in the per-protocol population.

To confirm and support the available data on the clinical effects of LeucoPatch in DFUs a large multinational RCT of estimated 250 type 1 and type 2 diabetic patients is ongoing at 20 sites in England, seven in Denmark, and three in Sweden (ClinicalTrials.gov, identifier NCT02224742). The trial is being coordinated by the Nottingham Clinical Trials Unit, Queen's Medical Center, Nottingham, United Kingdom. The selected inclusion and exclusion criteria are listed in Table 11.3. The primary endpoint is the number of DFUs healed within 20 weeks. Healing is defined as complete epithelialization without discharge that is maintained for 4 weeks and is

Table 11.3 Selected inclusion and exclusion criteria for the LeucoPatch II trial

Inclusion criteria	Exclusion criteria
• People aged 18 years and over who have diabetes complicated by one or more ulcers on a foot or both feet below the level of the malleoli, excluding ulcers confined to the interdigital cleft. • Eligible ulcers will be hard-to-heal, meaning that the cross-sectional area will decrease by less than 50% during a 4-week run-in period. • $HbA_{1c} \leq 12.0\%$ at screening. • The wound area of the index ulcer will be $50–1000\,mm^2$ at the end of the 4-week run-in period. • At randomization, the index ulcer will be clinically noninfected, according to Infectious Diseases Society of America criteria • Either the ankle–brachial index in the affected limb will be 0.50–1.40 or the dorsalis pedis pulse and/or tibialis posterior pulse will be palpable.	• Hemoglobin concentration <6.5 mmol/L at screening. • Dialysis or an estimated glomerular filtration rate $<20\,mL/min/1.73\,m^2$. • Revascularization procedure in the affected limb planned or undertaken within the 4 weeks prior to screening. • Treatment of foot ulcers with growth factors, stem cells, or equivalent preparation within the 8 weeks prior to screening. • The need for continued use of negative pressure wound therapy. • Likely inability to comply with the need for weekly visits because of planned activity. • Participation in another interventional clinical foot ulcer healing trial within the 4 weeks prior to screening.

confirmed by an observer blinded to the randomization group. Secondary endpoints are time to healing, change in ulcer area, incidence of secondary infection, quality of life outcomes, and health economy.

11.8 Conclusions

DFUs are difficult to treat (Boulton et al., 2005). One important pathogenic mechanism for the nonhealing DFU phenotype seems to be the inability of these patients to mobilize an adequate immune response at the wound site due to microvascular and neurological pathologies. This impairs the clearance of infections and wound healing. LeucoPatch is the first blood-derived autologous product that delivers immune and repair cells to the wound. In vitro studies have shown that many of these cells are functional with respect to bacterial phagocytosis and growth factor/cytokine production. Results from the initial clinical studies indicate that LeucoPatch may promote the healing of DFUs of various severities. These preliminary findings are currently being verified in a large RCT on 250 patients with hard-to-heal DFUs.

References

Ågren MS, Eaglstein WH, Ferguson MWJ, Harding KG, Moore K, Saarialho-Kere UK, et al. Causes and effects of the chronic inflammation in venous leg ulcers. Acta Derm Venereol 2000;(Suppl. 210):3–17.

Ågren MS, Rasmussen K, Pakkenberg B, Jørgensen B. Growth factor and proteinase profile of Vivostat® platelet-rich fibrin linked to tissue repair. Vox Sang 2014;107:37–43.

Aller MA, Arias JI, Arraez-Aybar LA, Gilsanz C, Arias J. Wound healing reaction: a switch from gestation to senescence. World J Exp Med 2014;4:16–26.

Beckman J, Creager M, Libby P. Diabetes and atherosclerosis: epidemiology, pathophysiology, and management. JAMA 2002;287:2570–81.

Boulton AJ, Vileikyte L, Ragnarson-Tennvall G, Apelqvist J. The global burden of diabetic foot disease. Lancet 2005;366:1719–24.

Braverman IM. The cutaneous microcirculation. J Invest Dermatol Symp Proc 2000;5:3–9.

Del Corso M, Dohan DM. Letters to the Editor - Choukroun's platelet-rich fibrin membranes in periodontal surgery: understanding the biomaterial or believing in the magic of growth factors? J Periodontol 2009;80:1694–7.

Dohan Ehrenfest DM, Andia I, Zumstein MA, Zhang CQ, Pinto NR, Bielecki T. Classification of platelet concentrates (Platelet-Rich Plasma-PRP, Platelet-Rich Fibrin-PRF) for topical and infiltrative use in orthopedic and sports medicine: current consensus, clinical implications and perspectives. Muscles Ligaments Tendons J 2014;4:3–9.

Driver VR, Hanft J, Fylling CP, Beriou JM. A prospective, randomized, controlled trial of autologous platelet-rich plasma gel for the treatment of diabetic foot ulcers. Ostomy Wound Manage 2006;52:68–87.

Fife C, Carter M, Walker D, Thomson B. Wound care outcomes and associated cost among patients treated in US outpatient wound centers: data from the US wound registry. Wounds 2012;24:10–7.

Galkowska H, Olszewski WL, Wojewodzka U, Rosinski G, Karnafel W. Neurogenic factors in the impaired healing of diabetic foot ulcers. J Surg Res 2006;134:252–8.
Galkowska H, Wojewodzka U, Olszewski WL. Low recruitment of immune cells with increased expression of endothelial adhesion molecules in margins of the chronic diabetic foot ulcers. Wound Repair Regen 2005;13:248–54.
Jørgensen B, Karlsmark T, Vogensen H, Haase L, Lundquist R. A pilot study to evaluate the safety and clinical performance of Leucopatch, an autologous, additive-free, platelet-rich fibrin for the treatment of recalcitrant chronic wounds. Int J Low Extrem Wounds 2011;10:218–23.
Jörneskog G. Why critical limb ischemia criteria are not applicable to diabetic foot and what the consequences are. Scand J Surg 2012;101:114–8.
Kalani M, Brismar K, Fagrell B, Ostergren J, Jörneskog G. Transcutaneous oxygen tension and toe blood pressure as predictors for outcome of diabetic foot ulcers. Diabetes Care 1999;22:147–51.
Knighton D, Ciresi K, Fiegel V, Austin L, Butler E. Classification and treatment of chronic nonhealing wounds. Successful treatment with autologous platelet-derived wound healing factors (PDWHF). Ann Intern Med 1986;204:322–9.
Kushida S, Kakudo N, Morimoto N, Hara T, Ogawa T, Mitsui T, et al. Platelet and growth factor concentrations in activated platelet-rich plasma: a comparison of seven commercial separation systems. J Artif Organs 2014;17:186–92.
Lavery LA, Davis KE, Berriman SJ, Braun L, Nichols A, Kim PJ, et al. WHS guidelines update: diabetic foot ulcer treatment guidelines. Wound Repair Regen 2016;24:112–26.
Lindberger M1, Schröder HD, Schultzberg M, Kristensson K, Persson A, Ostman J, et al. Nerve fibre studies in skin biopsies in peripheral neuropathies. I. Immunohistochemical analysis of neuropeptides in diabetes mellitus. J Neurol Sci 1989;93:289–96.
Löndahl M, Tarnow L, Karlsmark T, Lundquist R, Nielsen AM, Michelsen M, et al. Use of an autologous leucocyte and platelet-rich fibrin patch on hard-to-heal DFUs: a pilot study. J Wound Care 2015;24:172–4, 176–8.
Lundquist R, Dziegiel MH, Ågren MS. Bioactivity and stability of endogenous fibrogenic factors in platelet-rich fibrin. Wound Repair Regen 2008;16:356–63.
Lundquist R, Holm N. Multilayered blood product. 2010. WO Patent WO2010020254.
Lundquist R, Holmstrøm K, Clausen C, Jørgensen B, Karlsmark T. Characteristics of an autologous leukocyte and platelet-rich fibrin patch intended for the treatment of recalcitrant wounds. Wound Repair Regen 2013;21:66–76.
Magalon J, Bausset O, Serratrice N, Giraudo L, Aboudou H, Veran J, et al. Characterization and comparison of 5 platelet-rich plasma preparations in a single-donor model. Arthroscopy 2014;30:629–38.
Margolis DJ, Kantor J, Santanna J, Strom BL, Berlin JA. Effectiveness of platelet releasate for the treatment of diabetic neuropathic foot ulcers. Diabetes Care 2001;24:483–8.
Marston WA. Dermagraft, a bioengineered human dermal equivalent for the treatment of chronic nonhealing diabetic foot ulcer. Expert Rev Med Devices 2004;1:21–31.
Martínez CE, Smith PC, Palma Alvarado VA. The influence of platelet-derived products on angiogenesis and tissue repair: a concise update. Front Physiol 2015;6:290.
Martínez-Zapata MJ, Martí-Carvajal A, Solà I, Bolibar I, Angel Expósito J, Rodriguez L, et al. Efficacy and safety of the use of autologous plasma rich in platelets for tissue regeneration: a systematic review. Transfusion 2009;49:44–56.
Martinez-Zapata MJ, Martí-Carvajal AJ, Solà I, Expósito JA, Bolíbar I, Rodríguez L, et al. Autologous platelet-rich plasma for treating chronic wounds. Cochrane Database Syst Rev 2012;10:CD006899.

Mazzucco L, Balbo V, Cattana E, Guaschino R, Borzini P. Not every PRP-gel is born equal. Evaluation of growth factor availability for tissues through four PRP-gel preparations: Fibrinet, RegenPRP-Kit, Plateltex and one manual procedure. Vox Sang 2009;97:110–8.
McInnes AD. Diabetic foot disease in the United Kingdom: about time to put feet first. Foot Ankle Res 2012;5:26.
Mosser DM, Edwards JP. Exploring the full spectrum of macrophage activation. Nat Rev Immunol 2008;8:958–69.
de Oliveira AL, Moore Z. Treatment of the diabetic foot by offloading: a systematic review. J Wound Care 2015;24:560, 562–70.
Pradhan L, Nabzdyk C, Andersen ND, LoGerfo FW, Veves A. Inflammation and neuropeptides: the connection in diabetic wound healing. Expert Rev Mol Med 2009;11:e2.
Prompers L, Huijberts M, Apelqvist J, Jude E, Piaggesi A, Bakker K, et al. High prevalence of ischaemia, infection and serious comorbidity in patients with diabetic foot disease in Europe. Baseline results from the Eurodiale study. Diabetologia 2007;50:18–25.
Rendell MS, Milliken BK, Finnegan MF, Finney DA, Healy JC. The skin blood flow response in wound healing. Microvasc Res 1997;53:222–34.
Renner R, Sticherling M, Rüger R, Simon J. Persistence of bacteria like *Pseudomonas aeruginosa* in non-healing venous ulcers. Eur J Dermatol 2012;22:751–7.
Sen C, Gordillo G, Roy S. Human skin wounds: a major and snowballing threat to public health and the economy. Wound Repair Regen 2009;17:763–71.
Sun PC, Kuo CD, Chi LY, Lin HD, Wei SH, Chen CS. Microcirculatory vasomotor changes are associated with severity of peripheral neuropathy in patients with type 2 diabetes. Diab Vasc Dis Res 2013;10:270–6.
Thomsen K, Trøstrup H, Christophersen L, Lundquist R, Høiby N, Moser C. The phagocytic fitness of leucopatches may impact the healing of chronic wounds. Clin Exp Immunol 2016 [Epub ahead of print].
Veves A, Falanga V. Graftskin, a human skin equivalent, is effective in the management of non-infected neuropathic diabetic foot ulcers. Diabetes Care 2001;24:290–5.
Villela DL, Santos VL. Evidence on the use of platelet-rich plasma for diabetic ulcer: a systematic review. Growth Factors 2010;28:111–6.
Wieman TJ, Smiell JM, Su Y. Efficacy and safety of a topical gel formulation of recombinant human platelet-derived growth factor-BB (becaplermin) in patients with chronic neuropathic diabetic ulcers. A phase III randomized placebo-controlled double-blind study. Diabetes Care 1998;21:822–7.

Index

'*Note*: Page numbers followed by "f" indicate figures, "t" indicate tables.'

M

N

R

S

Edwards Brothers Malloy
Ann Arbor MI. USA
July 19, 2016